Acute and Chronic Changes in Neural Excitability During Physical Activity in Non-Pathological States

Acute and Chronic Changes in Neural Excitability During Physical Activity in Non-Pathological States

Special Issue Editor

Kevin Power

MDPI • Basel • Beijing • Wuhan • Barcelona • Belgrade • Manchester • Tokyo • Cluj • Tianjin

Special Issue Editor
Kevin Power
Faculty of Medicine Memorial University of Newfoundland
Canada

Editorial Office
MDPI
St. Alban-Anlage 66
4052 Basel, Switzerland

This is a reprint of articles from the Special Issue published online in the open access journal *Brain Sciences* (ISSN 2076-3425) (available at: https://www.mdpi.com/journal/brainsci/special_issues/Neural_Excitability).

For citation purposes, cite each article independently as indicated on the article page online and as indicated below:

LastName, A.A.; LastName, B.B.; LastName, C.C. Article Title. *Journal Name* **Year**, *Article Number*, Page Range.

ISBN 978-3-03928-796-3 (Pbk)
ISBN 978-3-03928-797-0 (PDF)

Contents

About the Special Issue Editor

Kevin Power (PhD in Neurophysiology) is a Full Professor in the School of Human Kinetics and Recreation at Memorial University of Newfoundland. He received his bachelor's (2001) and master's degrees (2003) from the School of Human Kinetics and Recreation at Memorial University and his PhD in neuroscience from the University of Manitoba (2009).

Dr. Power is the Co-Director of the Human Neurophysiology Lab at Memorial University with his colleague, Dr. Duane Button. Together with his team of students, Dr. Power studies the neural control of arm cycling, a rhythmic motor output that has similar neural control strategies to locomotion. Using this model, Dr. Power examines how the motor cortex and spinal motoneurones produce movement and adapt to exercise interventions, the findings of which have implications for guiding neurorehabilitation programs for persons following stroke or traumatic brain injury. Dr. Power has served as an Associate Editor for Applied Physiology Nutrition and Metabolism since 2017 and is the Vice-Chair of Research with the Canadian Society for Exercise Physiology (CSEP) and is himself a Certified Exercise Physiologist® and course examiner for CSEP.

brain sciences

MDPI

Article

Neuromuscular Mechanisms Underlying Changes in Force Production during an Attentional Focus Task

Shawn Wiseman [1], Shahab Alizadeh [1], Israel Halperin [2,3], Behzad Lahouti [1], Nicholas J. Snow [1,4], Kevin E. Power [1,4] and Duane C. Button [1,4,*]

1 School of Human Kinetics and Recreation, Memorial University of Newfoundland, St. John's, NL A1C 5S7, Canada; saw072@mun.ca (S.W.); salizadeh@mun.ca (S.A.); blahouti@mun.ca (B.L.); njsnow@mun.ca (N.J.S.); kevinp@mun.ca (K.E.P.)

2 School of Public Health, Sackler Faculty of Medicine, Tel-Aviv University, Tel-Aviv 6997801, Israel; ihalperin@tauex.tau.ac.il

3 Sylvan Adams Sports Institute, Tel Aviv University, Tel-Aviv 6997801, Israel

4 Faculty of Medicine, Memorial University of Newfoundland, St. John's, NL A1C 5S7, Canada

* Correspondence: dbutton@mun.ca; Tel.: +1-709-864-4886

Received: 25 November 2019; Accepted: 3 January 2020; Published: 7 January 2020

Abstract: We examined the effects of attentional focus cues on maximal voluntary force output of the elbow flexors and the underlying physiological mechanisms. Eleven males participated in two randomized experimental sessions. In each session, four randomized blocks of three maximal voluntary contractions (MVC) were performed. The blocks consisted of two externally and two internally attentional focus cued blocks. In one of the sessions, corticospinal excitability (CSE) was measured. During the stimulation session transcranial magnetic, transmastoid and Erb's point stimulations were used to induce motor evoked potentials (MEPs), cervicomedullary MEP (CMEPs) and maximal muscle action potential (Mmax), respectively in the biceps brachii. Across both sessions forces were lower ($p = 0.024$) under the internal (282.4 ± 60.3 N) compared to the external condition (310.7 ± 11.3 N). Muscle co-activation was greater ($p = 0.016$) under the internal (26.3 ± 11.5%) compared with the external condition (21.5 ± 9.4%). There was no change in CSE. Across both sessions, force measurements were lower ($p = 0.033$) during the stimulation (279.0 ± 47.1 N) compared with the no-stimulation session (314.1 ± 57.5 N). In conclusion, external focus increased force, likely due to reduced co-activation. Stimulating the corticospinal pathway may confound attentional focus. The stimulations may distract participants from the cues and/or disrupt areas of the cortex responsible for attention and focus.

Keywords: corticospinal excitability; attentional focus; co-contraction

1. Introduction

The effects of attentional focus instructions on motor learning and performance have been extensively studied in the past 20 years [1,2]. Specifically, two types of instructions have been compared: those that elicit an internal focus (IF) and external focus (EF) of attention [1,2]. EF leads one to focus on the intended effects of movements on the environment. For example, focusing on the bull's eye during a dart throwing task. Conversely, IF leads one to focus on a body part or muscle group. For example, focusing on wrist movement during a dart throwing task [3]. Many studies report that EF enhances motor learning and physical performance compared to IF [1,4–7]. This includes tasks that require accuracy, balance, strength and speed. The effects are consistent across children, adults, older adults, and those suffering from mental disease [8–10]. These effects are arguably some of the most established ones identified in human movement science.

Despite the impressive number of studies comparing attentional focus strategies across tasks and populations, little is known about the underpinning neuromuscular mechanisms that can explain the observed effects. A handful of studies examined if attentional focus strategies lead to different brain activation patterns using electroencephalography (EEG) and functional magnetic resonance imaging (fMRI) [11–14]. EEG alpha power is generally lower during EF and associated with more ideal alpha frequencies [15]. An fMRI study observed higher activation in the motor cortex during EF compared to IF. Thus, some evidence shows that various parts of the brain are activated differently between EF and IF which may account for enhanced neuromuscular performance during EF. The most commonly used tool to shed light on the mechanistic pathway explaining the superiority of EF is surface electromyography (EMG), which is a general measure of central nervous system excitation. A repeated—although not consistent—pattern is that EF leads to decreased muscle EMG activity from both the agonist and antagonist muscle groups involved in the task execution compared to IF [16]. The reduction in EMG activity during EF may promote effective and efficient movement patterns [17]. However, EEG, fMRI and EMG alone cannot pinpoint the motor pathways leading to the enhanced movement patterns associated with EF. Indeed, EF can promote superior motor performance by eliciting greater nervous system excitation, less inhibition, or a combination thereof from the brain to the spinal cord.

Nervous system excitation and inhibition can be examined through measuring corticospinal excitability via transcranial magnetic stimulation (TMS), and transmastoid electrical stimulation (TMES) [18,19]. TMS elicits a motor evoked potential (MEP) in a muscle of interest, while TMES elicits a cervicomedullary MEP. TMS-evoked MEP amplitudes are used to quantify corticospinal excitability (CSE) [20]. Alterations in CSE could occur anywhere along the corticospinal pathway (i.e., from cortex to motoneuron). The combined use of the aforementioned techniques is used to determine whether the modulation of CSE is predominantly supraspinal or spinal [18]. The corticospinal tract is examined due to its importance in the organization of single and multi-jointed movements. The corticospinal fibers control the spinal motoneurons that innervate the muscles of the trunk and limbs [21]. Many modulators have been shown to influence CSE from caffeine indigestion [22] to arousal imagery [23]. It is possible that EF may increase corticospinal excitability, decrease corticospinal inhibition, or a combination of both which would account, in part, for the increase in motor performance seen. This would further our understanding of the pathways and underlying mechanisms to address the changes in motor performance and learning seen with attentional focus feedback.

In view of the insight that can be gained using corticospinal excitability measurement techniques, and the limited knowledge accumulated to date on the pathways accounting for attentional focus instructions effects, combining the two in a single experiment is a worthwhile endeavor. Therefore, the aim of this study was to (1) compare CSE to the biceps brachii between EF and IF cued maximal voluntary contractions (MVC) of the elbow flexors and (2) compare co-activation patterns of the biceps brachii and triceps brachii between EF and IF cued MVC of the elbow flexors. We hypothesized that (1) CSE would be modulated differently between EF and IF cued condition and (2) co-activation would be greater with an IF cued condition.

2. Materials and Methods

Eleven resistance-trained males (1.77 ± 0.02 m, 84.32 ± 3.22 kg, 23.8 ± 2.36 years) participated in the experimental study. Resistance-trained status was determined as meeting the Canadian Society of Exercise Physiology guidelines of two hours a week of resistance training for at least a year. We chose to recruit only resistance-trained participants because corticospinal excitability is training dependent [24–27]. Participants completed a magnetic stimulation safety checklist prior to participation in order to screen for potential contraindications with magnetic stimulation procedures [28]. Verbal description about the procedures was given to the participants and if accepted they gave their informed written consent. The study was approved by The Memorial University of Newfoundland

Interdisciplinary Committee on Ethics in Human Research and was in accordance with the Tri-Council guidelines in Canada with full disclosure of potential risks to participants (HK-20190008).

2.1. Elbow Flexor Force

Participants were seated in a custom-built chair (Technical Services, Memorial University of Newfoundland, St. John's, NL, Canada) in an upright position, with chest and head strapped in place to minimize movement, with hips and knees flexed at 90°. The forearm was held horizontal, positioned supine with the shoulders resting against the back of the chair, and placed in a custom-made orthosis that was connected to a load cell (S-beam hanging load cells, model number LC101-500, Omegadyne, Inc., Sunbury, OH, USA) which was calibrated prior to the measurement. The load cell detected force output, which was amplified (×1000) (CED 1902, Cambridge Electronic Design Ltd., Cambridge, UK) and displayed on a computer screen. Data was sampled at 5000 Hz. Participants were instructed to maintain an upright position with their head in a neutral position during contractions. Visual feedback was given to all participants during all contractions prior to and during the conditions as a line on a computer screen in front of them showing when to begin and end the contraction. Participants were only able to view the amount of their force production and biceps brachii EMG activity.

2.2. Electromyography

Electromyography (EMG) activity was recorded from the dominant arm's biceps brachii and triceps brachii using surface EMG (10 mm MediTrace Ag-AgCl, Graphic Controls Ltd., Buffalo, NY, USA). Electrodes were placed 2 cm apart (center to center) over the midpoint of the muscle belly of the participant's biceps brachii and triceps brachii lateral head. A ground electrode was placed over the lateral epicondyle of the dominant knee. Skin preparation for all recording electrodes included shaving to remove excess hair and cleaning with an isopropyl alcohol swab to remove dry epithelial cells. An inter-electrode impedance of <5 kΩ was obtained prior to recording to ensure an adequate signal-to-noise ratio. EMG signals were amplified (×1000) (CED 1902) and filtered using a 3-pole Butterworth filter with cut-off frequencies of 10–1000 Hz. All signals were analog-digitally converted at a sampling rate of 5 kHz using a CED 1401 (Cambridge Electronic Design Ltd., Cambridge, UK) interface.

2.3. Stimulation Conditions

Motor responses from the bicep brachii were elicited via (1) transcranial magnetic stimulation (TMS), (2) transmastoid electrical stimulation (TMES) and (3) brachial plexus electrical stimulation at Erb's point. Stimulation intensities used for TMS and TMES were adjusted similar to that of Pearcey et al. (2014) so that the evoked potentials produced by each, evoked motor evoked potentials (MEPs), and CMEPs, respectively, were of similar amplitude and normalized to a maximal M-wave (Mmax) [26]. Stimulation intensities were then set during an isometric elbow flexion contraction equal to 5% of MVC.

2.3.1. Transcranial Magnetic Stimulation (TMS)

TMS- (MEPs) were used to measure corticospinal excitability. A TMS (Magstim 200, maximal output 2.0 Tesla) circular coil (13 cm outside diameter) was placed directly over the vertex of the head to induce MEPs in the active (5% MVC) biceps brachii muscle. The vertex was located by marking the measured halfway points between the nasion to inion and tragus to tragus. The coil was flipped to ensure the induced current flow was anterior to posterior in the target motor cortex (A side up for right side, B side up for left) to activate the dominant biceps brachii [29–31]. The optimum stimulation site was determined by initially placing the coil over the vertex and slightly moving the coil towards the non-dominant hemisphere in three directions, i.e., anteriorly, posteriorly, and laterally. Once the highest MEP response was elicited, that position was determined as the optimum stimulation site. Stimulation intensity was set to elicit a MEP 10%–20% of Mmax taken as an average of eight trials in the biceps brachii during a 5% MVC.

2.3.2. Transmastoid Electrical Stimulation (TMES)

Stimulation (Model DS7AH, Digitimer Ltd., Welwyn Garden City, Hertfordshire, UK) was applied via surface electrodes placed over the mastoid processes and current was passed between them (200 μs duration, 80–200 mA). Stimulation intensity was adjusted to prevent ventral root activation by closely monitoring CMEP responses for any decrease in onset latency (~2 ms), which shows cervical ventral root activation [32]. Stimulation intensity was adjusted to elicit a response that matched the size of MEP amplitude, taken as an average of eight trials, in the biceps brachii during a 5% MVC.

2.3.3. Brachial Plexus Stimulation

Stimulation (Model DS7AH, Digitimer Ltd., Welwyn Garden City, Hertfordshire, UK) of the brachial plexus was used to measure maximal compound muscle action potential (Mmax). Erb's point was electrically stimulated via a cathode and anode placed on the skin over the supraclavicular fossa and the acromion process, respectively. Current pulses were delivered as a singlet (200 μs duration, 90–185 mA, 400 V). The electrical current was gradually increased until Mmax of the biceps brachii at a 5% MVC was observed.

2.4. *Experimental Protocol*

Participants completed a familiarization session and two experimental sessions that were randomized. Each session took place on separate days with 48 h between each session.

2.4.1. Familiarization Session

Participants performed two 5 s MVCs of the dominant elbow flexors, with 2 min of rest between contractions. If the force from the two MVCs differed by greater than 5%, a third MVC was performed. Following completion of the MVCs, participants practiced holding the 5% MVC contraction for 10 s at each position. Participants then received the three different types of stimulations at various intensities to ensure that they were comfortable to endure the stimulation paradigm involved in each experimental session.

2.4.2. Stimulation Session

Upon arrival, the participants were prepared for EMG and asked to perform two elbow flexor MVCs. A 10 min rest period was then issued to ensure no effect of the MVC on the CSE measurements [33]. Following the rest period, the experimental procedures began and the stimulation intensities for the Mmax, MEP, and CMEP of the biceps brachii during 5% MVC were determined. Participants then moved on to perform a semi-randomized protocol where they completed four blocks of 3 MVCs of the elbow flexors with 3 min of rest between MVCs. Five minutes of rest was given between each block of conditions. A total of 12 MVCs were performed. Participants were verbally encouraged with the same attentional focus cue provided immediately before each contraction in each block of conditions. Participants were either asked to "focus on pulling up on the handle as hard and as quickly as you possibly can" (external cue) or to "focus on contracting your biceps as hard and as quickly as you possibly can" (internal cue). In total participants were EF cued six times or IF cued six times. These cues were countered balanced between sets. During each condition participants received counter-balanced TMSs and TMESs at 1.5 and 3 s and an M-wave was given at the 4.5 s mark. See Figures 1 and 2 for the experimental set-up.

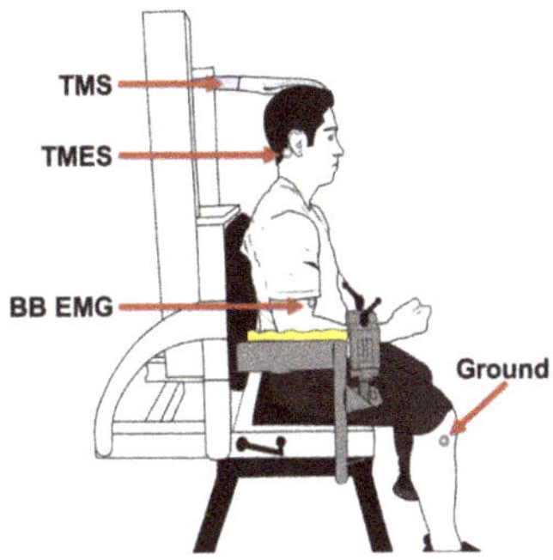

Figure 1. Participant's experimental set-up. Participants were positioned up right in an elevated chair with shoulders at 0 degrees and elbows at 90 degrees. TMS: transcranial magnetic stimulation, TMES: transmastoid electrical stimulation, BB EMG: biceps brachii electromyography. Ground: ground electrode.

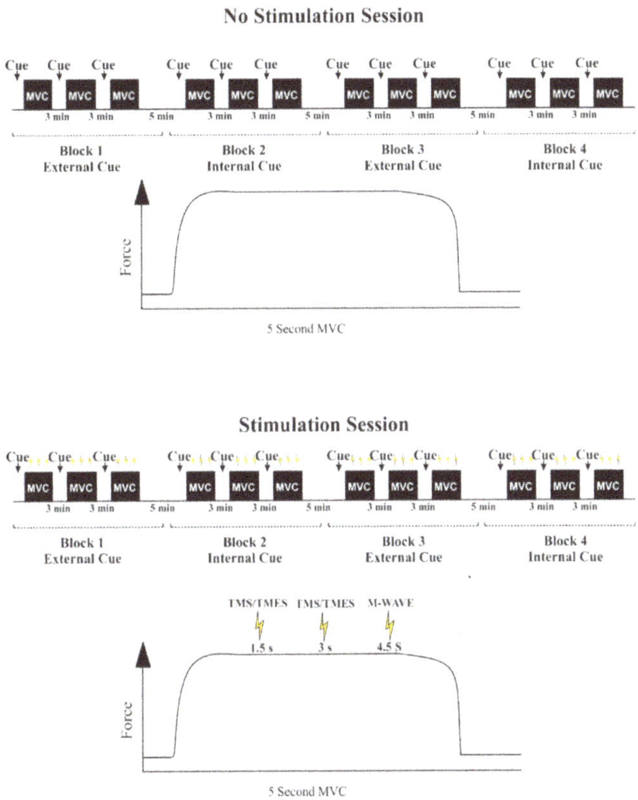

Figure 2. Experimental protocol. Each participant completed two experimental sessions (no stimulation session above, stimulation session below) which were randomized. Within each session, participants completed two blocks of three externally cued condition and two blocks of internally cued condition which were also randomized. Maximal voluntary contractions (MVC) were held for 5 s, beginning and ending at 2 and 7 s respectively, and during the stimulation session a transcranial magnetic stimulation (TMS) and transmastoid electrical stimulation (TMES) pulse was randomly delivered at 1.5 and 3.0 s marks with an M-Wave delivered each time at the 4.5 s mark. The traces underneath the sequence blocks represents the timing of the no-stimulation/stimulation during a 5 s MVC contraction.

2.4.3. Non-Stimulation Session

If the stimulation session was completed first, then the non-stimulation session was completed 48 h after the first session. This session was identical to the stimulation session except no stimulations were used. This session was included in the study to examine if stimulations impact a participant's ability to perform an MVC and their ability to focus on the attentional focus cues.

2.5. Data Analysis

Force, EMG, and CSE data were measured offline using Signal 4.0 software (Cambridge Electronic Design Ltd., Cambridge, Hertfordshire, UK). All offline computations were conducted using Microsoft Office Excel 2016 (Microsoft Corporation, Redmond, WA, USA).

Maximum voluntary isometric force. Peak elbow flexor's force was obtained from all six MVCs under each condition (external cued condition and internal cued condition) during both no-stimulation and stimulation sessions. MVC force output was measured as the peak amplitude from no force to maximum force.

Electromyography (EMG). Root mean square EMG (rmsEMG) from the biceps brachii and triceps brachii muscles was calculated from $t = 1$ s to $t = 2$ s interval of each MVC once the force reached its peak output under each condition (external cued condition and internal cued condition) and during each session (no-stimulation and stimulation). Additionally, muscle co-activation was quantified by computing the percentage of triceps brachii rmsEMG/biceps brachii rmsEMG [34]. To illustrate the amount of co-activation per unit of force production and illustrate the relationship between force and coactivation, the percentage ratio of muscle co-activation per Newton of force was calculated for MVCs from both the external and internal cued conditions during both no-stimulation and stimulation sessions.

Corticospinal excitability (CSE). During the stimulation session only, biceps brachii MEP, CMEP, and Mmax peak-to-peak amplitudes (mV) were extracted during all six MVCs under each condition (external cued condition and internal cued condition). See Figure 3 for raw data for MEP, CMEP and Mmax responses during a MVC. Since amplitudes and areas give similar results, we used MEP, CMEP, and Mmax amplitudes for comparisons [35]. MEP and CMEP peak-to-peak amplitudes were normalized to Mmax amplitudes (%Mmax), given Mmax is a stable measure of muscle activity during maximal muscle fibre recruitment [36]. Simply put, the MEP and CMEP values were divided by the M-wave produced from the same trial. As well, ratios of normalized MEP/CMEP amplitude were calculated [37]. The MEP/CMEP would indicate the level of supraspinal or spinal excitability. Levels above 1 are indicative of higher supraspinal excitability compared to spinal excitability [30].

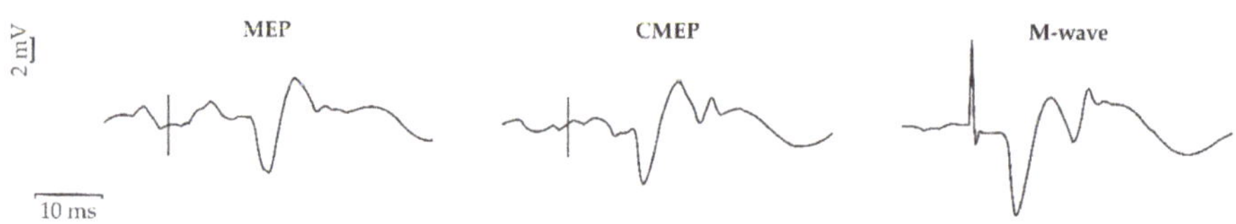

Figure 3. Raw data sample for MEP, CMEP and Mmax responses during an MVC from a single participant.

2.6. Statistical Analysis

Prior to statistical analyses all data underwent quality control checks in Microsoft Office Excel 2016 (Microsoft Corporation, Redmond, WA, USA) for missing data points and outliers. In terms of missing data, only one participant was unable to complete the stimulation session (P09). This participant was not included in CSE analyses; however, their MVC peak force and rmsEMG data (trial 1 to trial 6) for both the external condition and internal conditions (12 trials) were subsequently imputed for the stimulation session to enable groupwise comparisons across sessions. Additionally, two participants (P10, P11) were missing force data for one MVC trial each (trial 4), under both the internal and external conditions, for the stimulation session alone (four trials). The missing datapoints were in a time series where data for both sessions (i.e., stimulation and non-stimulation sessions) were not available. In total, 16 datapoints were missing for MVC peak force (6.1%) and 12 datapoints each were missing for rmsEMG of both biceps brachii (4.5%) and triceps brachii (4.5%). Missing data were imputed by determining the series average for the entire sample, including both conditions (external condition, internal condition), at their respective timepoints and sessions using the Missing Values Analysis and Transform functions in SPSS (V26.0, IBM Corporation, Armonk, NY, USA). Outliers were considered datapoints that exceeded the sample mean by ± three standard deviations (SD). No outliers were identified.

Statistical analyses were completed using SPSS (V26.0, IBM Corporation, Armonk, NY, USA). Assumptions of normality (Shapiro–Wilk test), sphericity (Mauchly's test), and homogeneity of variances (Levene's test) were tested for all outcome measures where appropriate. For the Shapiro–Wilk test, statistical significance was set at $p < 0.001$ [38]. In the event of a violation of the assumption of sphericity, p-values were adjusted using the Greenhouse–Geisser correction. If the assumption of homogeneity of variances was violated, p-values were adjusted (equal variances not assumed).

To rule out whether measures of MVC peak force, rmsEMG, or CSE changed over subsequent trials (trial 1 to trial 6), separate one-way repeated-measures analyses of variance (ANOVAs) with the factor TRIAL (6 levels) were conducted on all data independently for internal and external conditions, as well as stimulation and no-stimulation sessions. This test was used to guide subsequent analyses in terms of whether trials were pooled or tested separately. For MVC peak force, the main effect of TRIAL was statistically significant in all cases ($F_{(5, 50)} \geq 3.982$, $p \leq 0.022$). Similarly, with reference to rmsEMG data, the main effect of TRIAL was statistically significant in most cases ($F_{(5, 50)} \leq 6.690$, $p \geq 0001$). However, regarding CSE, the main effect of TRIAL was not statistically significant in any case ($F_{(5, 45)} \leq 2.137$, $p \geq 0.150$). Consequently, in main statistical tests, TRIAL was considered a separate factor for MVC peak force and rmsEMG data, whereas all levels of the factor TRIAL were pooled for CSE.

For main statistical tests, repeated-measures ANOVAs and paired-samples t-tests were used, with designs depending on the result of the above one-way repeated-measures ANOVAs. Peak force measurements from MVCs were compared across trials (trial 1 to trial 6), conditions (external cued condition and internal cued condition), and sessions (no-stimulation and stimulation) using a $6 \times 2 \times 2$ three-way repeated-measures ANOVA with the factors TRIAL, CONDITION, and SESSION, respectively. Due to variability in EMG recordings, raw rmsEMG values for biceps brachii and triceps brachii were examined separately for each session (no-stimulation and stimulation) across trials (trial 1 to trial 6) and conditions (external cued condition and internal cued condition) using 2×2 two-way repeated measures ANOVAs with the factors TRIAL and CONDITION, respectively, given they were not normalized [33]. Because triceps brachii/biceps brachii co-activation values were normalized, they were compared as square root transformed values across trials (trial 1 to trial 6), conditions (external cued condition and internal cued condition), and sessions (no-stimulation and stimulation) using separate $6 \times 2 \times 2$ three-way repeated measures ANOVAs with the factors TRIAL, CONDITION, and SESSION, respectively [33]. For CSE, average values across all trials (trial 1 to trial 6) for Mmax amplitude (mV), as well as MEP/Mmax, CMEP/Mmax, and square root transformed CMEP/MEP ratios, were compared across conditions (external cued condition and internal cued condition) using separate paired-samples t-tests. Finally, to investigate the relationship between changes in peak force and co-activation across stimulation conditions, two analyses were performed. First, a 2×2 two-way repeated-measures ANOVA with the factors CONDITION and SESSION was conducted on the percentage ratios of muscle co-activation per Newton of force calculated from MVCs from both the external and internal conditions during both no-stimulation and stimulation sessions. Last, simple bivariate correlations (Pearson's r) were calculated between changes in MVC peak force and triceps brachii/biceps brachii co-activation from external to internal cued condition in the no-stimulation and stimulation sessions separately. Strength of the correlation coefficients (r) was interpreted as <0.3 (negligible), 0.3–0.5 (weak), 0.5–0.7 (moderate), 0.7–0.9 (strong), and > 0.9 (very strong) [39].

Statistical significance for main tests was set at $p \leq 0.05$. In the event of a statistically significant ANOVA outcome, pairwise comparisons were completed post hoc using the Bonferroni correction. In the text, tables' and figures' data are reported as mean ± SD.

3. Results

3.1. Data Distribution

All data were normally distributed (MVC: $W_{(11)} = 0.821$–0.966, $p = 0.018$–0.848; rmsEMG: $W_{(11)} = 0.699$–0.982, $p = 0.001$–0.976; CSE: $W_{(10)} = 0.684$–0.934, $p = 0.001$–0.490), with the exception of MEP/CMEP ratio values under the internal condition of the stimulation session ($W_{(10)} = 0.628$, $p = 0.0001$) and muscle co-activation (% triceps/biceps brachii rmsEMG) under the internal condition of the stimulation session ($W_{(11)} = 0.639$–0.858, $p = 0.0001$–0.054). Thus, all MEP/CMEP ratio and muscle co-activation values were square root transformed using the Transform function in SPSS, resulting in normal distributions (MEP/CMEP: $W_{(10)} = 0.740$–0.879, $p = 0.003$–0.126; co-activation: $W_{(11)} = 0.745$–0.956, $p = 0.002$–0.715).

3.2. Peak Force

MVC peak force are shown in Figure 4, Tables 1 and 2. The three-way repeated-measures ANOVA on peak force measurements from elbow flexor MVCs revealed three statistically significant main effects. First, a statistically significant main effect of CONDITION ($F_{(1, 10)} = 7.033$, $p = 0.024$) showed that force was significantly less under the internal conditions (282.4 ± 60.3 N) versus the external condition (310.7 ± 11.3 N) (Figure 4A). Next, a statistically significant main effect of SESSION ($F_{(1, 10)} = 6.076$, $p = 0.033$) demonstrated that force measurements were significantly smaller during the stimulation session (279.0 ± 47.1 N) than the no-stimulation session (314.1 ± 57.5 N) (Figure 4B). Finally, there was a statistically significant main effect of TRIAL ($F_{(5, 50)} = 14.262$, $p = 0.00001$) (see Table 2 for multiple comparisons). Neither the TRIAL × CONDITION ($F_{(5, 50)} = 1.701$, $p = 0.152$), TRIAL × SESSION ($F_{(5, 50)} = 0.211$, $p = 0.891$), CONDITION × SESSION ($F_{(5, 50)} = 1.365$, $p = 0.270$), nor TRIAL × CONDITION × SESSION interactions ($F_{(5, 50)} = 1.344$, $p = 0.281$) were statistically significant.

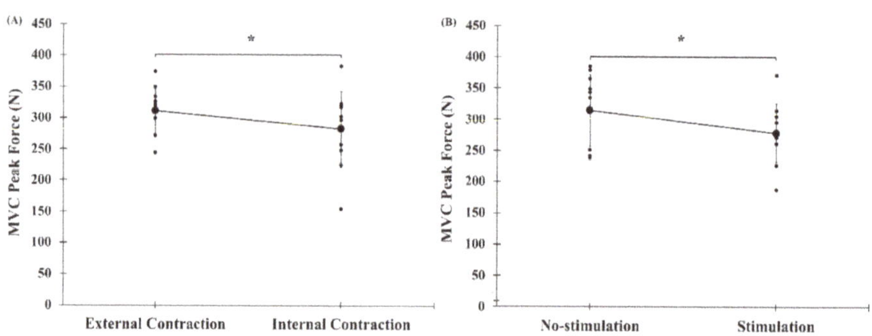

Figure 4. Peak force values for MVCs, measured in Newtons (N). Smaller points represent individual participant data, larger points represent mean, and error bars represent one standard deviation. (**A**) Peak force values for external versus internal conditions, collapsed across all trials (trial 1 to trial 6) and sessions (no-stimulation, stimulation), demonstrating the significant main effect of CONDITION. (**B**) Peak force values for no-stimulation versus stimulation session, collapsed across all trials (trial 1 to trial 6) and conditions (external condition, internal condition), signifying the significant main effect of SESSION. *, statistically significant at $p < 0.05$.

Table 1. Mean ± SD of MVC force and electromyographic (EMG) data, collapsed across trials (trial 1 to trial 6), presented for conditions (external cue, internal cue) during both no-stimulation and stimulation sessions.

	No-Stimulation		Stimulation	
	External Cue (Range)	Internal Cue (Range)	External Cue (Range)	Internal Cue (Range)
MVC Peak Force (N)	333.5 ± 43.7, (242.6–375.5)	294.8 ± 76.7, (184.2–398.6)	287.9 ± 38.7, (245.3–376.0)	270.1 ± 62.4, (124.3–367.1)
Biceps Brachii rmsEMG	0.73 ± 0.51, (0.27–1.88)	0.60 ± 0.38, (0.16–1.47)	0.59 ± 0.33, (0.26–1.41)	0.53 ± 0.25, (0.21–1.04)
Triceps Brachii rmsEMG	0.12 ± 0.03, (0.07–0.18)	0.15 ± 0.09, (0.05–0.34)	0.11 ± 0.04, (0.05–0.17)	0.12 ± 0.04, (0.05–0.17)
Co-activation (% Triceps/Biceps rmsEMG)	22.2 ± 13.8, (8.6–49.6)	31.5 ± 19.5, (12.1–77.2)	24.3 ± 15.7, (11.3–63.0)	26.8 ± 17.5, (12.7–74.4)
% Co-activation per Newton Force	0.06 ± 0.04, (0.03–0.14)	0.12 ± 0.08, (0.03–0.24)	0.09 ± 0.06, (0.03–0.23)	0.11 ± 0.08, (0.03–0.29)

N: Newton, rmsEMG: Root mean square of EMG signal.

Table 2. Mean ± SD of MVC force and electromyographic data, collapsed across sessions (no-stimulation, stimulation) and conditions (external condition, internal condition), for MVC trial 1 to trial 6.

		1 (Range)	2 (Range)	3 (Range)	4 (Range)	5 (Range)	6 (Range)
MVC Peak Force (N)		313.8 ± 45.8, (229.9–373.8) ^{d,e}	304.7 ± 42.1, (224.4–372.2) ^{d,e}	308.1 ± 36.2, (258.9–377.7) ^{d,e}	276.0 ± 34.1, (227.5–336.1) ^{a,b,c}	284.0 ± 30.4, (23.4–320.9) ^{a,b,c}	288.6 ± 34.8, (219.4–335.6)
Biceps Brachii rmsEMG	S	0.72 ± 0.46, (0.25–1.83)	0.70 ± 0.46, (0.22–1.76) ^{e}	0.69 ± 0.45, (0.21–1.71) ^{d}	0.64 ± 0.43, (0.21–1.59) ^{c}	0.64 ± 0.45, (0.23–1.67) ^{b}	0.61 ± 0.40, (0.17–1.50)
	NS	0.63 ± 0.28, (0.31–1.10)	0.57 ± 0.31, (0.18–1.27)	0.58 ± 0.33, (0.20–1.38)	0.53 ± 0.33, (0.18–1.35)	0.52 ± 0.26, (0.21–1.14)	0.53 ± 0.27, (0.20–1.15)
Triceps Brachii rmsEMG (mV·s)	S	0.14 ± 0.05, (0.07–0.23)	0.15 ± 0.06, (0.07–0.26)	0.14 ± 0.06, (0.06–0.25)	0.14 ± 0.08, (0.06–0.35)	0.13 ± 0.06 (0.06–0.25)	0.12 ± 0.05, (0.06–0.20)
	NS	0.12 ± 0.04, (0.05–0.16)	0.11 ± 0.04, (0.05–0.17)	0.12 ± 0.04, (0.06–0.17)	0.11 ± 0.04, (0.05–0.18)	0.12 ± 0.05, (0.05–0.19)	0.11 ± 0.04, (0.05–0.16)
Co-activation (%Triceps/Biceps rmsEMG) (mV·s)		23.9 ± 9.8, (12.8–40.5)	27.0 ± 13.4, (13.1–53.9)	25.5 ± 12.1, (12.3–48.7)	27.6 ± 12.6, (12.1–55.8)	26.8 ± 12.9, (12.5–54.5)	26.3 ± 14.2, (11.8–55.9)
%Co-activation per Newton Force		0.08 ± 0.03, (0.03–0.13)	0.09 ± 0.04, (0.04–0.16)	0.08 ± 0.04, (0.03–0.15)	0.10 ± 0.04, (0.04–0.18)	0.09 ± 0.04, (0.04–0.17)	0.09 ± 0.05, (0.04–0.18)

N: Newton, rmsEMG: Root mean square of EMG signal, S: Stimulation, NS: No stimulation. ^{a}, statistically significant difference versus trial 1, $p < 0.05$. ^{b}, statistically significant difference versus trial 2, $p < 0.05$. ^{c}, statistically significant difference versus trial 3, $p < 0.05$. ^{d}, statistically significant difference versus trial 4, $p < 0.05$. ^{e}, statistically significant difference versus trial 5, $p < 0.05$.

3.3. *Electromyography (EMG)*

Biceps brachii and triceps brachii rmsEMG data are displayed in Tables 1 and 2.

3.3.1. Biceps Brachii

No-stimulation session. For biceps brachii rmsEMG during the no-stimulation session there was a statistically significant main effect of TRIAL ($F_{(5, 50)} = 7.341$, $p = 0.001$) (see Table 2 for multiple comparisons). The main effect of CONDITION trended towards significance ($F_{(1, 10)} = 3.958$, $p = 0.075$) and indicated that biceps brachii rmsEMG tended to be greater under the external condition (0.73 ± 0.51% MVC) compared to internal condition (0.60 ± 0.38% MVC). The TRIAL × CONDITION interaction effect was not statistically significant ($F_{(5, 50)} = 1.83$, $p = 0.133$) (Table 2).

Stimulation session. During the stimulation session, the main effect of TRIAL trended towards significance ($F_{(5, 50)} = 3.317$, $p = 0.068$) and suggested that rmsEMG tended to be greater under trial 3 (0.58 ± 0.33% MVC) versus trial 4 (0.53 ± 0.33% MVC) (see Table 2 for multiple comparisons). Otherwise, there was neither a statistically significant main effect of CONDITION ($F_{(1, 10)} = 2.407$, $p = 0.152$) nor TRIAL × CONDITION interaction effect ($F_{(5, 50)} = 0.506$, $p = 0.565$) (Table 2).

3.3.2. Triceps Brachii

No-stimulation session. With reference to triceps brachii rmsEMG throughout the no-stimulation session, there were no statistically significant main effects of TRIAL ($F_{(5, 50)}$ = 1.722, p = 0.210) or CONDITION ($F_{(1, 10)}$ = 2.178, p = 0.171), nor a two-way TRIAL × CONDITION interaction effect ($F_{(5, 50)}$ = 0.510, p = 0.528).

Stimulation session. In the stimulation session, there were no statistically significant effects of TRIAL ($F_{(5, 50)}$ = 1.443, p = 0.226), CONDITION ($F_{(1, 10)}$ = 0.141, p = 0.716), or TRIAL × CONDITION ($F_{(5, 50)}$ = 0.642, p = 0.583), for triceps brachii rmsEMG.

3.3.3. Co-Activation

Muscle co-activation data (expressed as % triceps/biceps rmsEMG) are shown in Figure 5A and Tables 1 and 2. The three-way repeated-measures ANOVA on percentage values of co-activation demonstrated a statistically significant main effect of CONDITION ($F_{(1, 10)}$ = 8.438, p = 0.016), whereby muscle co-activation was significantly greater under the internal condition (26.3 ± 11.5%) versus external contraction condition (21.5 ± 9.4%) (Figure 5A).

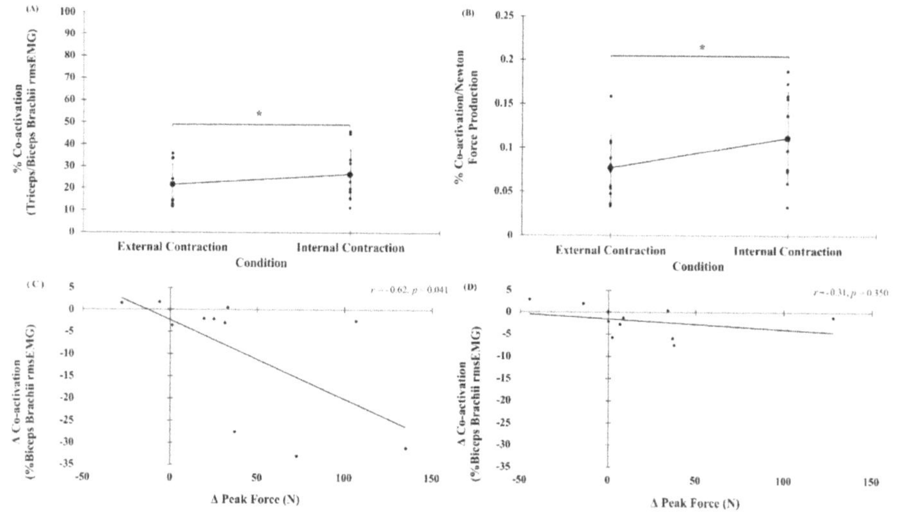

Figure 5. Data expressing the relationship between muscle co-activation and MVC peak force. In panels A–B, smaller points represent individual participant data, larger points represent mean, and error bars represent one standard deviation. In panels D–E, points represent individual data. (**A**) Muscle co-activation values for external and internal conditions, collapsed across trials (trial 1 to trial 6) and sessions (no-stimulation, stimulation), demonstrating the significant main effect of CONDITION. (**B**) Percentage of muscle co-activation/MVC peak force (co-activation per Newton force production) for external versus internal conditions, collapsed across session (no-stimulation, stimulation), illustrating the significant main effect of CONDITION. (C-D) Scatterplots demonstrating relationship between changes in MVC peak force and triceps brachii/biceps brachii co-activation across external and internal conditions during the (**C**) no-stimulation and (**D**) stimulation sessions. *, statistically significant at $p < 0.05$.

The main effects of TRIAL ($F_{(5, 50)}$ = 2.123, p = 0.136) and SESSION ($F_{(1, 10)}$ = 0.029, p = 0.869) were not statistically significant. Likewise, neither the TRIAL × CONDITION ($F_{(5, 50)}$ = 0.175, p = 0.971), TRIAL × SESSION ($F_{(5, 50)}$ = 0.419, p = 0.833), CONDITION × SESSION ($F_{(5, 50)}$ = 1.969, p = 0.191), nor TRIAL × CONDITION × SESSION ($F_{(5, 50)}$ = 2.072, p = 0.144) interaction effects were statistically significant.

3.4. Corticospinal Excitability (CSE)

CSE data are presented for each condition (external cue, internal cue), collapsed across trials (trial 1 to trial 6) in Table 3.

Table 3. Mean ± SD of the corticospinal excitability (CSE), collapsed across trials (trial 1 to trial 6), for external and internal contraction conditions. Data presented as mean (M), standard deviation (SD), and range.

	External Cue (Range)	Internal Cue (Range)
Mmax Amplitude (mV)	8.62 ± 4.97, (3.0–20.1)	8.48 ± 5.0, (1.63–17.47)
MEP Amplitude (Ratio of Mmax)	0.83 ± 0.36, (0.48–1.65)	1.01 ± 0.78, (0.48–2.95)
CMEP Amplitude (Ratio of Mmax)	0.64 ± 0.38, (0.30–1.76)	0.79 ± 0.71, (0.23–2.20)
MEP/CMEP Ratio	1.64 ± 0.99, (0.92–3.64)	1.97 ± 1.77, (0.82–6.44)

Mmax: Maximal compound motor unit action potential, MEP: motor evoked potential, CMEP: Cervicomedullary MEP.

There was no statistically significant difference ($t_{(9)} = -0.508$, $p = 0.624$; $t_{(9)} = 0.598$, $p = 0.565$; $t_{(9)} = 0.340$, $p = 0.742$; and $t_{(9)} = -1.215$, $p = 0.255$) in Mmax, MEP, or CMEP amplitudes or MEP/CMEP ratios, respectively across external cued condition and internal cued conditions. A sample of MEP, CMEP, and M-wave responses during stimulation is presented Figure 2.

3.5. Co-Activation/MVC Peak Force

There was a statistically significant main effect of CONDITION for ratios of co-activation/Newton force produced in MVCs ($F_{(1, 10)} = 11.307$, $p = 0.007$), which indicated that under the external condition (0.08 ± 0.04%) less muscle co-activation occurred per Newton of force production compared to the internal condition (0.11 ± 0.05%; $p = 0.007$) (Figure 5B). Neither the main effect of SESSION ($F_{(1, 10)} = 0.131$, $p = 0.725$) nor the CONDITION × SESSION two-way interaction effect ($F_{(5, 50)} = 1.333$, $p = 0.275$) reached statistical significance.

3.6. Relationship between Change in Peak Force and Co-Activation

Values of percent muscle co-activation per Newton of force production in MVCs, and correlations between changes in MVC peak force and triceps brachii/biceps brachii co-activation, are shown in Figure 5B–D and Tables 1 and 2.

Correlations

No-stimulation session. During the no-stimulation session there was a statistically significant negative correlation between changes in MVC peak force (38.8 ± 48.6 N) and triceps brachii/biceps brachii co-activation (−9.2 ± 13.9%) across external and internal conditions ($r_{(9)} = -0.623$, $p = 0.041$, moderate correlation), suggesting increased co-activation was related to reduced MVC force production in the internal condition (Figure 5C).

Stimulation session. In the stimulation session the relationship between changes across external and internal conditions in MVC peak force and triceps brachii/biceps brachii co-activation was not present ($r_{(9)} = -0.312$, $p = 0.350$, weak correlation) (Figure 5D).

4. Discussion

The purpose of this study was to examine if consistent superior motor performance observed with an external, compared with an internal focus instruction, is mediated by different corticospinal excitability processes. We observed three key findings. First, consistent with the literature, force production was greater with external focus instructions. Second, the greater force outputs were accompanied with lower co-contraction ratios between the biceps and triceps brachii (measured as rmsEMG Triceps Brachii/rmsEMG Biceps Brachii) under the external focus condition, leading to a more effective contractions strategy. Third, the neuromuscular strategy identified with the EMG patterns did not coincide with a change in corticospinal excitability. This finding likely stems from an interaction between the stimulation techniques for measuring CSE and attentional focus. We speculate that the stimulation negated the effect of an external focused cue. This assumption is supported by the higher forces produced under the external focus condition in the non-stimulation session.

Maximal elbow flexor force is affected by the type of attentional focus cue. Participants were able to produce more force when provided an external focus cue (310.7 ± 11.3 N) compared to internal cue (282.4 ± 60.3 N) condition during the non-stimulation session. This is consistent with previous research which showed enhanced force production when given an external cue over no cue and internal focus cues. For example, Marchant et al. (2009) found that during concentric elbow flexion completed at a set speed, an external cue led to a higher peak net torque (102.10 ± 2.42% MVC) than the internal condition (95.33 ± 2.08% MVC) [6]. Halperin et al. (2016) reiterated these results showing that when given an external focus cue during an isometric mid-thigh pull, trained athletes applied 9% more force compared to those that received an internal cue, and 5% more force than control [40]. This supports that external focus cues enhance force output compared to internal ones.

While there was an observed difference in force production between conditions in the non-stimulation session, there were no significant changes in force production between conditions during the stimulation session. This finding is not consistent with previous research as it is well documented that attentional focus alters force production [1,5–7,40], which can be accounted for by a number of possible reasons. First, the stimulation may have distracted the participants from focusing on the provided cue. For example, the stimulation may have caused discomfort, led to fear, or attracted interest, all of which could distract participants from the provided instruction. Additionally, the notion that the stimulations are provided at random intervals may channeled their attention to anticipation, amplifying the distraction from the cues. Second, the use of the stimulation techniques disrupted areas of the cortex responsible for attention. It is known that transcranial magnetic stimulation (TMS) can disrupt cortical function. For example, Ashbridge et al. (1997) suggested that TMS disrupts an area in the front parietal lobe responsible for the focal attention necessary for feature binding in a conjunction search task [41]. Another study showed that repetitive TMS of the intraparietal sulcus and the frontal eye fields during an auditory spatial attention task impaired visually cued auditory attention [42]. With each stimulation pulse, it is possible that more than just the cortical area of interest was being stimulated [43], and therefore it is possible that cortical areas involved in attention were unintentionally disrupted. It has been studied that attentional focus potentiate short-term plasticity in the motor cortex through the premotor-to-motor connection, which is why TMS stimulation of the motor cortex disrupts attentional focus [44]. Either of the mentioned reasons, or a combination thereof would confound the effects of attentional focus instructions on force production and explain the differences between sessions.

Moreover, force produced in the stimulation session (279.0 ± 47.1 N) was lower compared to the non-stimulation session (314.1 ± 57.5 N), collapsed across the two instruction conditions. This finding is aligned with Button and Behm [45], who showed that the expectation of an interpolated twitch stimulation reduced voluntary force production by 9.5%. However, to date there appears to be a lack of research showing how stimulation of the nervous system using TMS and TMES influences force production. This result should be replicated and expanded upon in future studies as it implies that the use of stimulations could confound a study involving force production

Mechanisms underlying changes in elbow flexor maximal force with attentional focus cues.

Electromyography. Our results showed greater co-activation with an internal compared to an external cue. This is consistent with a previous study by Lohse et al. (2011), who reported greater co-contraction between the lateral aspect of the soleus and the tibialis anterior with an internal focus cue during a submaximal plantar flexion task [46]. Greater co-activation of the agonist and antagonist musculature is a possible mechanism underpinning why maximal force production is lower during internal focused cues. Based on the current EMG findings, and aligned with Wulf's [1] "Constrained action hypothesis", it appears that force production is impaired with an internal cue due to disruption of natural automatized movement as supported by increases in co-activation by increasing the antagonist and decreasing the agonist muscle activity, whereas the greater EMG pattern of the agonists observed during the external cue could be a result of greater motor unit recruitment and/or rate coding. In contrast to some studies, we did not find a significant difference in EMG activity between focus cues. However, a large, albeit not statistically significant, effect was observed in which greater neuromuscular activity of the biceps brachii was associated with an external focus cue, suggesting that external cue leads to greater motor unit recruitment and/or rate coding.

Corticospinal Excitability. Corticospinal tract output can be altered by multiple variables, such as exercise, injury, disuse and disease and potentially attentional cues. As force output is increased, both supraspinal and spinal excitability also increased illustrating that changes in the excitability of cortical neurons and/or spinal motoneurons are occurring [26,27]. Measures of corticospinal excitability, specifically supraspinal and spinal excitability were used during one of the two sessions (stimulation session). This allowed us to determine whether or not the increase in maximal elbow flexor force with an external focus cue was due, in part, to enhanced corticospinal excitability at the supraspinal or spinal level(s) or combination thereof, of the biceps brachii. We expected to see an increase or change in corticospinal excitability at the supraspinal or spinal level(s) of the biceps brachii with an external focus cue as increased central drive is a well-known mechanism underlying increases in force production [26,27,34]. However, we were unable to support this possibility with the current study due to a lack of differences in corticospinal excitability responses of the biceps brachii during elbow flexor MVCs when receiving external versus internal focus cues. There are a couple technical considerations that must be noted in relation to the measurement of corticospinal excitability. First, because MEP [26,47] and CMEP [26] amplitudes are dependent on background EMG during isometric contractions and EMG was different between conditions (i.e., attentional cues), it is possible that the MEPs and CMEPs between conditions may not be comparable. Second, MEP and CMEP amplitudes were matched during a 5% MVC to equal 10%–20% of Mmax. The motor output and activation of cortical neurons and spinal motoneurons required to produce a 5% MVC is different than 100% MVC. Because the same stimulation intensities were used during a 5% and a 100% MVC, we cannot rule out that the corticospinal tract (TMS and TMES) during 100% MVC was not optimally stimulated. Thus, the MEP and CMEP responses may have been suboptimal. In fact, during 100% MVC the MEP and CMEP amplitudes were much higher than 10%–20% of Mmax amplitude. Nonetheless, force and EMG and MEP, CMEP and Mmax amplitudes of the biceps brachii were not significantly different between attentional focus cues. To our knowledge, this appears to be the only study examining corticospinal excitability and attentional focus feedback. In view of the possible reasons raised earlier that can account for the lack of effects under the stimulation condition, it seems like a worthwhile attempt to conceptually replicate the current study with participants who are more experienced with isometric contractions and corticospinal excitability techniques, which may reduce its distracting effects, and thus possibly lead to different results.

5. Conclusions

In conclusion, force production during a MVC of the elbow flexors followed the known pattern in which external cue leads to superior performance compared to internal cue. This finding was accompanied by greater co-activation of the triceps brachii and biceps brachii which appear to be an underlying mechanism for this impairment. Interestingly, and in contrast to the non-stimulation sessions, the use of stimulation techniques impaired attention by way of distraction or impairment of certain areas in the brain, which nullified the established effects of attentional focus instruction on maximal force production.

Author Contributions: Conceptualization, S.W., I.H., and D.C.B.; Methodology, S.W., I.H., B.L., K.E.P. and D.C.B.; Formal Analysis, S.W., S.A., I.H., B.L., N.J.S., K.E.P. and D.C.B.; Resources, D.C.B.; Data Curation, S.W., S.A. and

N.J.S.; Writing—Original Draft Preparation, S.W., S.A., I.H., B.L., N.J.S., K.E.P. and D.C.B.; Writing—Review & Editing, S.W., S.A., I.H. and D.C.B.; Visualization, S.W., S.A., I.H., N.J.S., K.E.P. and D.C.B.; Supervision, D.C.B.; Project Administration, D.C.B.; Funding Acquisition, D.C.B. All authors have read and agreed to the published version of the manuscript.

Funding: This research was funded by the Natural Sciences and Engineering Research Council of Canada, grant number (NSERC-#RGPIN-2018-03876).

Acknowledgments: This study was supported by MITACS ACELERATE funding to S.A. as well as NSERC Discovery Grant to D.C.B. We would like to thank Thamir Alkanani for technical support and the participants for volunteering their time.

Conflicts of Interest: The authors declare no conflict of interest.

References

1. Wulf, G. Attentional focus and motor learning: A review of 15 years. *Int. Rev. Sport Exerc. Psychol.* **2013**, *6*, 77–104. [CrossRef]
2. Lauber, B.; Keller, M. Improving motor performance: Selected aspects of augmented feedback in exercise and health. *Eur. J. Sport Sci.* **2014**, *14*, 36–43. [CrossRef] [PubMed]
3. Porter, J.; Nolan, R.; Ostrowski, E.; Wulf, G. Directing attention externally enhances agility performance: A qualitative and quantitative analysis of the efficacy of using verbal instructions to focus attention. *Front. Psychol.* **2010**, *1*, 216–223. [CrossRef] [PubMed]
4. Wulf, G.; Lewthwaite, R. Optimizing performance through intrinsic motivation and attention for learning: The OPTIMAL theory of motor learning. *Psychon. Bull. Rev.* **2016**, *23*, 1382–1414. [CrossRef] [PubMed]
5. Marchant, D.C.; Greig, M.; Bullough, J.; Hitchen, D. Instructions to adopt an external focus enhance muscular endurance. *Res. Q. Exerc. Sport* **2011**, *82*, 466–473. [CrossRef] [PubMed]
6. Marchant, D.C.; Greig, M.; Scott, C. Attentional focusing instructions influence force production and muscular activity during isokinetic elbow flexions. *J. Strength Cond. Res.* **2009**, *23*, 2358–2366. [CrossRef] [PubMed]
7. Greig, M.; Marchant, D. Speed dependant influence of attentional focusing instructions on force production and muscular activity during isokinetic elbow flexions. *Hum. Mov. Sci.* **2014**, *33*, 135–148. [CrossRef]
8. Rotem-Lehrer, N.; Laufer, Y. Effect of focus of attention on transfer of a postural control task following an ankle sprain. *J. Orthop. Sports Phys. Ther.* **2007**, *37*, 564–569. [CrossRef]
9. Lessa, H.T.; Chiviacowsky, S. Self-controlled practice benefits motor learning in older adults. *Hum. Mov. Sci.* **2015**, *40*, 372–380. [CrossRef]
10. Halperin, I.; Chapman, D.W.; Martin, D.T.; Abbiss, C. The effects of attentional focus instructions on punching velocity and impact forces among trained combat athletes. *J. Sports Sci.* **2017**, *35*, 500–507. [CrossRef]
11. Zentgraf, K.; Lorey, B.; Bischoff, M.; Zimmermann, K.; Stark, R.; Munzert, J. Neural correlates of attentional focusing during finger movements: A fMRI study. *J. Mot. Behav.* **2009**, *41*, 535–541. [CrossRef] [PubMed]
12. Van der Lubbe, R.H.; Bundt, C.; Abrahamse, E.L. Internal and external spatial attention examined with lateralized EEG power spectra. *Brain Res.* **2014**, *1583*, 179–192. [CrossRef] [PubMed]
13. Crews, D.J.; Landers, D.M. Electroencephalographic measures of attentional patterns prior to the golf putt. *Med. Sci. Sports Exerc.* **1993**, *25*, 116–126. [CrossRef] [PubMed]
14. Benedek, M.; Schickel, R.J.; Jauk, E.; Fink, A.; Neubauer, A.C. Alpha power increases in right parietal cortex reflects focused internal attention. *Neuropsychologia* **2014**, *56*, 393–400. [CrossRef] [PubMed]
15. Ray, W.J.; Cole, H.W. EEG alpha activity reflects attentional demands, and beta activity reflects emotional and cognitive processes. *Science* **1985**, *228*, 750–752. [CrossRef]
16. Zachry, T.; Wulf, G.; Mercer, J.; Bezodis, N. Increased movement accuracy and reduced EMG activity as the result of adopting an external focus of attention. *Brain Res. Bull.* **2005**, *67*, 304–309. [CrossRef]
17. Wulf, G.; Dufek, J.S.; Lozano, L.; Pettigrew, C. Increased jump height and reduced EMG activity with an external focus. *Hum. Mov. Sci.* **2010**, *29*, 440–448. [CrossRef]
18. McNeil, C.J.; Butler, J.E.; Taylor, J.L.; Gandevia, S.C. Testing the excitability of human motoneurons. *Front. Hum. Neurosci.* **2013**, *7*, 152–161. [CrossRef]
19. Rossini, P.M.; Rossi, S. Clinical applications of motor evoked potentials. *Electroencephalogr. Clin. Neurophysiol.* **1998**, *106*, 180–194. [CrossRef]
20. Kobayashi, M.; Pascual-Leone, A. Transcranial magnetic stimulation in neurology. *Lancet Neurol.* **2003**, *2*, 145–156. [CrossRef]

21. Kandel, E.R.; Schwartz, J.H.; Jessell, T.M.; Biochemistry, D.o.; Jessell, M.B.T.; Siegelbaum, S.; Hudspeth, A. *Principles of Neural Science*, 5th ed.; McGraw Hill Professional: New York, NY, USA, 2012; Volume 4, p. 1760.
22. Specterman, M.; Bhuiya, A.; Kuppuswamy, A.; Strutton, P.; Catley, M.; Davey, N. The effect of an energy drink containing glucose and caffeine on human corticospinal excitability. *Physiol. Behav.* **2005**, *83*, 723–728. [CrossRef] [PubMed]
23. Fadiga, L.; Buccino, G.; Craighero, L.; Fogassi, L.; Gallese, V.; Pavesi, G. Corticospinal excitability is specifically modulated by motor imagery: A magnetic stimulation study. *Neuropsychologia* **1998**, *37*, 147–158. [CrossRef]
24. Carroll, T.J.; Riek, S.; Carson, R.G. The sites of neural adaptation induced by resistance training in humans. *J. Physiol.* **2002**, *544*, 641–652. [CrossRef] [PubMed]
25. Falvo, M.J.; Sirevaag, E.J.; Rohrbaugh, J.W.; Earhart, G.M. Resistance training induces supraspinal adaptations: Evidence from movement-related cortical potentials. *Eur. J. Appl. Physiol.* **2010**, *109*, 923–933. [CrossRef] [PubMed]
26. Pearcey, G.E.; Power, K.E.; Button, D.C. Differences in supraspinal and spinal excitability during various force outputs of the biceps brachii in chronic-and non-resistance trained individuals. *PLoS ONE* **2014**, *9*, e98468. [CrossRef] [PubMed]
27. Philpott, D.T.; Pearcey, G.E.; Forman, D.; Power, K.E.; Button, D.C. Chronic resistance training enhances the spinal excitability of the biceps brachii in the non-dominant arm at moderate contraction intensities. *Neurosci. Lett.* **2015**, *585*, 12–16. [CrossRef]
28. Rossi, S.; Hallett, M.; Rossini, P.M.; Pascual-Leone, A.; Safety of TMS Consensus Group. Safety, ethical considerations, and application guidelines for the use of transcranial magnetic stimulation in clinical practice and research. *Clin. Neurophysiol.* **2009**, *120*, 2008–2039. [CrossRef]
29. Lahouti, B.; Lockyer, E.J.; Wiseman, S.; Power, K.E.; Button, D.C. Short-interval intracortical inhibition of the biceps brachii in chronic-resistance versus non-resistance-trained individuals. *Exp. Brain Res.* **2019**, *237*, 3023–3032. [CrossRef]
30. Lockyer, E.J.; Hosel, K.; Nippard, A.P.; Button, D.C.; Power, K.E. Corticospinal-Evoked Responses from the Biceps Brachii during Arm Cycling across Multiple Power Outputs. *Brain Sci.* **2019**, *9*, 205. [CrossRef]
31. Stefanelli, L.; Lockyer, E.J.; Collins, B.W.; Snow, N.J.; Crocker, J.; Kent, C.; Power, K.E.; Button, D.C. Delayed-Onset Muscle Soreness and Topical Analgesic Alter Corticospinal Excitability of the Biceps Brachii. *Med. Sci. Sports Exerc.* **2019**, *51*, 2344–2356. [CrossRef]
32. Taylor, J.L.; Todd, G.; Gandevia, S.C. Evidence for a supraspinal contribution to human muscle fatigue. *Clin. Exp. Pharmacol. Physiol.* **2006**, *33*, 400–405. [CrossRef]
33. Cadigan, E.W.; Collins, B.W.; Philpott, D.T.; Kippenhuck, G.; Brenton, M.; Button, D.C. Maximal voluntary activation of the elbow flexors is under predicted by transcranial magnetic stimulation compared to motor point stimulation prior to and following muscle fatigue. *Front. Physiol.* **2017**, *8*, 707–718. [CrossRef] [PubMed]
34. Levenez, M.; Garland, S.J.; Klass, M.; Duchateau, J. Cortical and spinal modulation of antagonist coactivation during a submaximal fatiguing contraction in humans. *J. Neurophysiol.* **2008**, *99*, 554–563. [CrossRef] [PubMed]
35. Calder, K.M.; Hall, L.-A.; Lester, S.M.; Inglis, J.G.; Gabriel, D.A. Reliability of the biceps brachii M-wave. *J. Neuroeng. Rehabil.* **2005**, *2*, 33–41. [CrossRef] [PubMed]
36. Gandevia, S.; Petersen, N.; Butler, J.; Taylor, J. Impaired response of human motoneurones to corticospinal stimulation after voluntary exercise. *J. Physiol.* **1999**, *521*, 749–759. [CrossRef]
37. Gamst, G.; Meyers, L.S.; Guarino, A.J. *Analysis of Variance Designs: A Conceptual and Computational Approach with SPSS and SAS*; Cambridge University Press: New York, NY, USA, 2008; p. 594.
38. Halaki, M.; Ginn, K.A. Normalization of EMG Signals: To Normalize or Not to Normalize and What to Normalize to? In *Computational Intelligence in Electromyography Analysis: A Perspective on Current Applications and Future Challenges*; InTech: Rijeka, Crotia, 2012; pp. 175–194.
39. Mukaka, M.M. A guide to appropriate use of correlation coefficient in medical research. *Malawi Med. J.* **2012**, *24*, 69–71. [PubMed]
40. Halperin, I.; Williams, K.J.; Martin, D.T.; Chapman, D.W. The effects of attentional focusing instructions on force production during the isometric midthigh pull. *J. Strength Cond. Res.* **2016**, *30*, 919–923. [CrossRef]
41. Ashbridge, E.; Walsh, V.; Cowey, A. Temporal aspects of visual search studied by transcranial magnetic stimulation. *Neuropsychologia* **1997**, *35*, 1121–1131. [CrossRef]

42. Smith, D.T.; Jackson, S.R.; Rorden, C. Repetitive transcranial magnetic stimulation over frontal eye fields disrupts visually cued auditory attention. *Brain Stimul.* **2009**, *2*, 81–87. [CrossRef]
43. Bolognini, N.; Ro, T. Transcranial magnetic stimulation: Disrupting neural activity to alter and assess brain function. *J. Neurosci.* **2010**, *30*, 9647–9650. [CrossRef]
44. Conte, A.; Gilio, F.; Iezzi, E.; Frasca, V.; Inghilleri, M.; Berardelli, A. Attention influences the excitability of cortical motor areas in healthy humans. *Exp. Brain Res.* **2007**, *182*, 109–117. [CrossRef] [PubMed]
45. Button, D.C.; Behm, D.G. The effect of stimulus anticipation on the interpolated twitch technique. *J. Sports Sci. Med.* **2008**, *7*, 520–524.
46. Lohse, K.R.; Sherwood, D.E.; Healy, A.F. Neuromuscular effects of shifting the focus of attention in a simple force production task. *J. Mot. Behav.* **2011**, *43*, 173–184. [CrossRef] [PubMed]
47. Hasegawa, Y.; Kasai, T.; Tsuji, T.; Yahagi, S. Further insight into the task-dependent excitability of motor evoked potentials in first dorsal interosseous muscle in humans. *Exp. Brain Res.* **2001**, *140*, 387–396. [CrossRef] [PubMed]

brain sciences

MDPI

Article

The Inhibitory Tendon-Evoked Reflex Is Increased in the Torque-Enhanced State Following Active Lengthening Compared to a Purely Isometric Contraction

Vincenzo S. Contento [1], Brian H. Dalton [2] and Geoffrey A. Power [1,*]

1 Department of Human Health and Nutritional Sciences, College of Biological Sciences, University of Guelph, Guelph, ON N1G 2W1, Canada; vcontent@uoguelph.ca

2 School of Health and Exercise Science, University of British Columbia, Kelowna, BC V1V 1V7, Canada; brian.dalton@ubc.ca

* Correspondence: gapower@uoguelph.ca

Received: 22 November 2019; Accepted: 18 December 2019; Published: 23 December 2019

Abstract: Residual torque enhancement (rTE) is a history-dependent property of muscle, which results in an increase in steady-state isometric torque production following an active lengthening contraction as compared to a purely isometric (ISO) contraction at the same muscle length and level of activation. Once thought to be only an intrinsic property of muscle, recent evidence during voluntary contractions indicates a neuromechanical coupling between motor neuron excitability and the contractile state of the muscle. However, the mechanism by which this occurs has yet to be elucidated. The purpose of this study was to investigate inhibition arising from tendon-mediated feedback (e.g., Golgi tendon organ; GTO) through tendon electrical stimulation (TStim) in the ISO and rTE states during activation-matching and torque-matching tasks. Fourteen male participants (22 ± 2 years) performed 10 activation-matching contractions at 40% of their maximum tibialis anterior electromyography amplitude (5 ISO/5 rTE) and 10 torque-matching contractions at 40% of their maximum dorsiflexion torque (5 ISO/5 rTE). During both tasks, 10 TStim were delivered during the isometric steady state of all contractions, and the resulting tendon-evoked inhibitory reflexes were averaged and analyzed. Reflex amplitude increased by ~23% in the rTE state compared to the ISO state for the activation-matching task, and no differences were detected for the torque-matching task. The current data indicate an important relationship between afferent feedback in the torque-enhanced state and voluntary control of submaximal contractions. The history-dependent properties of muscle is likely to alter motor neuron excitability through modifications in tension- or torque-mediated afferent feedback arising from the tendon.

Keywords: electromyography; history dependence of force; residual force enhancement; eccentric; golgi tendon organ; afferent

---

1. Introduction

Residual torque enhancement (rTE) is a history-dependent muscle property that results in an increase in steady-state isometric (ISO) torque production following an active lengthening contraction as compared to a purely isometric contraction at the same muscle length and level of activation [1–3]. The presence of rTE has been demonstrated in vitro, from the level of the sarcomere [4] to whole muscle preparations [5], and at the single muscle fiber level [6] to the whole muscle level, via electrical stimulation [7,8] and submaximal and maximal voluntary isometric contractions (MVCs) in humans [2,9,10]. Although numerous reports have focused on the basic mechanical mechanisms of rTE [5,11–14], the neural consequences of rTE have been largely underappreciated. Recent studies have reported that, during voluntary contractions, rTE may be linked to excitability modifications within the corticospinal pathway [9,15]. For example, rTE led to an increase in corticospinal excitability during plantar flexion MVCs [15] and a decrease in spinal excitability during submaximal steady-state dorsiflexion contractions [15]. However, it is not clear what factors may be modulating these changes in excitability.

Sypkes et al. [15] proposed that reduced spinal excitability during a condition of enhanced muscle force production capacity may be corresponding to greater inhibition of the agonist motor neuron pool arising from tendon-mediated feedback or, more specifically, the Golgi tendon organ (GTO) [15]. The GTOs are located in series with extrafusal muscle fibers at the interface of the musculotendinous junction (MTJ) and act to sense tension produced by its corresponding activated muscle [16–18]. Rising muscle tension increases the firing of the Ib afferent neurons that project from the GTO to a variety of targets within the central nervous system [17,19], including inhibitory interneurons synapsing onto the agonist motor neuron pool [20,21]. An effective mode to assess the efficacy of inhibitory pathways onto the motor neuron pool is using tendon electrical stimulation (TStim) to elicit a short latency reflex [22,23]. Applying TStim during times of increased muscle tension (i.e., rTE state) may elicit increased Ib inhibitory reflex parameters [22,24,25]. Recently, using this technique, we have shown that the tendon-evoked inhibitory reflex is reduced in the shortening-induced residual torque depressed state [23]. The opposite may be true for a condition of enhanced torque production (e.g., rTE). For a review on residual torque depression, please see Chen et al. [26].

Residual torque enhancement can be measured as the increase in torque following active lengthening in the isometric steady-state phase compared to purely ISO contractions during activation-matched tasks (i.e., electromyography), and it has been observed during both submaximal and maximal voluntary activation [27–29]. It is proposed that rTE occurs due to a stiffening and shortening of titin's free spring length in the presence of calcium and cross-bridge cycling [27], effectively increasing the contribution of passive force to total force following active lengthening. Therefore, while matching torque, the increase in passive tension associated with rTE [5,30] requires less muscle activation as compared with a purely ISO contraction [28,29]. This activation reduction is regularly measured through a decrease in electromyography (EMG) and has been observed in both submaximally and maximally activated contractions [3,7,28,31]. These two methods of evaluating rTE—activation matching and torque matching—provide a unique approach to investigate the role of tension-mediated afferents on previously observed reductions in spinal excitability.

The purpose of this study was to determine how the tendon-evoked inhibitory reflex is modified between the rTE and ISO states and to discuss its implications in the reduction of motor neuron pool excitability in the rTE state. During activation-matched ISO and rTE contractions, muscle tension in the isometric steady state is expected to be greater, whereas muscle tension is expected to be equivalent during torque-matched ISO and rTE contractions. If previously observed reductions in agonist motor neuron pool excitability during rTE contractions are in fact modulated from tendon-mediated feedback, we hypothesize that increases in TStim-elucidated reflex parameters (such as amplitude) should be apparent in rTE contractions during isometric steady state compared to purely ISO contractions for the activation-matching trials but not the torque-matching ones.

2. Materials and Methods

2.1. Participants

Fourteen healthy male participants with a mean age of 22 ± 2 years, height of 180 ± 6 cm, and mass of 81.8 ± 13.2 kg with no prior history of neuromuscular disease or ankle joint injuries were recruited from the university population. Participants gave written informed consent prior to testing. All procedures were approved by the Human Research Ethics Board of the University of Guelph (REB: 15NV008).

2.2. Experimental Set Up

A HUMAC NORM dynamometer (CSMi Medical Solutions, Stoughton, MA, USA) was used to record torque, angular velocity, and position. Each participant sat with their right hip and knee angles set at 110° and 140° (180° = full extension), respectively. Joint angles were measured using a goniometer. The right knee was immobilized just proximal to the patella with the dynamometer's leg restraint preventing hip flexion and a cushion positioned beneath the distal hamstrings preventing hip extension, while movement at the torso was restricted with a four-point seatbelt harness. The right foot was fixed to the dorsi/plantar flexor adaptor with one inelastic strap secured over the ankle and another across the mid-distal portion of the metatarsals. The dynamometer's maximum ankle dorsi- and plantar flexion angles were set to 0° and 40° plantar flexion (PF; 0° = neutral), respectively, allowing for 40° of ankle excursion.

Locations for the surface EMG electrodes (Ag/AgCl, 1.5 × 1 cm: Kendall, Mansfield, MA, USA) were prepared by shaving and cleaning the skin with alcohol swabs. The active electrode was placed over the tibialis anterior (TA) approximately 7 cm inferior and 2 cm lateral to the tibial tuberosity, and a reference electrode was placed inferiorly, adjacent to the active electrode in line with the muscle fibers. To record antagonist activity, the active electrode was placed on the soleus, along the midline of the leg approximately 2 cm inferior to the border of the heads of the gastrocnemii, and a reference electrode was placed inferiorly, bordering the active electrode. A single ground electrode was positioned over the patella.

Surface EMG, torque, angular velocity, dynamometer position, and stimulus trigger data were digitized using a 12-bit analog-to-digital converter (PowerLab System 16/35, ADInstruments, Bella Vista, Australia) and analyzed with Labchart software (Labchart, Pro Modules 2014, version 8). Torque angular velocity and position as well as EMG data were recorded at a sampling rate of 1000 and 2000 Hz, respectively. The EMG data were bandpass filtered using a digital filter (10–1000 Hz).

2.3. Peripheral Nerve Stimulation

To test the voluntary activation of the dorsiflexors (see next section) and obtain compound muscle action potentials (M-waves) from the TA and soleus, peripheral nerve stimulation was delivered transcutaneously with a standard clinical bar electrode (Empi, St. Paul, MN, USA) coated in conductive gel. The deep fibular nerve, innervating the dorsiflexor muscles, was located by palpating the head of the fibula and moving posteroinferiorly until the nerve was identified. Stimulation distal to the bifurcation of the common fibular nerve was ensured in order to limit activation of the peroneal muscles. The tibial nerve was stimulated via a bar electrode positioned within the popliteal fossa to maximize the M-wave. All stimulations were delivered as a single square-wave pulse from a constant-current, high-voltage stimulator (model DS7AH, Digitimer, Welwyn Garden City, Hertfordshire, UK). Voltage and pulse width were set to a maximum of 400 V and 200 μs, respectively. The current was increased incrementally until a plateau was reached for the peak-to-peak amplitude of the resting M-wave (Mmax). To ensure consistent activation of all motor neurons throughout the experiment, the current was then increased to a supramaximal level, equivalent to 110% of that required to generate Mmax (range: 20–200 mA and 25–250 mA for deep fibular and tibial nerves, respectively). An overview of all experimental procedures is provided in Figure 1A.

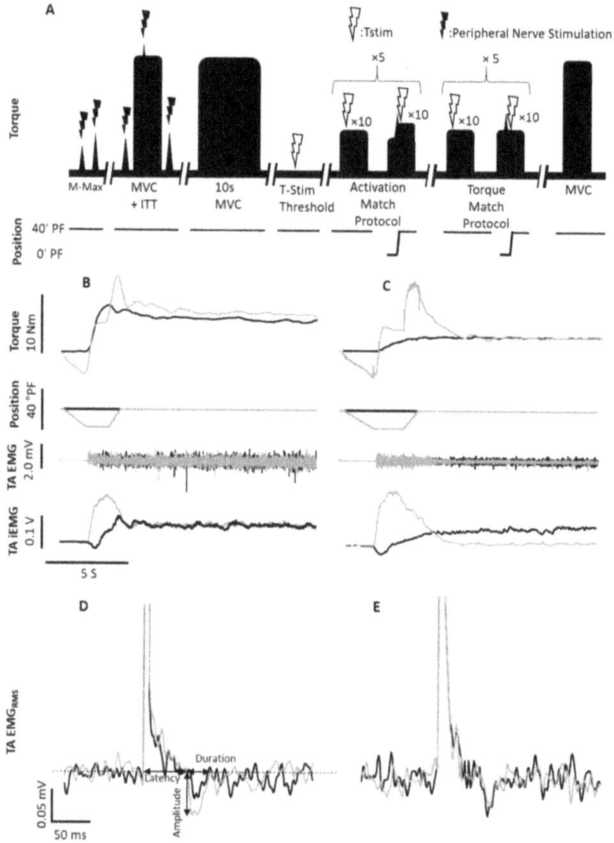

Figure 1. Timeline and raw data traces. Schematic timeline of experimental procedures (**A**). To establish maximum compound muscle action potentials (Mmax) for the tibialis anterior (TA) and soleus, peripheral nerve stimulation was performed on the deep branch of the common fibular and tibial nerves, respectively. A 10 s maximum voluntary contraction (MVC) was followed by an initial MVC performed with the interpolated twitch technique (ITT) to assess maximum torque and voluntary activation. Tendon electrical stimulation (TStim) was applied at increasing currents until perceptual threshold (PT) was found, which was defined as the minimum current intensity that induced a tingling or tapping sensation that was detectable to the participant. Stimulation intensity was then increased to 6 × PT, a current at which participants reported a muscular sensation, including a tugging or pulling at the muscular insertion, a deep tingling near the cathode, or a muscle twitch. If a visible muscle twitch was induced, stimulation intensity was reduced to the maximum current that failed to produce a visible muscle twitch. This stimulation intensity was used for all subsequent trials. Participants then performed 10 pairs of activation-matched dorsiflexion trials (**B**) consisting of an isometric (ISO) trial (black trace) followed by a residual torque-enhanced (rTE) trial (gray trace). These contractions were performed at 40 ± 5% of the participant's maximum TA activation, and 10 TStims were applied during the isometric steady state of each contraction. Participants then performed 10 pairs of torque-matched dorsiflexion trials (**C**) consisting of an ISO trial (black trace) followed by an rTE trial (gray trace). These contractions were performed at 40% of the participants maximum dorsiflexion torque, and 10 TStims were applied during the isometric steady state of each contraction to evoke an inhibitory reflex (activation-matched trial (**D**), torque-matched trial (**E**)). Lastly, participants were instructed to perform a final MVC to assess for fatigue. A resting period of 5 min was given to participants after all contractions, and each contraction was performed with visual feedback as well as verbal encouragement.

2.4. Maximum Voluntary Contraction and Voluntary Activation

Voluntary activation of the dorsiflexors was assessed during brief MVCs (~5 s) performed twice, separated by 4 min of rest both prior to and following the experimental trials. The interpolated twitch technique was used to evaluate voluntary activation [32]. The torque resulting from peripheral nerve stimulation delivered during the plateau phase of the MVC was compared to a resting twitch evoked 1–2 s after relaxation. The level of voluntary activation was calculated as follows: voluntary activation (%) = [1 −(interpolated twitch torque/resting twitch torque)] × 100%. Participants were encouraged verbally and provided visual feedback of torque output during all MVC attempts [33]. All participants were required to reach a minimum of 95% voluntary activation in order to be included in the study and were given five minutes of rest following the qualifying MVCs before continuing with the experiment.

2.5. Determining Submaximal Muscle Activation

To determine the submaximal integrated EMG (iEMG) and torque targets, participants were instructed to perform a 10 s dorsiflexion MVC at an ankle angle of 40° PF. The average iEMG and torque collected between 6 and 8 s was then used to determine the 40% submaximal iEMG and torque targets [23] (Figure 1B). For activation-matched contractions, a ± 5% window was calculated about the 40% iEMG target, and participants were instructed to maintain their iEMG amplitude within set guidelines marking this target window [23].

2.6. Tendon Electrical Stimulation

Percutaneous TStim was used to induce tendon-evoked inhibitory agonist reflexes. This technique involved percutaneous square-wave electrical stimulation of the tendon near the MTJ and evoked a short-latency (<50 ms) reflexive inhibition in the agonist muscle. This reflex has been demonstrated for several upper and lower limb muscles [22–25] and is thought to be mediated via Ib spinal pathways owing to the short latency and the polarity of the response, which is consistent with Ib autogenic inhibition [22,23]. However, contributions from other sources, such as muscle or tendon type III afferents, cannot be ruled out completely [25]. Still, non-GTO origins of TStim, specifically cutaneous receptors overlying the tendon or muscle stimulation via current spread, were excluded in prior studies [24]. Previous reports [22,34] have demonstrated that indwelling electrical stimulation of the tendon evokes an inhibitory response with similar characteristics as percutaneous stimulation, indicating that the percutaneous tendon-evoked reflex technique used here is most likely tension-mediated via GTO afferents. Ag/AgCl electrodes (1.5 × 1 cm: Kendall, Mansfield, MA, USA) were used for TStim in order to generate tendon-evoked inhibitory reflexes. The cathode was placed near the MTJ of the TA, and the anode was placed over the distal tendon at the level of the malleoli [23]. Single stimuli were presented with a constant-current, high-voltage stimulator (DS7AH). Voltage was set to a maximum of 400 V and pulse width to 200 μs. The stimulation protocol was initiated with the detection of perceptual threshold (PT), which was defined as the minimum current intensity that induced a tingling or tapping sensation that was detectable to the participant (6.63 ± 2.13 mA) [23]. Stimulation intensity was then increased to 6 × PT, a current at which participants have reported a muscular sensation, including a tugging or pulling at the muscular insertion, a deep tingling near the cathode, or a muscle twitch [9,23]. If a visible muscle twitch was induced, stimulation intensity was reduced to the maximum current that failed to produce a visible muscle twitch [23]. This stimulation intensity was used for all subsequent trials (5.2 ± 0.90 × PT).

2.7. Experimental Procedures

Each rTE trial was preceded by an ISO trial, and protocol A was followed by protocol B. Five rTE trials and five ISO trials were performed for each of the two protocols for a total of 20 contractions. Participants were provided visual feedback of the iEMG and torque amplitudes on a computer monitor and were verbally encouraged to match the target as closely as possible during all submaximal contractions. Four minutes of rest separated all submaximal contractions.

2.8. Protocol "A": Activation-Matching Condition

For each rTE trial, the protocol consisted of a 40% iEMG contraction involving a 2 s isometric phase at an ankle angle of 90°, a 1 s isokinetic lengthening phase (angular velocity: 40°/s) and ~20 s isometric phase at 40° PF (Figure 1B). A series of 10 TStim pulses were delivered at random 1–4 s intervals [9,23] during the isometric phase at 40° PF. During the ISO trials, an isometric dorsiflexion contraction corresponding to 40% iEMG was performed for ~23 s at an ankle angle of 40° PF, with a similar pattern of stimuli delivered as described for the rTE trials.

2.9. Protocol "B": Torque-Matching Condition

For each rTE and ISO trial, participants were instructed to maintain 40% MVC torque. The movement and stimulation protocols were identical to those of protocol A (Figure 1C).

2.10. Data Analysis and Statistics

Mean torque from each protocol contraction was calculated from 500 ms prior to the first stimulation to the end of each contraction [23]. Root mean squared EMG (EMG_{RMS}) amplitude was calculated in a 500 ms window that occurred between 6 and 8 s after contraction initiation following the achievement of an isometric steady state. It was ensured that the window selected was matched between corresponding rTE and ISO trials [23]. The EMG_{RMS} of the resting Mmax recorded at the TA and soleus was used to normalize the voluntary TA and soleus EMG, respectively. A paired *t*-test was performed to compare the torque and EMG data between rTE and ISO trials to validate the presence of rTE and activation reduction.

To obtain the reflex parameters, a stimulus-triggered average of the raw TA EMG_{RMS} was generated using Labchart software (Labchart, Pro Modules 2014, version 8). For each protocol, separate rTE and ISO averages were constructed from all stimuli delivered in each condition; therefore, each average was composed of 50 stimuli delivered over 5 contractions (Figure 1D,E). A paired *t*-test was performed to compare reflex characteristics between the rTE and ISO states, including latency, duration, amplitude, and change in average EMG_{RMS} from baseline, in order to characterize changes in the tendon-evoked inhibitory reflex in the rTE and AR state [23]. Baseline was measured as the average EMG_{RMS} in a 300–500 ms window occurring before TStim, and the onset of the stimulation artifact was defined as 0 s. Latency was measured as the time from the initiation of the stimulus artifact to the sharp decrease in baseline EMG_{RMS} occurring at the start of the reflex. Duration was measured as the time occurring from the initial decrease in EMG_{RMS} from baseline at the start of the reflex to when baseline EMG_{RMS} was once again reached. Amplitude was measured as the magnitude of EMG_{RMS} from baseline to the lowest trough in the reflex [23].

Change in average EMG_{RMS} from baseline was measured as the difference between baseline EMG_{RMS} and the average EMG_{RMS} from the duration of the reflex. Paired *t*-tests were also used to detect any differences in the torque produced during MVCs performed before and after the experiment in order to assess any effects of fatigue during the experimental protocol. Descriptive data found in text are reported as means ± standard deviation, while data presented in figures are reported as means ± standard error of the mean. Significance was determined based on $\alpha < 0.05$.

3. Results

3.1. Maximum Voluntary Contraction and Voluntary Activation

The mean pretrial MVC torque was 27.6 ± 6.4 Nm, and all participants were capable of achieving near-maximal values for voluntary activation (98.9 ± 1.5%). Following the 20 submaximal contractions, MVC torque was not different from the pretrial values (27.1 ± 6.6 Nm).

3.2. *Dorsiflexion Torque and Muscle Activity*

3.2.1. Activation Matching

Normalized EMG_{RMS} of both TA (p = 0.8) and soleus (p = 0.6) were not different between rTE and ISO contractions (Figure 2B,C). Following active lengthening, steady-state isometric torque was 15.0 ± 9.7% (p < 0.0001) greater than that produced during the purely isometric contractions at the corresponding muscle length and level of activation (Figure 2A). Participants successfully maintained the EMG target level such that iEMG of the TA did not differ in the rTE and ISO contractions (p = 0.1; Figure 1B). This indicates indirectly that motor neuron output was similar in both the rTE and ISO states. For the tendon-evoked inhibitory reflex, reflex latency was not significantly different between the rTE and ISO states, with the onset of inhibition occurring at 47.8 ± 7.8 ms following TStim in rTE trials and 48.6 ± 4.8 ms following TStim in ISO trials (p = 0.6; Figure 2E). Further, reflex duration (p = 0.5; Figure 2F) and reduction in average reflex EMG_{RMS} from baseline were not different when rTE and ISO contractions were compared (p = 0.2). However, inhibitory reflex amplitude differed by 22.6 ± 41.8% (p < 0.05; Figure 2D) in the rTE state compared to the ISO state.

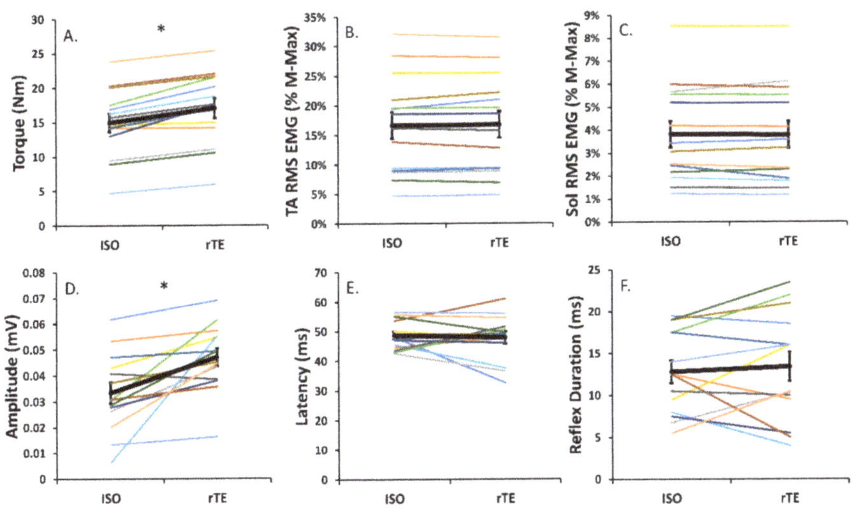

Figure 2. Activation-matching trial. Mean values for each participant across measures (colored lines) and the group mean (black line; error bars indicate standard error of the mean) in the rTE and ISO states. For the activation-matching trial, in the rTE state as compared to the ISO state, there was a 15.0% increase in torque (**A**) and a 22.6% increase in reflex amplitude (**D**) (* p < 0.05). There was no significant difference in EMG_{RMS} collected from the tibialis anterior (**B**) or soleus (**C**) reflex duration (**F**) or reflex latency (**E**) between the two states.

3.2.2. Torque Matching

Steady-state isometric torque was not different following active lengthening when the rTE and ISO states were compared (p = 0.2 Figure 3A). In the rTE state, however, there was a significant 14.4 ± 13.3% decrease in normalized EMG_{RMS} (p < 0.01; Figure 3B) and a 26.9 ± 23.7% decrease in iEMG (p < 0.01) for the TA compared to ISO contractions, with no change in antagonist coactivation (Figure 3C). For the tendon-evoked inhibitory reflex, reflex latency was not significantly different between the rTE and ISO states, with the onset of inhibition occurring at 48.0 ± 6.8 ms following TStim in rTE trials and 50.6 ± 5.6 ms following TStim in ISO trials (p = 0.1; Figure 3E). Further, reflex duration (p = 0.2; Figure 3F), reflex amplitude (p = 0.2; Figure 3D), and reduction in average reflex EMG_{RMS} from baseline (p = 0.2) were not different when rTE and ISO contractions were compared.

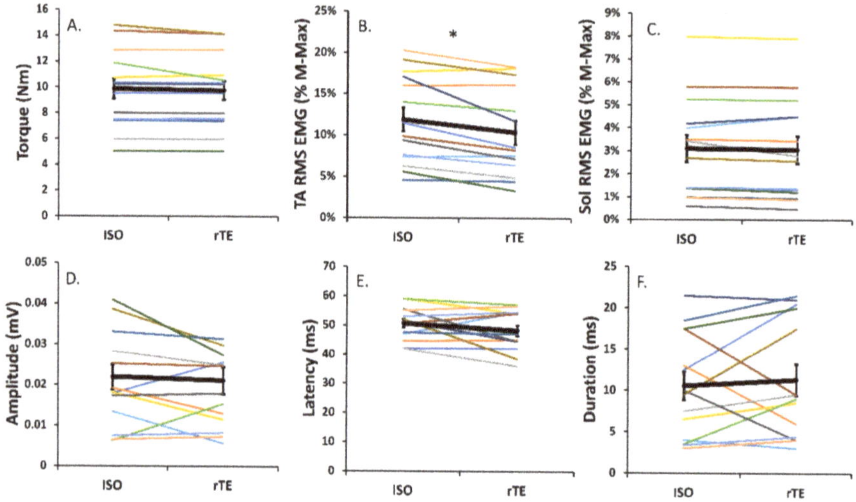

Figure 3. Torque-matching trial. Mean values for each participant across measures (colored lines) and the group mean (black line; error bars indicate standard error of the mean) in rTE and ISO states. For the torque-matching trial, in the rTE state as compared to the ISO state, as expected, there was a 14.4% decrease in TA activation (**B**) (*$p < 0.05$). There were no differences in torque (**A**), antagonist coactivation (**C**), or any reflex parameter (**D–F**) between the two states.

4. Discussion

The purpose of this study was to determine how the tendon-mediated inhibitory reflex is modified between the rTE and ISO states through TStim of the tibialis anterior during submaximal activation-matched and torque-matched dorsiflexions. The activation-matched task successfully elicited an ~15% increase in torque in the rTE state compared to ISO, while the torque-matching task resulted in an ~14% decrease in EMG. Our hypothesis of an increased tendon-mediated inhibitory reflex in the rTE state was supported by a 23% increase in inhibitory reflex magnitude in the rTE state compared to the ISO state during the activation-matching task but not the torque-matching task. Therefore, these results support an underlying tension-mediated factor as a plausible explanation for the previously reported decrease in agonist motor neuron pool excitability in the rTE state [15].

In the force-enhanced isomeric steady state, there is greater relative contribution of passive force to total force production, possibly owing to stiffening of the giant molecular spring titin [27]. This increase in passive tension can manifest in rTE contractions in two ways when compared to purely ISO contractions. The first is an activation reduction in torque-matching tasks in which no change in muscle tension occurs [3,29]. The second is an increase in torque, and consequently muscle tension, during activation matching [29,35]. Critical to the present study was the use of both paradigms: activation matching and torque matching. Previous reports of decreases in agonist motor neuron pool [15] and increases in corticospinal [9] excitability have been observed for submaximal and maximal rTE contractions compared to purely ISO contractions. Given the aforementioned investigations [9,15] had similar motor neuron outputs for both ISO and rTE contractions, the alteration in spinal excitability was most likely owing to peripheral sensory inputs. One source of peripheral input upon the motor neuron pool are muscle spindles. While activating their respective Ia afferents via Achilles tendon vibration, no modulatory effect on rTE was observed [35]. Therefore, a tension-mediated factor was speculated as the most likely factor driving the previously reported results [15]. Golgi tendon organs increase Ib afferent firing during periods of increased muscle tension [16] and excite inhibitory interneurons within the spinal cord [17,19]. Excitation of these inhibitory interneurons results in decreased neural output of the agonist motor neuron pool, which can be observed within the surface

EMG signal [24]. Through single pulse electrical stimulation of the TA tendon, an inhibitory reflex can be observed in the EMG_{RMS} ~50 ms following stimulation [22,24,25]. Following active lengthening, during an activation-matching task, rTE torque increased by ~15% compared to ISO, which resulted in an ~23% increase in tendon-evoked inhibitory reflex magnitude. The torque-matching task served as a control, i.e., with no changes in torque or muscle tension between rTE and ISO states, no changes in tendon-evoked reflexes should occur. This is particularly important because it ensures any changes observed in the activation-matching trials are indeed due to GTO-mediated reflexes. While there was an ~14% reduction in TA activation (normalized EMG_{RMS}), indicating the presence of rTE, negligible differences in muscle tension were detected, and as expected, there was no significant differences in the tendon-evoked inhibitory reflex across the rTE and ISO tasks. The lack of changes in reflex characteristics during the torque-matching task indicates our findings were due to a tension-mediated factor. Therefore, the tension-dependent GTO and the Ib afferent is most likely a key contributor in modulating agonist motor neuron excitability during voluntary control of submaximal contractions in the rTE state.

The results of our investigation are similar to previous reports on rTE and activation reduction. While we found ~15% increase in torque across all 40% activation-matched rTE trials in comparison to ISO, Pinniger and Cresswell [36] found a 12% increase in torque at 25% of maximally activated TA dorsiflexion. The 40% MVC torque-matching task induced an ~14% reduction in activation of the tibialis anterior EMG_{RMS}. Other studies reported a similar activation reduction ranging from 5% to 20% [28,29,31]. Additionally, the aim of this study was to elucidate a measurable change in tendon-evoked reflex parameters between ISO and rTE states, where we hypothesized an increase in inhibitory reflex parameters in the rTE state compared to the ISO during the activation-matching protocol but not the torque-matching one. In support of this hypothesis, we found an ~23% increase in inhibitory reflex magnitude in the rTE state compared to the ISO state during the activation-matching task but not the torque-matching one. Under similar premises—decreased torque production in the torque-depressed state eliciting a decrease in tendon-evoked reflex parameters—Sypkes et al. [23] found a 16% reduction in reflex magnitude. The symmetry in the findings of Sypkes et al. [23] and the present study demonstrate a clear relationship between these history-dependent properties of force and the GTO mechanoreceptor. The resulting neural reflex pathway consisting of Ib afferent neurons and spinal inhibitory interneurons provide a plausible explanation to decreased motor neuron pool excitability reported previously [15].

A limitation to the present study was the lack of randomization for the order of protocols A and B. This had the potential to change perceptual threshold and reduce stimulus efficacy throughout the testing session. However, when protocol A ISO was compared to protocol B ISO, there were no differences in reflex parameters (Reflex Latency, Reflex Duration, Reflex Magnitude; Change in TA EMG_{RMS} from Baseline). Thus, the lack of randomization does not appear to influence the results.

5. Conclusions

Residual torque enhancement, a history-dependent property of muscle, was present during submaximal activation- and torque-matching tasks. In the isometric steady state following an active lengthening contraction, there was an increase in tendon-evoked inhibitory reflex magnitude compared to purely isometric contractions for the activation-matching task but not the torque-matching task. This observation likely characterizes a tension-dependent increased contribution of tendon-mediated inhibitory feedback on the agonist motor neuron pool and as such may explain—at least partially—the documented decrease in agonist motor neuron excitability during an rTE state [15]. This study provides novel insight into the peripheral contributions of the history dependence of force and bridges the gap between the central nervous system and a property that was once thought to be purely intrinsic to muscle.

Author Contributions: Conceptualization, V.S.C., B.H.D., and G.A.P.; data curation, V.S.C.; formal analysis, V.S.C. and G.A.P.; funding acquisition, G.A.P.; investigation, V.S.C. and G.A.P.; methodology, V.S.C., B.H.D., and G.A.P.;

supervision, B.H.D. and G.A.P.; writing—original draft, V.S.C., B.H.D., and G.A.P.; writing—review & editing, V.S.C., B.H.D., and G.A.P. All authors have read and agreed to the published version of the manuscript.

Funding: This research was funded by the Natural Sciences and Engineering Research Council of Canada (NSERC).

Acknowledgments: We would like to thank all of the participants in this study.

Conflicts of Interest: The authors declare no conflict of interest.

References

1. Abbott, B.C.; Aubert, X.M. The force exerted by active striated muscle during and after change of length. *J. Physiol.* **1952**, *117*, 77. [PubMed]
2. Power, G.A.; Rice, C.L.; Vandervoort, A.A. Residual force enhancement following eccentric induced muscle damage. *J. Biomech.* **2012**, *45*, 1835. [CrossRef] [PubMed]
3. Seiberl, W.; Power, G.A.; Hahn, D. Residual force enhancement in humans: Current evidence and unresolved issues. *J. Electromyogr. Kinesiol.* **2015**, *25*, 571–580. [CrossRef] [PubMed]
4. Joumaa, V.; Fitzowich, A.; Herzog, W. Energy cost of isometric force production after active shortening in skinned muscle fibres. *J. Exp. Biol.* **2017**, *220*, 1509. [CrossRef]
5. Herzog, W.; Schappacher, G.; DuVall, M.; Leonard, T.R.; Herzog, G.A. Residual Force Enhancement Following Eccentric Contractions: A New Mechanism Involving Titin. *Am. Physiol. Soc.* **2016**, *31*, 300–312. [CrossRef]
6. Pinnell, R.A.M.; Mashouri, P.; Mazara, N.; Weersink, E.; Brown, S.H.M.; Power, G.A. Residual force enhancement and force depression in human single muscle fibres. *J. Biomech.* **2019**, *91*, 164–169. [CrossRef]
7. Lee, H.D.; Herzog, W. Force enhancement following muscle stretch of electrically stimulated and voluntarily activated human adductor pollicis. *J. Physiol.* **2002**, *545*, 321. [CrossRef]
8. Seiberl, W.; Power, G.A.; Herzog, W.; Hahn, D. The stretch-shortening cycle (SSC) revisited: Residual force enhancement contributes to increased performance during fast SSCs of human m. adductor pollicis. *Physiol. Rep.* **2015**, *3*, e12401. [CrossRef]
9. Hahn, D.; Hoffman, B.W.; Carroll, T.J.; Cresswell, A.G. Cortical and spinal excitability during and after lengthening contractions of the human plantar flexor muscles performed with maximal voluntary effort. *PLoS ONE* **2012**, *7*, e49907. [CrossRef]
10. Chapman, N.; Whitting, J.; Broadbent, S.; Crowley-McHattan, Z.; Meir, R. Residual force enhancement in humans: A systematic review. *J. Appl. Biomech.* **2018**, *34*, 240. [CrossRef]
11. Lee, E.J.; Joumaa, V.; Herzog, W. New insights into the passive force enhancement in skeletal muscles. *J. Biomech.* **2007**, *40*, 719. [CrossRef] [PubMed]
12. Joumaa, V.; Rassier, D.E.; Leonard, T.R.; Herzog, W. The origin of passive force enhancement in skeletal muscle. *Am. J. Physiol. Cell Physiol.* **2008**, *294*, C74. [CrossRef] [PubMed]
13. Koppes, R.A.; Herzog, W.; Corr, D.T. Force enhancement in lengthening contractions of cat soleus muscle in situ: Transient and steady-state aspects. *Physiol. Repor.* **2013**, *1*. [CrossRef] [PubMed]
14. Power, G.A.; Makrakos, D.P.; Rice, C.L.; Vandervoot, A.A. Enhanced force production in old age is not a far stretch: An investigation of residual force enhancement and muscle architecture. *Physiol. Rep.* **2013**, *1*. [CrossRef] [PubMed]
15. Sypkes, C.T.; Kozlowski, B.J.; Grant, J.; Bent, L.R.; McNeil, C.J.; Power, G.A. The influence of residual force enhancement on spinal and supraspinal excitability. *PeerJ.* **2018**, *6*. [CrossRef] [PubMed]
16. Houk, J.; Henneman, E. Responses of Golgi tendon organs to active contractions of the soleus muscle of the cat. *J. Neurophysiol.* **1967**, *30*, 466–481. [CrossRef]
17. Jami, L. Golgi tendon organs in mammalian skeletal muscle: Functional properties and central actions. *Physiol. Rev.* **1992**, *72*, 623–666. [CrossRef]
18. Pearson, K.G.; Gordon, J. Spinal reflexes. In *Principles of Neural Science*, 4th ed.; Kandel, E.R., Schwartz, J.H., Jessell, T.M., Eds.; McGraw-Hill: New York, NY, USA, 2000; pp. 713–736.
19. Jankowska, E. Interneuronal relay in spinal pathways from proprioceptors. *Prog. Neurobiol.* **1992**, *38*, 335–378. [CrossRef]
20. Conway, B.A.; Hultborn, H.; Kiehn, O. Proprioceptive input resets central locomotor rhythm in the spinal cat. *Exper. Brain Res.* **1987**, *68*, 643. [CrossRef]

21. McCrea, D.A.; Shefchyk, S.J.; Stephens, M.J.; Pearson, K.G. Disynaptic group I excitation of synergist ankle extensor motoneurones during fictive locomotion in the cat. *J. Physiol.* **1995**, *487*, 527. [CrossRef]
22. Horslen, B.C.; Inglis, J.T.; Blouin, J.S.; Carpenter, M.G. Both standing and postural threat decrease Achilles' tendon reflex inhibition from tendon electrical stimulation. *J. Physiol.* **2017**, *595*, 4493–4506. [CrossRef] [PubMed]
23. Sypkes, C.T.; Dalton, B.H.; Stuart, J.; Power, G.A. Inhibitory tendon-evoked reflex is attenuated in the torque depressed isometric steady-state following active lengthening. *Appl. Physiol. Nutr. Metab.* **2019**, in press.
24. Burne, J.A.; Lippold, O.C. Reflex inhibition following electrical stimulation over muscle tendons in man. *Brain.* **1996**, *119*, 1107–1114. [CrossRef] [PubMed]
25. Chen, J.; Hahn, D.; Power, G.A. Shortening-induced residual force depression in humans. *J. Appl. Physiol.* **2019**, *126*, 1066–1073. [CrossRef] [PubMed]
26. Jones, A.A.; Power, G.A.; Herzog, W. History dependence of the electromyogram: Implications for isometric steady-state EMG parameters following a lengthening or shortening contraction. *J. Electromyogr. Kinesiol.* **2016**, *27*, 30–38. [CrossRef] [PubMed]
27. Priori, A.; Berardelli, A.; Inghilleri, M.; Pedace, F.; Giovannelli, M.; Manfredi, M. Electrical stimulation over muscle tendons in humans. Evidence favouring presynaptic inhibition of Ia fibres due to the activation of group III tendon afferents. *Brain* **1998**, *121*, 373–380. [CrossRef]
28. Oskouei, A.E.; Herzog, W. Observations on force enhancement in submaximal voluntary contractions of human adductor pollicis muscle. *J. App. Physiol.* **2005**, *98*, 2087. [CrossRef]
29. Paquin, J.; Power, G.A. History dependence of the EMG-torque relationship. *J. Electromyogr. Kinesiol.* **2018**, *41*, 109–115. [CrossRef]
30. Freundt, J.K.; Linke, W.A. Titin as a force-generating protein under regulatory control. *J. Appl. Physiol.* **2019**, *126*, 1474–1482. [CrossRef]
31. Herzog, W. The multiple roles of titin in muscle contraction and force production. *Biophys. Rev.* **2018**, *10*, 1187–1199. [CrossRef]
32. Belanger, A.Y.; McComas, A.J. Extent of motor unit activation during effort. *J. Appl. Physiol.* **1981**, *51*, 1131. [CrossRef]
33. Gandevia, S.C. Spinal and supraspinal factors in human muscle fatigue. *Physiol. Rev.* **2001**, *81*, 1725–1789. [CrossRef] [PubMed]
34. Rogasch, N.C.; Burne, J.A.; Türker, K.S. Comparison of the inhibitory response to tendon and cutaneous afferent stimulation in the human lower limb. *J. Neurophysiol.* **2012**, *107*, 564–572. [CrossRef] [PubMed]
35. Dalton, B.H.; Contento, V.C.; Power, G.A. Residual force enhancement during submaximal and maximal effort contractions of the plantar flexors across knee angle. *J. Biomech.* **2018**, *78*, 70–76. [CrossRef] [PubMed]
36. Pinniger, G.J.; Cresswell, A.G. Residual force enhancement after lengthening is present during submaximal plantar flexion and dorsiflexion actions in humans. *J. Appl. Physiol.* **2007**, *102*, 18–25. [CrossRef] [PubMed]

brain sciences

MDPI

Article

The Task at Hand: Fatigue-Associated Changes in Cortical Excitability During Writing

Kezia T. M. Cinelli, Lara A. Green and Jayne M. Kalmar *

Department of Kinesiology, Wilfrid Laurier University, Waterloo, ON N2L 3C5, Canada; cine2150@mylaurier.ca (K.T.M.C.); lgreen@wlu.ca (L.A.G.)

* Correspondence: jkalmar@wlu.ca

Received: 11 October 2019; Accepted: 29 November 2019; Published: 2 December 2019

Abstract: Measures of corticospinal excitability (CSE) made via transcranial magnetic stimulation (TMS) depend on the task performed during stimulation. Our purpose was to determine whether fatigue-induced changes in CSE made during a conventional laboratory task (isometric finger abduction) reflect the changes measured during a natural motor task (writing). We assessed single-and paired-pulse motor evoked potentials (MEPs) recorded from the first dorsal interosseous (FDI) of 19 participants before and after a fatigue protocol (submaximal isometric contractions) on two randomized days. The fatigue protocol was identical on the two days, but the tasks used to assess CSE before and after fatigue differed. Specifically, MEPs were evoked during a writing task on one day and during isometric finger abduction to a low-level target that matched muscle activation during writing on the other day. There was greater variability in MEP amplitude ($F(1,18) = 13.55$, $p < 0.01$) during writing compared to abduction. When participants were divided into groups according to writing style (printers, $n = 8$; cursive writers, $n = 8$), a task x fatigue x style interaction was revealed for intracortical facilitation ($F(1,14) = 9.90$, $p < 0.01$), which increased by 28% after fatigue in printers but did not change in cursive writers nor during the abduction task. This study is the first to assess CSE during hand-writing. Our finding that fatigue-induced changes in intracortical facilitation depend on the motor task used during TMS, highlights the need to consider the task-dependent nature of CSE when applying results to movement outside of the laboratory.

Keywords: corticospinal excitability; task-dependent; transcranial magnetic stimulation; fatigue; writing

---

1. Introduction

Single-pulse transcranial magnetic stimulation (TMS) elicits motor evoked potentials (MEPs) that are recorded from the muscle of interest using surface electromyography (EMG). MEPs are used as a measure of corticospinal excitability (CSE) that reflect the excitability of the pathway from the site of stimulation to the site of recording, such that both cortical and spinal mechanisms contribute to changes in the MEP evoked using single-pulse TMS. Intracortical mechanisms that may contribute to CSE are assessed using conditioned MEPs elicited via paired-pulse TMS. For example, paired-pulses with a brief interpulse interval (1–5 msec) provide a measure of short interval intracortical inhibition (SICI), whereas longer interpulse intervals (10–15 msec) provide a measure of intracortical facilitation (ICF) [1]. One challenge associated with using single- and paired-pulse TMS to assess CSE is the variability in MEP amplitude, both within and between participants. To minimize variability, most TMS protocols are conducted with the muscle at rest or during submaximal isometric muscle contractions to standardize levels of muscle activation and control for joint position, muscle length, movement, and other factors. Although these laboratory tasks are intended to minimize variability, they do not reflect activities of daily living to which the results may be extrapolated. This is problematic because

CSE depends on the net excitability of the targeted brain region and net excitatory and inhibitory input to spinal motor neurons at the time of stimulation, all of which vary with different motor tasks and states [2–4]. Thus, the conclusions drawn from TMS studies depend on the motor task employed during stimulation and therefore may not translate to movement outside of laboratory settings.

Muscles of the hand (e.g., first dorsal interosseous (FDI) and adductor pollicis (AP)) are frequently used in studies of CSE because of the ease with which MEPs are elicited in distal muscles of the upper extremity. In such studies, MEPs are typically evoked with the hand at rest or during submaximal isometric contractions. For example, FDI MEPs are often assessed with the hand pronated with the wrist and hand secured to allow movement only at the metacarpal phalangeal joint of the index finger. Outside the lab, however, we use intrinsic muscles of the hands in a variety of positions, including power and precision grips. Measures of CSE differ between these grips. For example, FDI MEP amplitude is greater during conventional abduction tasks compared to power, pincer, or grasping tasks [5–7], whereas APB MEPs are greater during pincer tasks compared to during power tasks or at rest [8]. More direct measures made in the monkey reveal that corticospinal neurons are more active during a precision grip compared to a power grip, despite increased EMG activity in the latter [9]. This suggests that CSE is not simply related to the force of contraction, but also to the specific task. Pearce and Kidgell (2010) demonstrate that CSE is modulated when the precision required for a given task is increased [10] Furthermore, more complex tasks will require the involvement of proximal muscles for stability during finger movements. Flament and colleagues [11] compared index finger abduction to simple manual tasks including precision grip, power grip, grasping of a petri dish, and rotation of a bottle cap. Compared to isolated finger abduction, which was restricted to FDI use, all other tasks required activation of at least one additional muscle, which was speculated to contribute to the increased CSE found during the complex tasks [11].

Conventional laboratory tasks, such as matching force output to a static target displayed on a computer monitor, require very different cognitive demands compared to more natural motor tasks such as hand writing, use of a keyboard, or object manipulation. Nonetheless, there is a tendency to attribute the TMS results solely to the changes within or downstream to the primary motor cortex, and to overlook the influence of upstream cognitive processes transmitted to the primary motor cortex [12]. For example, internally guided movements, such as writing or drawing, have been shown to generate greater activation in the pre-supplementary motor area and dorsal premotor cortex as compared to externally guided movements, such as tracing [13]. Using the hand to convey language introduces additional cognitive influences on the motor system. In fact, mere observation of letters and words can alter CSE measured in the FDI when at rest [14,15]. Interestingly, this effect is specific to handwritten text, including handwritten "non-words", and is not observed with typed text [14,15], indicating that the recognition of hand-writing represents a unique cognitive demand that is conveyed to the motor system, likely via the mirror neuron system [14]. Accordingly, it is unlikely that estimates of CSE obtained during simple, isometric laboratory tasks would translate directly to CSE during more complex tasks outside the lab. In this study, we compared a hand-writing task, which represents an internally generated and complex task that is familiar and relevant outside the laboratory, to a conventional externally guided, isometric force-matching task often used in TMS research.

We wanted to determine whether changes in CSE in response to a well-studied intervention (neuromuscular fatigue) would depend on the motor task that is used to produce background muscle activity during the delivery of TMS. TMS has been used for many years to understand the role of the central nervous system in neuromuscular fatigue (for review, see [16,17]). Such studies have revealed that CSE is briefly facilitated and then undergoes a more prolonged period of depression following fatigue when MEPs are evoked in resting muscles [18–21]. Post-exercise depression is also observed after a fatigue protocol when MEPs are evoked at rest just prior to a voluntary contraction [22]. On the other hand, MEPs recover much more quickly when assessed during a strong muscle contraction after fatigue [23,24], suggesting that reductions in CSE after fatigue are overcome with sufficient voluntary drive. Given that measures of CSE depend on limb position, muscle activation level, and other aspects of the motor task, such as different cognitive demands, we speculate that fatigue-induced changes in CSE made during a conventional laboratory task may not represent fatigue-induced changes in CSE made during the more relevant and complex task of writing. Therefore, the purpose of this study is to elicit neuromuscular fatigue and then compare fatigue-associated changes in CSE during writing to the same measures made during a conventional isometric finger abduction task. We hypothesized that CSE would be greater during the writing task compared to the simple isometric abduction task. Furthermore, we hypothesized that the effect of fatigue on CSE would depend on the motor task employed during stimulation, even when the fatigue task itself was the same.

2. Materials and Methods

2.1. Participants

Twenty participants with a mean age of 22.6 ± 1.1 were recruited from the Wilfrid Laurier University student population. This sample size was based on a power calculation made using data from an earlier study of intracortical mechanisms of fatigue from our lab [22]. All participants were right-handed as determined by the Annett Handedness Questionnaire [25]. One participant was excluded from analysis due to an inability to complete the full testing protocol, leaving 19 participants (six male). Exclusion criteria included any neurological conditions, and orthopaedic conditions or pain of the hand, wrist, or arm. The study was approved by the Wilfrid Laurier University Research Ethics Board (REB #5381), and all participants provided written informed consent prior to participating.

2.2. Experimental Design

In this repeated-measures study design, each participant attended two testing sessions occurring no less than 48 h apart and was tested at the same time on both days to decrease between session variability [26]. On one day, participants completed a writing task and on the other day, participants completed an abduction task. The order of the two days was randomized and counterbalanced between participants. On each day participants began with a short familiarization to each task. Following familiarization, participants completed 90 pre-fatigue trials of the selected task (writing or abduction), a fatigue protocol of intermittent isometric abduction contractions, and finally 90 post-fatigue trials of the same task (writing or abduction) (Figure 1).

On each testing day participants completed a familiarization protocol including the writing task, the abduction task, and maximal voluntary contractions (MVCs). The writing task included 20 trials where the participant repetitively wrote the word "name" on a iPad tablet with a Adonit Pro3 precision disc stylus pen. The word "name" was specifically selected for the writing task because it is familiar, it is a short enough word to be written within the 5-s recording frame, and because the letters n, a, m, and e do not contain ascenders or descenders that would require greater finger movement. The tablet screen presented a blank 7 × 2 cm rectangle on a series of Powerpoint slides that were refreshed every 5 s. The rectangle served as the boundaries within which the participant was asked to write the word. These writing boundaries and the refreshing of the writing "page" on the tablet before each trial, allowed the participants to maintain a constant hand position and to minimize wrist deviations between trials. The participants were allowed to place the tablet in any position and orientation on the table in front of them. This position was marked and maintained between trials, blocks, and testing days. Participants were instructed to maintain the same self-selected grip of the pen (e.g., dynamic tripod, lateral tripod, quadrupod, etc.) for the duration of the protocol. In the abduction task, TMS was applied during isometric abduction of the right index finger against a force transducer for 3 s with a 2 s rest. The target level of contraction during the abduction task was set to the average RMS amplitude of FDI EMG activity over the 20 familiarization writing trials so that the contraction intensity was matched to activity required during writing for each participant (Figure 2). This level of activity was marked with horizontal cursors (±2.5%) to allow the participant to match it using a real-time smoothed and rectified EMG biofeedback channel. The location of the force transducer was adjusted to be in line with the self-selected graphic tablet location, such that shoulder and elbow angle were maintained between the writing and abduction tasks (Figure 3). The hand was in a pronated position for abduction with the third finger secured by a strap and wooden dowels placed on either side of the wrist. Participants then performed three isometric MVCs from which the highest was taken as maximal finger abduction force.

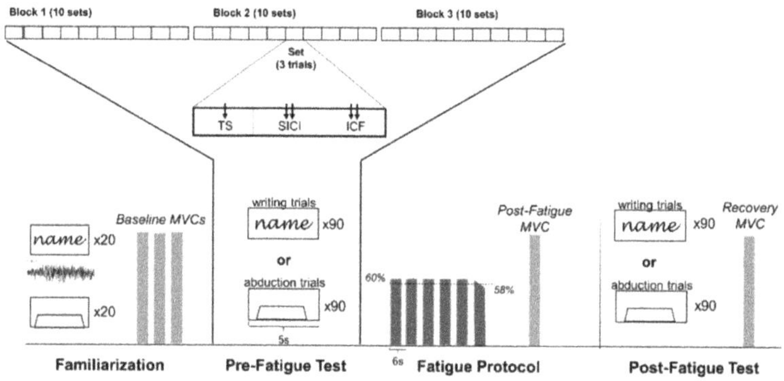

Figure 1. Each day began with a familiarization period that included 20 repetitions of each task to record muscle activation during writing, familiarizing participants with the force-tracing task, and an assessment of baseline maximal voluntary contraction (MVC). Participants then completed the Pre-Fatigue Test, which included 90 trials (abduction or writing task), followed by a Fatigue Protocol that was comprised of submaximal (60% MVC target) abduction contractions until failure (force fell below 58% MVC force for >3 s during attempts to hold the 60% target), followed by a post-fatigue MVC. The protocol ended with the Post-Fatigue Test that was identical to the Pre-Fatigue Test followed by a recovery MVC. The 90 pre-fatigue and post-fatigue trials were separated into 3 blocks of 10 sets each. Each set contained 3 randomized trials: (1) test MEP (TS), (2) short-interval intracortical inhibition (SICI) paired-pulse, and (3) intracortical facilitation (ICF) paired-pulse stimulation, resulting in pseudo-randomization of TS, SICI, and ICF across each block.

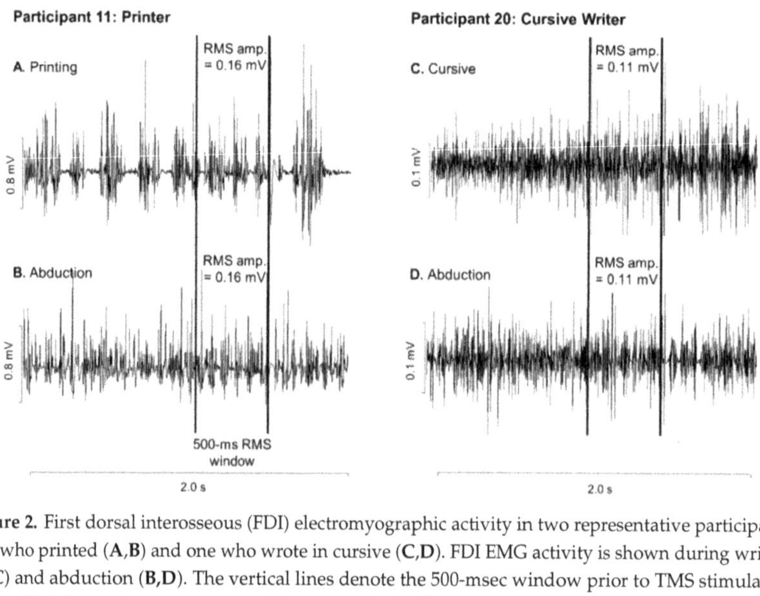

Figure 2. First dorsal interosseous (FDI) electromyographic activity in two representative participants: one who printed (**A**,**B**) and one who wrote in cursive (**C**,**D**). FDI EMG activity is shown during writing (**A**,**C**) and abduction (**B**,**D**). The vertical lines denote the 500-msec window prior to TMS stimulation from which the root-mean-square activity was calculated.

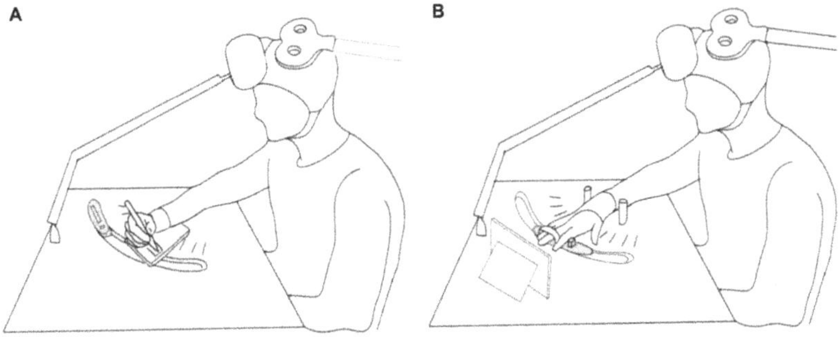

Figure 3. Experimental set-up of participant positioning during the writing (**A**) and abduction (**B**) tasks. The table design allowed the position of the force transducer to match the self-selected writing position of the graphic tablet.

Following familiarization, the participant was set up for TMS. The pre-fatigue and post-fatigue tests included a total of 90 trials, which consisted of 30 pulses each of TMS stimulation type, including single-pulse test MEP (TS), and paired-pulse stimuli to assess short-interval intracortical inhibition (SICI), and intracortical facilitation (ICF). The 90 trials were performed in 3 blocks, with 10 sets per block and 3 trials per set to pseudo-randomize the TMS stimulation such that each set would include a single-pulse TS, SICI, and ICF evoked in random order (Figure 1). Each trial was 5 seconds in duration. On the writing day, the timing of the screen refresh was set to ensure that the stimulus was delivered as the participant wrote the letter "a" or "m". Timing between pulses (5 s) was never changed. On the abduction day, the TMS stimulation occurred 2 seconds into the 5-second trial to ensure the participant was in the plateau portion of the submaximal, isometric, finger abduction contraction.

The fatigue protocol consisted of repeated 4-second isometric finger abduction contractions at 60% of the participant's MVC, with two seconds between each contraction. Task failure was defined as the point at which force fell below 58% MVC for more than three seconds despite encouragement to meet and hold the 60% target. Immediately following task failure, participants completed a post-fatigue MVC before beginning the post-fatigue test (Figure 1). It is important to note that on both days, regardless of the condition (abduction or writing), participants performed the same isometric finger abduction fatigue protocol. In this way, we examined the task-dependent nature of TMS measures of CSE rather than the task-dependent nature of neuromuscular fatigue.

2.3. *Experimental Set-Up and Recordings*

2.3.1. Electromyography and Force Recordings

The skin over the right FDI muscle was cleaned with isopropyl alcohol and two Ag/AgCl EMG electrodes were affixed in a bipolar configuration (0.5 cm recording surface, 1 cm interelectrode distance) over the muscle belly. A ground electrode was placed on the dorsal aspect of the right hand after being shaven and cleaned with alcohol. To allow the hand to rest comfortably on the graphic tablet, a glove covering the hand and the fourth and fifth digits was worn by participants. The skin over the extensor carpi radialis (ECR) and the flexor carpi radialis (FCR) was prepared for EMG by shaving, abrading, and cleansing with alcohol. Parallel bar surface EMG electrodes (10 × 1 mm Ag contacts, 1 cm interelectrode distance, DE-2.1 DELSYS Inc., Natick, MA, USA) were affixed over the muscle belly of the ECR and FCR, and a ground electrode was placed on the elbow. EMG and force signals were digitized at 2000 Hz using the Micro 1401-3 data acquisition unit and Signal 6.0 waveform acquisition software (Cambridge Electronics Design, Cambridge, UK). The FDI EMG signal was pre-amplified 300× and band-pass filtered from 15 to 450 Hz (Motion Lab Systems, Inc. Los Angeles, CA, USA). The

ECR and FCR EMG data was amplified 1000× and band-pass filtered from 20 Hz to 450 Hz (Bagnoli-16, DELSYS Inc., Natick, MA, USA).

2.3.2. Transcranial Magnetic Stimulation

Participants were seated at a table with an adjustable headrest situated in front of the forehead to allow the body and head to be supported in a comfortable writing position (Figure 3). A figure-eight magnetic stimulating coil (D70 coil, Magstim Company Ltd., Whitland, UK) was positioned over the primary motor cortex with the handle positioned posteriorly 45° to the midsagittal line and the induced current in a posterior to anterior direction and held in place using a lighting support arm and clamp (Manfrotto Supports, Cassola, Italy) with additional support and position maintenance by the investigator. The TMS coil was moved in small increments in order to determine the optimal site for generating a motor evoked potential (MEP) in the FDI via suprathreshold stimulations from the BiStim2 system (Magstim Company Ltd., Whitland, UK). Once located, this spot was marked on the cap worn by participants. Stimulator output was then adjusted to find the active motor threshold (AMT), determined as the minimum intensity that elicited a 200 μV MEP in 5 out of 10 of trials while the participant maintained a target matched to the average EMG over 20 writing trials recorded during the familiarization period at the beginning of each experimental day. Pulses were delivered at intervals of at least 5 s during threshold hunting. Paired-pulse TMS was used to assess SICI and ICF with a conditioning stimulus of 80% AMT preceding the test MEP set at 120% AMT. To elicit SICI, the conditioning stimulus preceded the test MEP by 3 ms To elicit ICF, the conditioning stimulus preceded the test MEP by 12 ms [27].

2.3.3. Motor Evoked Potentials (MEPs)

Motor evoked potential (MEP) peak-to-peak amplitudes were measured offline using Signal 6.0 (Cambridge Electronics Design, Cambridge, UK). Trials were excluded if (a) the participant responded incorrectly (e.g., no contraction), (b) TMS stimulation occurred between EMG bursts during writing, and (c) no MEP was elicited (MEP < 200 μV) [28]. Exclusion criteria a, b, and c were based on trial by trial inspection of the waveform data. SICI and ICF MEPs were normalized to the corresponding test MEP within the same set (each series of three randomized trials that include TS, SICI, and ICF). If a test MEP was excluded, MEPs were normalized to the test MEP in the next closest set such that the conditioned MEP was always normalized to an unconditioned MEP that was no further than three trials away. Test MEP peak-to-peak amplitudes are reported in millivolts (mV). The coefficient of variation of the test MEP amplitude was calculated for each participant to determine the consistency of the measure. Conditioned MEP peak-to-peak amplitudes (SICI and ICF) were normalized to the test MEP amplitude and are therefore reported as ratio values where <1 would indicate inhibition and >1 would indicate facilitation. Sample data from one participant are shown in Figure 4.

2.3.4. Cortical Silent Period

The duration of the test MEP CSP was measured offline using a custom script in MATLAB (MathWorks, Inc., Natick, MA, USA). CSP duration was calculated from the point of the test pulse stimulus delivery to the point at which EMG activity returned to prestimulus activity (i.e., average RMS amplitude calculated over a 300-ms period prior to stimulation). CSP was calculated as the average for the 30 test pulse trials only (CSP following paired-pulses were not included).

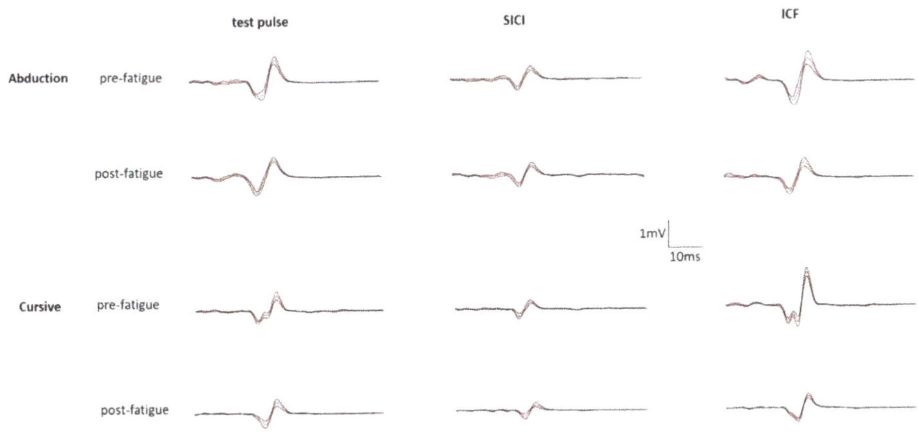

Figure 4. Block averages (red line) and error (black lines) of evoked potentials for one participant. Each block included 10 single test pulses, 10 SICI paired-pulses, and 10 ICF paired-pulses. This participant completed the writing task in cursive. Pre-fatigue averages are shown above the post-fatigue averages.

2.4. Statistical Analysis

Analysis was performed using Statistica 13.2 (TIBCO Software Inc., Palo Alto, CA, USA). For TMS pulses, trial outliers were removed if they fell >2 standard deviations from the average within each individual participant. Two participants were excluded from the SICI analysis due to a large number of outliers occurring within a single block of trials. One participant was removed from the CSP analysis due to an inability return to the target level of EMG activity following stimulation. Assumptions for the analysis of variance (ANOVA) were tested for each variable using the Shapiro–Wilk test for normality. Test MEP amplitudes were not normally distributed, thus a 1/sqrt transformation was applied for statistical analysis. Preliminary analysis was performed to ensure that there was no effect of block (i.e., the 3 blocks of 30 trials pre- and post-fatigue). No block main effect was present for any variable, therefore all pre-fatigue trials and all post-fatigue trials were averaged for each participant.

To examine potential differences between testing days a dependent samples t-test was conducted on baseline measures (see Table 1). To determine the effect of task on dependent measures, a one-way repeated measures ANOVA was performed comparing pre-fatigue values between writing and the abduction task. To determine the effect of task on the dependent measures during fatigue, a 2 (task: abduction versus writing) × 2 (fatigue: pre versus post) repeated measures ANOVA was performed for each dependent variable. The number of participants included in each analysis is noted in Tables 2 and 3.

Table 1. Baseline measures made during the familiarization period on the abduction and writing days. These measures included maximal index finger abduction force and first dorsal interosseous (FDI) activity during MVCs, the level of FDI activity during 20 writing trials, the number of fatiguing sets completed, and active motor threshold (AMT) for TMS (Mean ± SD).

	Abduction Day	Writing Day
Maximal Force (N)	19.49 ± 5.80	24.05 ± 11.38
Maximal FDI RMS amplitude (mV)	0.91 ± 0.38	0.94 ± 0.40
Writing EMG (% max)	18.7 ± 6.88	20.5 ± 9.62
AMT (%MSO)	39.1 ± 8.20	41.0 ± 7.30
# of Fatiguing Trials	43.50 ± 26.34	48.83 ± 45.34

FDI: first dorsal interosseous; MVC: maximal voluntary contraction; AMT: active motor threshold; RMS: root mean squared; EMG: electromyography; MSO: maximum stimulator output.

Table 2. Motor evoked potentials recorded from the FDI muscle and results of the repeated measures ANOVAs comparing Task (abduction versus writing) and Task × Fatigue conducted with all 19 participants.

	Abduction Task			Writing Task			Effects	
Measure	**Pre-Fatigue**	**Post-Fatigue**	**% Change Effect Size**	**Pre-Ftg**	**Post-Ftg**	**% Change Effect Size**	***F* Value § (Task)**	***F* Value (Task × Fatigue)**
Test MEP (n = 19)	1.64 ± 1.11	1.54 ± 1.13	−5.9% d = 0.09	1.85 ± 1.53	1.64 ± 1.65	−11.2% d = 0.13	F(1,18) = 0.60, p = 0.45, η_p^2 = 0.03	F(1,18) = 2.22, p = 0.15, η_p^2 = 0.11
SICI (n = 17)	0.95 ± 0.22	0.94 ± 0.30	−0.7% d = 0.03	0.82 ± 0.22	0.86 ± 0.21	4.1% d = 0.16	F(1,16) = 4.40, p = 0.052, η_p^2 = 0.22	F(1,16) = 0.36, p = 0.56, η_p^2 = 0.02
ICF (n = 19)	1.25 ± 0.33	1.19 ± 0.30	−4.2% d = 0.16	1.13 ± 0.28	1.26 ± 0.33	11% d = 0.41	F(1,18) = 1.69, p = 0.21, η_p^2 = 0.09	F(1,18) = 3.60, p = 0.07, η_p^2 = 0.17

§ Calculated on the pre-fatigued values only. % Change is calculated as ((post-pre)/pre) × 100 from the unrounded data. Cohen's d is calculated for the pre- to post-fatigue means. Abbreviations: SICI: short-interval intracortical inhibition; ICF: intracortical facilitation.

Table 3. Results of the ANOVA comparing Task × Style (printing versus cursive) and Task × Fatigue × Style conducted with the subset of participants (n = 16) that could be classified as printers (bold font) or cursive writers.

		Abduction Task			Writing Task			Effects	
Measure	**Method (n)**	**Pre-Fatigue**	**Post-Fatigue**	**% Change Effect Size**	**Pre-Fatigue**	**Post-Fatigue**	**% Change Effect Size**	***F* Value § (Task × Style)**	***F* Value (Task × Fatigue × Style)**
Test MEP	**Printers (8)**	**1.49 ± 1.16**	**1.54 ± 0.82**	**3.0% d = 0.05**	**1.96 ± 1.54**	**1.77 ± 1.50**	−10.0% d = 0.13	F(1,14) = 2.95, p = 0.11, η_p^2 = 0.17	F(1,14) = 0.77, p = 0.40, η_p^2 = 0.05
	Cursive writers (8)	1.93 ± 1.16	1.90 ± 1.43	−1.4% d = 0.02	2.09 ± 1.78	1.86 ± 2.08	−10.9% d = 0.12		
SICI	**Printers (7)**	**1.03 ± 0.28**	**1.05 ± 0.25**	**1.5% d = 0.06**	**0.71 ± 0.13**	**0.86 ± 0.16**	20.4% d = 1.02	$F_{(1,13)}$ = 8.00, p = 0.01, η_p^2 = 0.38	F(1,13) = 0.56, p = 0.47, η_p^2 =0.04
	Cursive writers (8)	0.87 ± 0.17	0.81 ± 0.33	−6.6% d = 0.22	0.86 ± 0.26	0.82 ± 0.26	−4.1% d = 0.13		
ICF	**Printers (8)**	**1.37 ± 0.35**	**1.21 ± 0.41**	**−12.0% d = 0.43**	**1.09 ± 0.29**	**1.40 ± 0.36**	28.1% d = 0.94	F(1,14) = 2.71, p = 0.12, η_p^2 = 0.16	F(1,14) = 9.90, p = 0.007, η_p^2 =0.41
	Cursive Writers (8)	1.15 ± 0.33	1.14 ± 0.21	−0.4% d = 0.02	1.19 ± 0.33	1.12 ± 0.31	−5.6% d = 0.21		

§ Calculated on the pre-fatigued values only. % Change is the (post − pre)/pre fatigue values. Cohen's d is calculated for the pre- to post-fatigue means. Abbreviations: SICI: short-interval intracortical inhibition; ICF: intracortical facilitation.

As the study progressed, we found that of the 19 participants, 8 wrote in cursive, 8 printed, and 3 used a combination of cursive and printing. Accordingly, we conducted an additional post hoc analysis of writing style. This separate-groups post hoc analysis included writing style (cursive vs. printing) as a categorical predictor to determine the baseline effect (task × writing style) and the effect of task and style on fatigue (task × fatigue × writing style) in a posteriori analysis. The three individuals who used a combination of printing and cursive writing strategies to complete the task were excluded from this analysis. Because this analysis of writing style was not planned a priori, it is important to note that it is underpowered.

Tukey's HSD testing was used for post hoc analysis where applicable. Partial eta-squared was calculated in Statistica for the ANOVA models. Cohen's d was calculated for the pre-fatigue to post-fatigue values where applicable [29].

3. Results

3.1. Baseline Measures

The baseline measures recorded during familiarization (force and FDI RMS amplitude during MVCs, FDI activity of writing, and AMT) were not different between the abduction and writing days. Additionally, the number of fatiguing trials performed was not different between the two days. The amount of fatigue elicited was not significantly different (Task × Style: $F(1,13) = 0.88$, $p = 0.37$) between printers on the writing day (−26.0% MVC force), cursive writers on the writing day (−18.6%), printers on the abduction day (−27.5%), and cursive writers on the abduction day (−22.6%).

3.2. Effect of Task

Motor Evoked Potentials

In the pre-fatigue trials, there was no effect of task (writing versus abduction) on the amplitude of the test MEP ($F(1,18) = 0.60$, $p = 0.45$, $\eta_p^2 = 0.03$, Figure 5A), ICF ($F(1,18) = 1.69$, $p = 0.21$, $\eta_p^2 = 0.09$, Figure 5C), or CSP ($F(1,17) = 1.14$, $p = 0.30$, $\eta_p^2 = 0.06$). However, the coefficient of variation of test MEP amplitude was 23% greater in the writing task compared to the abduction task ($F(1,18) = 13.55$, $p < 0.01$, $\eta_p^2 = 0.43$). The coefficient of variation of the test MEP was higher in both the printing (48.4 ± 7.3, $n = 8$) and cursive writing (42.4 ± 5.5, $n = 8$) styles, as compared to abduction in printers (38.3 ± 7.9, $n = 8$) and abduction in cursive writers (35.1 ± 8.2, $n = 8$). Although there appeared to be a trend toward a greater level of inhibition (SICI) during writing compared to abduction ($F(1,16) = 4.40$, $p = 0.052$, $\eta_p^2 = 0.22$, Table 2, Figure 5B), this analysis included only 17 participants and was not adequately powered. When writing style (printing vs. cursive writing) were factored in (task × style: $F(1,13) = 8.00$, $p = 0.01$, $\eta_p^2 = 0.38$, Table 3), post hoc analysis revealed a significant difference ($p < 0.01$) in SICI between printing (0.71 ± 0.13, $n = 7$) and abduction (1.03 ± 0.28, $n = 7$, Figure 6B) in printers, but no difference between cursive writing (0.86 ± 0.26, $n = 8$) and abduction (0.87 ± 0.17, $n = 8$) in cursive writers.

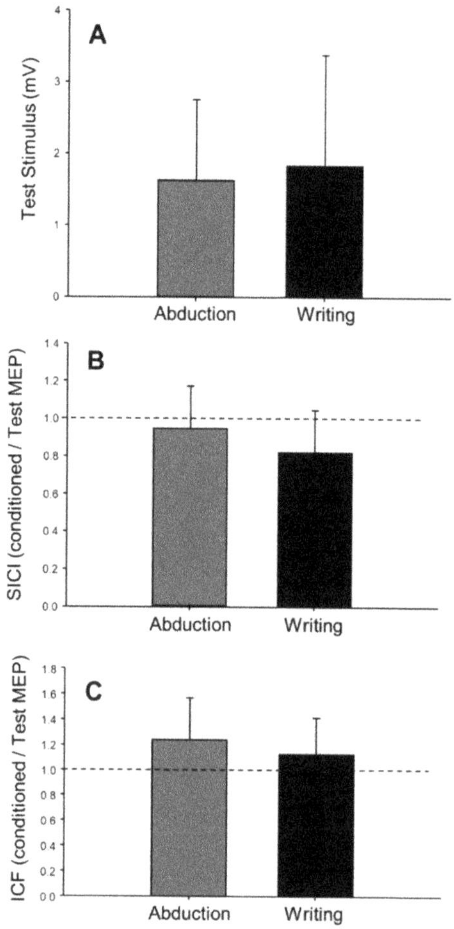

Figure 5. Pre-fatigue levels of test MEP (n = 19), intracortical facilitation (n = 19), and intracortical inhibition (n = 17) during finger abduction and writing (n = 19). There were no differences between test stimulus (**A**), short-interval intracortical inhibition (SICI; **B**), and intracortical facilitation (ICF; **C**) between the two tasks. In panels B and C, dashed lines represent no effect of the conditioning stimulus. Values below the line represent inhibition (conditioned MEP < test MEP), and values above the line represent facilitation (conditioned MEP > test MEP). Error bars represent standard deviation.

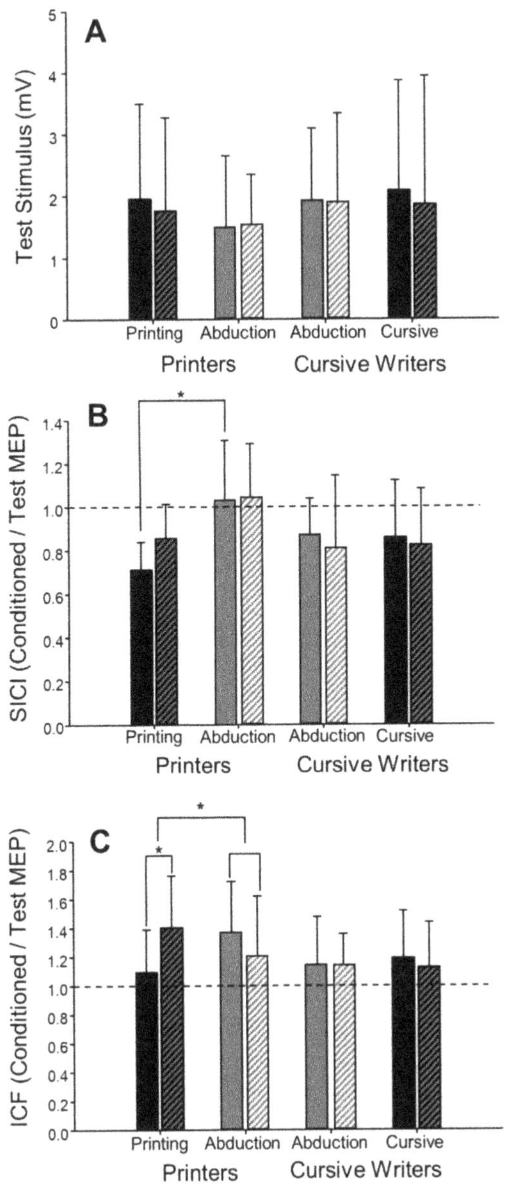

Figure 6. Motor evoked potentials during finger abduction and writing in printers ($n = 8$) and cursive writers ($n = 8$). There were no differences in single-pulse (test stimulus) MEPs between writers and printers or between tasks (test stimulus; **A**). Short-interval intracortical inhibition (SICI; **B**) was greater during printing than during abduction pre-fatigue (*, $p < 0.01$). In printers, there was a fatigue-associated increase in intracortical facilitation (ICF; **C**) when assessed during printing, but not during abduction (*, $p < 0.01$). Solid bars represent pre-fatigue values, and hatched bars represent post-fatigue values. In panels B and C, dashed lines represent no effect of the conditioning stimulus. Values below the line represent inhibition (conditioned MEP < test MEP), and values above the line represent facilitation (conditioned MEP > test MEP). Error bars represent standard deviation.

3.3. Muscle Activity

Task × fatigue repeated-measures ANOVAs were conducted for FDI, FCR, and ECR surface EMG measures of muscle activation. Level of FDI muscle activation was successfully maintained between the two tasks ($F(1,18) = 0.60$, $p = 0.45$, $\eta_p^2 = 0.03$). This was expected because average writing RMS amplitude was used as the muscle activation target during the abduction task. The abduction task elicited greater levels of FCR activity ($F(1,18) = 7.24$, $p = 0.02$, $\eta_p^2 = 0.29$) and lower levels of ECR activity ($F(1,18) = 7.73$, $p = 0.01$, $\eta_p^2 = 0.30$) compared to the writing task. Analysis of writing styles (task × style) revealed a significant interaction for both FCR activity ($F(1,14) = 4.70$, $p = 0.048$, $\eta_p^2 = 0.25$; Figure 7A) and ECR activity ($F(1,14) = 5.25$, $p = 0.04$, $\eta_p^2 = 0.27$; Figure 7B). Post hoc analysis identified that cursive writing required significantly less FCR activity (12.58 ± 7.42 μV) than abduction (29.79 ± 17.57 μV) in the cursive writing group. There was no significant difference between FCR activity during printing (18.36 ± 9.10 μV) or abduction (20.92 ± 13.07 μV) in the printing group. In the extensor muscle, post hoc analysis identified that printing (154.22 ± 71.53 μV) required significantly more ECR activity than abduction (70.36 ± 47.40 μV) in the printers. There was no significant difference between ECR activity during cursive writing (99.07 ± 27.63 μV) or abduction (92.84 ± 39.92 μV) in the cursive group.

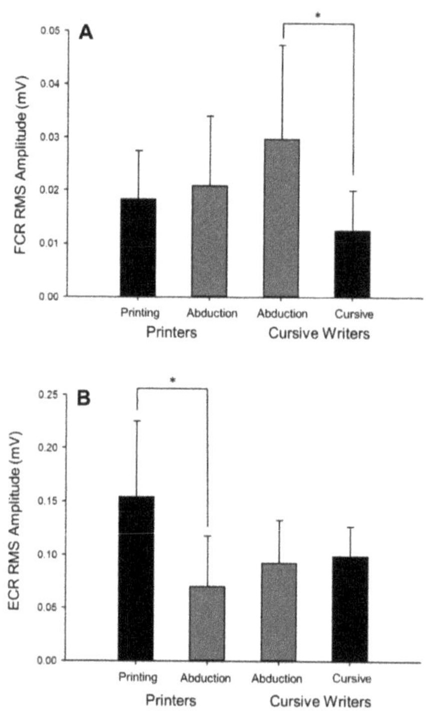

Figure 7. Pre-fatigue levels of wrist flexor (**A**) and wrist extensor (**B**) activity during the abduction and writing tasks in printers and cursive writers. Wrist flexor muscle activity was higher in the writing task than in the abduction task in cursive writers, whereas wrist extensor muscle activity was higher in the writing task than in the abduction task in the printers (*, $p < 0.05$). This analysis included participants who completed the writing task using a pure printing ($n = 8$) or cursive ($n = 8$) strategy. Error bars represent standard deviation.

3.4. Effect of Task on Fatigue

3.4.1. Motor Evoked Potentials

The task × fatigue repeated-measures ANOVA was not significant for any measure of CSE (see Table 2 for test MEP, SICI, and ICF results), including test MEP coefficient of variation (F(1,18) = 3.59, $p = 0.07$, $\eta_p^2 = 0.17$) and CSP (F(1,17) = 0.29, $p = 0.60$, $\eta_p^2 = 0.02$). However, when writing style was taken into consideration, a 3-way (task × fatigue × style) repeated-measures ANOVA revealed a significant 3-way interaction for ICF (F(1,14) = 9.90, $p < 0.01$, $\eta_p^2 = 0.41$) (Table 3). Post hoc analysis revealed a significant increase in ICF from pre-fatigue to post-fatigue (28.1%, d = 0.94, $p = 0.04$) in the printing group. Planned comparisons were conducted using percent change.

Scores (pre- to post-fatigue) to determine the task × style effect on fatigue. The increase in facilitation in printing (28.1%) was significantly different than the decrease in facilitation of printers performing the abduction task (−12.0%, $p < 0.01$, Figure 6C). There were no significant effects of writing style on the test MEP or SICI (see Table 3), test MEP coefficient of variation (F(1,14) = 0.02, $p = 0.88$, $\eta_p^2 = 0.002$), or CSP (F(1,13) = 0.30, $p = 0.60$, $\eta_p^2 = 0.02$).

3.4.2. Muscle Activity

The task × fatigue repeated measures ANOVA was not significant for FDI activity (F(1,18) = 1.31, $p = 0.27$, $\eta_p^2 = 0.07$), FCR activity (F(1,18) = 1.69, $p = 0.21$, $\eta_p^2 = 0.09$), or ECR activity (F(1,18) = 0.86, $p = 0.37$, $\eta_p^2 = 0.05$). Similarly, an analysis of writing style (task × fatigue × style) did not yield any significant interactions for FDI activity (F(1,14) = 0.57, $p = 0.46$, $\eta_p^2 = 0.04$), FCR activity (F(1,14) = 0.57, $p = 0.47$, $\eta_p^2 = 0.04$), or ECR activity (F(1,14) = 2.82, $p = 0.12$, $\eta_p^2 = 0.17$).

4. Discussion

Our study demonstrates that fatigue-associated changes in CSE depend on the motor task completed during single- and paired-pulse TMS stimulation. Not only did we find differences in fatigue-associated changes in cortical excitability assessed during hand-writing compared to isometric finger abduction, but we also found differences between participants who completed the writing task by printing compared to those who wrote in cursive. Specifically, we found that the same fatigue protocol elicited an increase in ICF during printing, but not during cursive writing. It is important to note that the fatigue protocol itself was the same on each day, and only the motor task during pre-fatigue and post-fatigue assessments of CSE differed. Furthermore, because the level of FDI muscle activation was held constant during the two tasks, the differences in CSE between writing and abduction days were not simply due to differences in voluntary drive to that muscle.

The present study examined CSE during voluntary movement. Because the level of muscle activation during voluntary movement affects levels of CSE [30–32], it was critical in the present study to match FDI EMG between tasks. We accomplished this by measuring FDI muscle activation during the writing task in a familiarization period each day, and then using biofeedback to set a target on the abduction day that matched muscle activation during writing for each participant. This approach to matching muscle activation between the two tasks (writing and abduction) was equally effective in printers and cursive writers. Therefore, differences in intracortical measures of excitability pre- and post-fatigue cannot be attributed to differing FDI EMG activity between the tasks.

During baseline (pre-fatigue) testing, the variability of single-pulse MEP amplitudes was greater in the writing task than the abduction task. This was expected because the abduction task is a single-joint, externally cued task frequently used in TMS studies for the purpose of reducing variability. The abduction task was maintained at a stable force level using visual feedback. To make the writing task more functionally-relevant, we did not control the style, or speed of each participant's writing. Therefore, it is not surprising that the writing task resulted in greater variability than the abduction task, both within and between subjects. Similar results have been found comparing test–retest reliability of

MEP amplitudes between static and dynamic tasks in the lower limb [33], where MEPs evoked during a static task (i.e., plateau at a target force level) were less variable than those evoked during a dynamic task (i.e., continuously increasing force level). The style of writing did not impact the MEP amplitude variability, as we found higher coefficient of variation values for both printing and cursive writing compared to abduction. A future challenge in TMS research will be to utilize laboratory tasks that balance variability with the relevance of the task to motor activities outside of the lab.

We hypothesized that CSE would be greater during writing compared to abduction due to the complexity of the writing task. However, single-pulse MEP amplitude was not different between tasks. Reports of differences in CSE between precision tasks and conventional abduction tasks are inconsistent. Original research in monkeys suggested that corticospinal neurons were more active during a precision task compared to a power task [9], a finding supported by several studies that found greater MEP amplitudes during complex, precision tasks compared to simple, power tasks [4,11,34]. Alternatively, larger MEP amplitudes have been reported during conventional abduction tasks compared to power, pincer, or grasping tasks [5–7]. The discrepancy between our findings and previous research may be that other studies have used precision tasks that were visually guided and externally controlled, whereas our writing task is a dynamic, internally generated task that is retrieved and implemented from memory [35,36]. Furthermore, writing involves higher levels of activation in the dorsal premotor cortex in comparison to simple finger contraction tasks, as well as unique activation in multiple brain regions (e.g., premotor cortex and anterior putamen) not activated during tapping or "zigzagging" finger actions [37]. This association of writing with higher cognitive demands would suggest that writing is a more complex task in comparison to simple finger contractions. On the other hand, because writing is learned and practiced from a young age into adulthood, it is also associated with a degree of automaticity [38]. Accordingly, our finding that writing was not associated with increased CSE compared to abduction, may have been due to either the increased complexity or the automaticity of writing compared to unpracticed, less natural, and externally guided precision-grip tasks used in previous studies.

Despite the fact that baseline (pre-fatigue) CSE did not differ between tasks, levels of intracortical inhibition (SICI) assessed using paired-pulse TMS trended ($p = 0.052$) toward greater inhibition during writing compared to abduction. When participants were subdivided into those who printed and those who wrote in cursive (excluding those who used a combination of printing and cursive within a single word), printers had the greatest intracortical inhibition during the writing task. We hypothesize that this may be due to the increased control required during the more intermittent task of writing distinct letters during printing, compared to the continuous nature of cursive writing (Figure 2). The differences in intracortical inhibition observed between writing styles may also be explained by the activity of the proximal forearm muscles. While FDI EMG activity between abduction and writing did not differ, extensor (ECR) activity was greater and flexor (FCR) activity was lower during the writing task compared to abduction. It has been suggested that activity and position of proximal muscles can have an effect on the corticospinal pathway leading to the distal muscle of interest [39–41]. This relationship has been shown between the FDI and proximal arm muscles (including the ECR, FCR, and deltoid muscles), where proximal muscle activity resulted in facilitation of distal muscle MEPs [39–41]. Similarly, forearm position (pronation vs. semi-supinated) is also known to alter CSE [42]. In our study, however, the forearm was semi-supinated for both printing and cursive writing, and therefore forearm position does not explain the differences in intracortical excitability that we observed when we compared the two different writing styles. The role that proximal muscle activation may play in the CSE of distal muscles highlights the importance of choosing laboratory tasks that more closely resemble natural movements outside of the laboratory to which we aim to extrapolate our results.

TMS is used to assess changes in corticospinal and intracortical excitability in response to many different interventions and perturbations (e.g., fatigue, strength training, skill training, disease, injury, aging, and pharmacological agents). We sought to determine whether changes in CSE following an intervention depend on the motor task used to elicit muscle activity during the delivery of TMS. To this end, we employed neuromuscular fatigue as an acute intervention, and measured changes in CSE during two different motor tasks before and after two identical fatigue protocols. It is well established that neuromuscular fatigue is associated with changes in CSE, specifically a reduction in unconditioned MEP (test MEP) amplitudes in the target muscle following neuromuscular fatigue [43,44]. This depression has been attributed to central mechanisms of fatigue [44,45]. Paired-pulse TMS techniques have been used to identify intracortical mechanisms of fatigue [22,46,47]. Following fatiguing contractions, SICI decreased [18,46,48]; however, the association between fatigue and ICF is not as clear. For example, ICF measured in the biceps brachii decreases during a sustained fatiguing contraction [48,49], whereas ICF measured in the FDI is elevated at the point of task failure when assessed at rest [22,46]. In other studies, ICF does not change with fatigue [47,49]. It has been suggested that decreased intracortical inhibition [18] and increased facilitation following fatigue [22] may serve as a compensatory mechanism to optimize motor output as fatigue develops.

In our study, there was no task-dependent effect of fatigue on any of the corticospinal and intracortical excitability measures between writing and abduction. However, most of our participants used one of two distinct writing styles: printing or cursive. Therefore, we completed additional a posteriori analyses to determine whether there was an effect of writing styles on fatigue-induced changes in CSE, and found that printers had a significant increase in ICF following fatigue. This is consistent with previous reports of increased ICF following fatigue that employ more conventional laboratory tasks [46,50]. However, this only occurred during printing, and was not found during cursive writing or the conventional finger abduction task. Previous research has suggested that increased ICF post-fatigue may be a mechanism to compensate for peripheral contractile failure or reduced upstream drive to the primary motor cortex [22]. However, this would not explain the increased ICF seen in printing but not in cursive writing or abduction. Possibly the differences in wrist stabilization during printing and cursive, reflected by differences in wrist extensor and wrist flexor muscle activity during the two tasks, had a greater impact on ICF than on SICI. Furthermore, it is possible that printing employs cortical circuits that are affected by isometric finger abduction fatigue task differently than the cortical circuits employed in cursive writing. Although there is very little research regarding the neural control of printing versus cursive writing, one clinical study reports impairment in cursive writing, but not printing, drawing, or the ability to draw continuous loops, following ischemic damage to the parietal lobe [51]. Another clinical case found that the ability to write in cursive was lost with the development of a large cranial tumor impinging on the left frontal lobe, while the ability to print remained intact. In this case, the patient regained the ability to write in cursive following tumor resection [52]. Case studies like these support the idea that printing and cursive writing have different cortical representations, and may therefore be associated with different inputs to the primary motor cortex. This novel finding certainly warrants further exploration, as printing and cursive writing appear to be two different tasks that further demonstrate the task-dependent nature of fatigue-associated changes in CSE.

We started this repeated-measures study with 20 participants based on a power analysis conducted using data from a previous study from our lab that also investigated intracortical mechanisms of fatigue [22]. Of the 20 participants we recruited, 1 participant was unable to complete the study, and 2 participants were excluded from the SICI analysis. Accordingly, this analysis is not robustly powered, and these data should be interpreted cautiously. The analysis of hand writing style was not planned in advance. From the start, we told participants to write as they preferred given that our aim was to assess corticospinal excitability during a natural motor task. As the study progressed, we found that many university students were unable to write in cursive, and we conducted the post hoc analysis of writing style. It should be noted that this a posteriori analysis of writing style (cursive vs. printing) is underpowered. Furthermore, in the present study, we assessed only three intracortical mechanisms (SICI, ICF, and cSP). Each of these measures has been associated with different types of neurotransmission (SICI with GABA-A receptors, ICF with glutamatergic transmission, and the cSP with GABA-B receptors) as reviewed by Reis [1]. It is possible that other intracortical (e.g., long interval intracortical inhibition) and interhemispheric pathways (e.g., interhemispheric inhibition) may reveal fatigue-associated changes that are dependent on the motor task that is conducted during stimulation. This is an area that requires further investigation

5. Conclusions

This study is the first to assess CSE and the effect of fatigue during a writing task. Although fatigue-associated changes in CSE have been well studied over the last three decades, this study highlights the importance of considering the task used during TMS measures of corticospinal and intracortical excitability. Although controlled laboratory tasks are required to reduce variability of motor evoked potentials to allow for reproducible results, it is important to note that CSE is task-dependent. Because of this, measures of CSE made during a laboratory task may not translate to motor tasks outside of the lab. Despite the fact that the fatigue task and the level of FDI muscle activation during TMS was the same in our study, hand-writing revealed fatigue-associated changes in CSE that were not evident during abduction. Therefore, the task-dependent nature of CSE emphasizes the need for experimental paradigms that better reflect relevant motor tasks or at least acknowledge the differences between the task employed in the laboratory and the movements outside of the lab to which results may be extrapolated. With this in mind, it is essential that studies of CSE consider the "task at hand".

Author Contributions: K.T.M.C.: pilot work, data collection, data analysis, and writing of manuscript, L.A.G.: supervision and instruction of techniques, data analysis, writing of manuscript, and revision of manuscript, J.M.K.: study conception, supervision, instruction of techniques, data analysis, and revision of manuscript.

Funding: This work was funded by a NSERC Discovery (386601) grant to J.M.K. and an NSERC PDF to L.A.G.

Acknowledgments: The authors would like to thank Ron Daniels, Laurier Science Maker Lab, for his assistance with the design and creation of custom-built TMS table used in this experiment.

Conflicts of Interest: The authors declare no conflicts of interest. The funders had no role in the design of the study; in the collection, analyses, or interpretation of data; in the writing of the manuscript; or in the decision to publish the results.

References

1. Reis, J.; Swayne, O.B.; Vandermeeren, Y.; Camus, M.; Dimyan, M.A.; Harris-Love, M.; Perez, M.A.; Ragert, P.; Rothwell, J.C.; Cohen, L.G. Contribution of Transcranial Magnetic Stimulation to the Understanding of Cortical Mechanisms Involved in Motor Control. *J. Physiol.* **2008**, *586*, 325–351. [CrossRef] [PubMed]
2. Kalmar, J.M. On Task: Considerations and Future Directions for Studies of Corticospinal Excitability in Exercise Neuroscience and Related Disciplines. *Appl. Physiol. Nutr. Metab.* **2018**, *43*, 1113–1121. [CrossRef] [PubMed]
3. Silvanto, J. State-Dependency of Transcranial Magnetic Stimulation. *Brain Topogr.* **2008**, *21*, 1–10. [CrossRef] [PubMed]
4. Tinazzi, M.; Farina, S.; Tamburin, S.; Facchini, S.; Fiaschi, A.; Restivo, D.; Berardelli, A. Task-Dependent Modulation of Excitatory and Inhibitory Functions within the Human Primary Motor Cortex. *Exp. Brain Res.* **2003**, *150*, 222–229. [CrossRef]
5. Datta, A.K.; Harrison, L.M.; Stephens, J.A. Task-Dependent Changes in the Size of Response to Magnetic Brain Stimulation in Human First Dorsal Interosseous Muscle. *J. Physiol.* **1989**, *418*, 13–23. [CrossRef]
6. Bunday, K.L.; Tazoe, T.; Rothwell, J.C.; Perez, M.A. Subcortical Control of Precision Grip after Human Spinal Cord Injury. *J. Neurosci.* **2014**, *34*, 7341–7350. [CrossRef]
7. Tazoe, T.; Perez, M.A. Cortical and Reticular Contributions to Human Precision and Power Grip. *J. Physiol.* **2017**, *595*, 2715–2730. [CrossRef]
8. Geevasinga, N.; Menon, P.; Kiernan, M.C.; Vucic, S. Motor Cortical Function and the Precision Grip. *Physiol. Rep.* **2014**, *2*, 1–10. [CrossRef]
9. Muir, R.B.; Lemon, R.N. Corticospinal Neurons with a Special Role in Precision Grip. *Brain Res.* **1983**, *261*, 312–316. [CrossRef]

10. Pearce, A.J.; Kidgell, D.J. Comparison of Corticomotor Excitability during Visuomotor Dynamic and Static Tasks. *J. Sci. Med. Sport* **2010**, *13*, 167–171. [CrossRef]
11. Flament, B.Y.D.; Goldsmith, P.; Buckley, C.J.; Lemon, R.N. Task Dependence of Responses in First Dorsal Interosseous Muscle to Magnetic Brain Stimulation in Man. *J. Physiol.* **1993**, *464*, 361–378. [CrossRef] [PubMed]
12. Bestmann, S.; Krakauer, J.W. The Uses and Interpretations of the Motor-Evoked Potential for Understanding Behaviour. *Exp. Brain Res.* **2015**, *233*, 679–689. [CrossRef] [PubMed]
13. Gowen, E.; Miall, R.C. Differentiation between External and Internal Cuing: An FMRI Study Comparing Tracing with Drawing. *Neuroimage* **2007**, *36*, 396–410. [CrossRef] [PubMed]
14. Gordon, C.L.; Spivey, M.J.; Balasubramaniam, R. Corticospinal Excitability during the Processing of Handwritten and Typed Words and Non-Words. *Neurosci. Lett.* **2017**, *651*, 232–236. [CrossRef] [PubMed]
15. Nakatsuka, M.; Thabit, M.N.; Koganemaru, S.; Nojima, I.; Fukuyama, H.; Mima, T. Writing's Shadow: Corticospinal Activation during Letter Observation. *J. Cogn. Neurosci.* **2012**, *24*, 1138–1148. [CrossRef] [PubMed]
16. Gruet, M.; Temesi, J.; Rupp, T.; Levy, P.; Millet, G.Y.; Verges, S. Stimulation of the Motor Cortex and Corticospinal Tract to Assess Human Muscle Fatigue. *Neuroscience* **2013**, *231*, 384–399. [CrossRef]
17. Taylor, J.L.; Gandevia, S.C. Transcranial Magnetic Stimulation and Human Muscle Fatigue. *Muscle Nerve* **2001**, *24*. [CrossRef]
18. Benwell, N.; Sacco, P.; Hammond, G.; Byrnes, M.L.; Mastaglia, F.L.; Thickbroom, G.W. Short-Interval Cortical Inhibition and Corticomotor Excitability with Fatiguing Hand Exercise: A Central Adaptation to Fatigue? *Exp. Brain Res.* **2006**, *170*, 191–198. [CrossRef]
19. Liepert, J.; Kotterba, S.; Tegenthoff, M.; Malin, J.-P. Central Fatigue Assessed By Transcranial Magnetic Stimulation. *Med. Sci. Sports Exerc.* **1996**, *19*, 1166–1175. [CrossRef]
20. Samii, A.; Wassermann, E.M.; Ikoma, K.; Mercuri, B.; George, M.S.; O'Fallon, A.; Dale, J.K.; Straus, S.E.; Hallett, M. Decreased Postexercise Facilitation of Motor Evoked Potentials in Patients with Chronic Fatigue Syndrome or Depression. *Neurology* **1996**, *47*, 1410–1414. [CrossRef]
21. Brasil-Neto, J.P.; Cohen, L.G.; Hallett, M. Central Fatigue as Revealed by Postexercise Decrement of Motor Evoked Potentials. *Muscle Nerve* **1994**, *17*, 713–719. [CrossRef] [PubMed]
22. Sharples, S.A.; Gould, J.A.; Vandenberk, M.S.; Kalmar, J.M. Cortical Mechanisms of Central Fatigue and Sense of Effort. *PLoS ONE* **2016**, 1–21. [CrossRef]
23. Taylor, J.L.; Butler, J.E.; Allen, G.M.; Gandevia, S.C. Changes in Motor Cortical Excitability during Human Muscle Fatigue. *J. Physiol.* **1996**, *490*, 519–528. [CrossRef] [PubMed]
24. Taylor, J.L.; Butler, J.E.; Gandevia, S.C. Altered Responses of Human Elbow Flexors to Peripheral-Nerve and Cortical Stimulation during a Sustained Maximal Voluntary Contraction. *Exp. Brain Res.* **1999**, *127*, 108–115. [CrossRef]
25. Dragovic, M.; Hammond, G. A Classification of Handedness Using the Annett Hand Preference Questionnaire. *Br. J. Psychol.* **2007**, *98*, 375–387. [CrossRef]
26. Matamala, J.M.; Howells, J.; Dharmadasa, T.; Trinh, T.; Ma, Y.; Lera, L.; Vucic, S.; Burke, D.; Kiernan, M.C. Inter-Session Reliability of Short-Interval Intracortical Inhibition Measured by Threshold Tracking TMS. *Neurosci. Lett.* **2018**, *674*, 18–23. [CrossRef]
27. Kujirai, T.; Caramia, M.D.; Rothwell, J.C.; Day, B.L.; Thompson, P.D.; Ferbert, A.; Wroe, S.; Asselman, P.; Marsden, C.D. Corticocortical Inhibition in Human Motor Cortex. *J. Physiol.* **1993**, *471*, 501–519. [CrossRef]
28. Van Campen, A.D.; Neubert, F.; Wery, P.; Van Den Wildenberg, M.; Ridderinkhof, K.R.; Mars, R.B. Paired-Pulse Transcranial Magnetic Stimulation Reveals Probability-Dependent Changes in Functional Connectivity between Right Inferior Frontal Cortex and Primary Motor Cortex during Go/No-Go Performance. *Front. Hum. Neurosci.* **2013**, *7*, 1–10. [CrossRef]
29. Cohen, J. *Statistical Power Analysis for the Behavioral Sciences*, 2nd ed.; Lawrence Erlbaum Associates, Publishers: Hillsdale, NJ, USA, 1988.
30. Darling, W.G.; Wolf, S.L.; Butler, A.J. Variability of Motor Potentials Evoked by Transcranial Magnetic Stimulation Depends on Muscle Activation. *Exp. Brain Res.* **2006**, *174*, 376–385. [CrossRef]
31. Devanne, H.; Lavoie, B.A.; Capaday, C. Input-Output Properties and Gain Changes in the Human Corticospinal Pathway. *Exp. Brain Res.* **1997**, *114*, 329–338. [CrossRef]

32. Kiers, L.; Cros, D.; Chiappa, K.H.; Fang, J. Variability of Motor Potentials Evoked by Transcranial Magnetic Stimulation. *Electroencephalogr. Clin. Neurophysiol. Evoked Potentials* **1993**, *89*, 415–423. [CrossRef]
33. Van Hedel, H.J.A.; Murer, C.; Dietz, V.; Curt, A. The Amplitude of Lower Leg Motor Evoked Potentials Is a Reliable Measure When Controlled for Torque and Motor Task. *J. Neurol.* **2007**, *254*, 1089–1098. [CrossRef] [PubMed]
34. Kouchtir-Devanne, N.; Capaday, C.; Cassim, F.; Derambure, P.; Devanne, H. Task-Dependent Changes of Motor Cortical Network Excitability during Precision Grip Compared to Isolated Finger Contraction. *J. Neurophysiol.* **2012**, *107*, 1522–1529. [CrossRef] [PubMed]
35. Elsinger, C.L.; Harrington, D.L.; Rao, S.M. From Preparation to Online Control: Reappraisal of Neural Circuitry Mediating Internally Generated and Externally Guided Actions. *Neuroimage* **2006**, *31*, 1177–1187. [CrossRef]
36. Debaere, F.; Wenderoth, N.; Sunaert, S.; Van Hecke, P.; Swinnen, S.P. Internal vs External Generation of Movements: Differential Neural Pathways Involved in Bimanual Coordination Performed in the Presence or Absence of Augmented Visual Feedback. *Neuroimage* **2003**, *19*, 764–776. [CrossRef]
37. Horovitz, S.G.; Gallea, C.; Najee-ullah, M.A.; Hallett, M. Functional Anatomy of Writing with the Dominant Hand. *PLoS ONE* **2013**, *8*, 1–10. [CrossRef]
38. Jones, D.; Christensen, C.A. Relationship between Automaticity in Handwriting and Students' Ability to Generate Written Text. *J. Educ. Psychol.* **1999**, *91*, 44–49. [CrossRef]
39. Devanne, H.; Cohen, L.G.; Kouchtir-Devanne, N.; Capaday, C. Integrated Motor Cortical Control of Task-Related Muscles during Pointing in Humans. *J. Neurophysiol.* **2002**, *87*, 3006–3017. [CrossRef]
40. Boyadjian, A.; Tyc, F.; Allam, N.; Brasil-Neto, J.P. Writer's Cramp: Cortical Excitability in Tasks Involving Proximo-Distal Coordination. *Acta Physiol. Scand.* **2011**, *203*, 321–330. [CrossRef]
41. Gagné, M.; Schneider, C. Dynamic Changes in Corticospinal Control of Precision Grip during Wrist Movements. *Brain Res.* **2007**, *1164*, 32–43. [CrossRef]
42. Perez, M.A.; Rothwell, J.C. Distinct Influence of Hand Posture on Cortical Activity during Human Grasping. *J. Neurosci.* **2015**, *35*, 4882–4889. [CrossRef] [PubMed]
43. Kalmar, J.M.; Cafarelli, E. Central Fatigue and Transcranial Magnetic Stimulation: Effect of Caffeine and the Confound of Peripheral Transmission Failure. *J. Neurosci. Methods* **2004**, *138*, 15–26. [CrossRef] [PubMed]
44. Brasil-Neto, J.P.; Pascual-Leone, A.; Valls-Solé, J.; Cammarota, A.; Cohen, L.G.; Hallett, M. Postexercise Depression of Motor Evoked Potentials: A Measure of Central Nervous System Fatigue. *Exp. Brain Res.* **1993**, *93*, 181–184. [CrossRef]
45. Gandevia, S.C.; Petersen, N.; Butler, J.E.; Taylor, J.L. Impaired Response of Human Motoneurones to Corticospinal Stimulation after Voluntary Exercise. *J. Physiol.* **1999**, *521*, 749–759. [CrossRef] [PubMed]
46. Maruyama, A.; Matsunaga, K.; Tanaka, N.; Rothwell, J.C. Muscle Fatigue Decreases Short-Interval Intracortical Inhibition after Exhaustive Intermittent Tasks. *Clin. Neurophysiol.* **2006**, *117*, 864–870. [CrossRef]
47. Latella, C.; Hendy, A.; Vanderwesthuizen, D.; Teo, W. The Modulation of Corticospinal Excitability and Inhibition Following Acute Resistance Exercise in Males and Females Acute Resistance Exercise in Males and Females. *Eur. J. Sport Sci.* **2018**, *1391*. [CrossRef]
48. Hunter, S.K.; McNeil, C.J.; Butler, J.E.; Gandevia, S.C.; Taylor, J.L. Short-Interval Cortical Inhibition and Intracortical Facilitation during Submaximal Voluntary Contractions Changes with Fatigue. *Exp. Brain Res.* **2016**, *234*, 2541–2551. [CrossRef]
49. Williams, P.S.; Hoffman, R.L.; Clark, B.C. Cortical and Spinal Mechanisms of Task Failure of Sustained Submaximal Fatiguing Contractions. *PLoS ONE* **2014**, *9*. [CrossRef]
50. Sharples, S.A.; Kalmar, J.M. Modulation of Cortical Excitability and Interhemispheric Inhibition Prior to Rhythmic Unimanual Contractions. *J. Neurosci. Methods* **2012**, *210*, 178–186. [CrossRef]
51. Popescu, I.M.; Vaidya, N.A. Isolated Inability to Write Cursively after Transient Ischemic Attack (TIA). *Cogn. Behav. Neurol.* **2007**, *20*, 131–135. [CrossRef]
52. Matiello, M.; Zimmerman, E.; Caplan, D.; Cohen, A.B. Reversible Cursive Agraphia. *Neurology* **2015**, *85*, 95–96. [CrossRef] [PubMed]

brain sciences

MDPI

Article

The Impact of Glucose on Corticospinal and Intracortical Excitability

Stephen L. Toepp, Claudia V. Turco, Mitchell B. Locke, Chiara Nicolini, Roshni Ravi and Aimee J. Nelson *

Department of Kinesiology, McMaster University, Hamilton, ON L8S 4K1, Canada; toeppsl@mcmaster.ca (S.L.T.); turcocv@mcmaster.ca (C.V.T.); lockemb@mcmaster.ca (M.B.L.); nicolinichiaratn@gmail.com (C.N.); roshniravi67@gmail.com (R.R.)

* Correspondence: nelsonaj@mcmaster.ca

Received: 21 October 2019; Accepted: 22 November 2019; Published: 25 November 2019

Abstract: Neurotransmission is highly dependent on the availability of glucose-derived energy, although it is unclear how glucose availability modulates corticospinal and intracortical excitability as assessed via transcranial magnetic stimulation (TMS). In this double-blinded placebo-controlled study, we tested the effect of acute glucose intake on motor-evoked potential (MEP) recruitment curves, short-interval intracortical inhibition (SICI), short-latency afferent inhibition (SAI) and long-latency afferent inhibition (LAI). Eighteen healthy males participated in four sessions. Session 1 involved acquisition of an individualized blood glucose response curve. This allowed measurements to be time-locked to an individualized glucose peak after consuming one of three drinks during the subsequent three sessions. Participants were administered a 300 mL concealed solution containing 75 g of glucose, sucralose, or water in separate sessions. Dependent measures were assessed at baseline and twice after drinking the solution. Secondary measures included blood glucose and mean arterial pressure. Corticospinal excitability and blood pressure increased following the drink across all treatments. No changes were observed in SICI, SAI or LAI. There was no rise in corticospinal excitability that was specific to the glucose drink, suggesting that acute changes in glucose levels do not necessarily alter TMS measures of corticospinal or intracortical excitability.

Keywords: glucose; SICI; SAI; LAI

1. Introduction

Glucose is the brain's primary energy substrate and provides the main carbon source for de novo synthesis of large compounds required for essential ranging processes, from neurotransmission to the management of oxidative stress [1–4]. Although the substantial influence of glucose is evident in several clinical contexts [5–7], a comprehensive neurophysiological profile has yet to be compiled in healthy humans during periods of fasting versus high-circulating glucose following feeding.

A small number of studies have used non-invasive transcranial magnetic stimulation (TMS) to compare neurophysiological measures in hyperglycemic, normoglycemic or fasting conditions [8–10]. Specterman et al. [10] reported a 3-fold increase in the size of motor-evoked potentials (MEPs) 60 min after ingestion of 68 g of glucose, such that greater increases in MEPs were correlated with greater increases in blood glucose levels. Badawy and colleagues [9] observed greater long interval intracortical inhibition (LICI) in epileptic and healthy individuals when in a fed (i.e., two hours after a meal) compared to fasted state (i.e., 12 h overnight). However, not all studies have detected an effect of glucose on TMS measures. Andersen and colleagues [8] manipulated glucose levels in type 1 diabetics via an intravenous glucose pump and did not observe any change in cortical motor thresholds, a common measure of corticomotor excitability. TMS can also be used to probe the sensorimotor system with measures of short- and long-latency afferent (SAI, LAI). To date, no studies have investigated the influence of glucose on afferent inhibition. SAI and LAI are impaired in populations with Alzheimer's and Parkinson's disease [11], both of which display altered central glucose metabolism [12,13].

Most research to date that directly tests the effects of glucose on the motor system provides information regarding a limited variety of neurophysiological measurements. For example, the study conducted by Specterman and colleagues [10] measured MEPs using a time-efficient protocol involving the delivery of only 15 suprathreshold TMS pulses to obtain a measure of average MEP size. This approach allows for relatively high temporal resolution but contrasts with more comprehensive tests which could probe underlying mechanisms. For example, acquisition of MEP recruitment curves, while more time consuming, provides information regarding cortical glutamate levels [14]. Investigating a proxy measure of glutamate levels is useful, since glutamatergic neurotransmission has been linked with glucose in in vitro [15,16] and in situ [17] neurobiological studies. It may also be worthwhile to probe gamma-aminobutyric acid (GABA)-mediated intracortical inhibition using paired-pulse TMS. Although Badawy and colleagues found no significant effect of glucose on short interval intracortical inhibition (SICI), all significant and non-significant differences between fasted and fed participants were in the direction of increased inhibition or decreased facilitation after feeding as opposed to after fasting [9]. This is counterintuitive, given the apparent increase in MEP size after glucose ingestion [10], and warrants further investigation.

The goal of this study was to test the effect of glucose ingestion on the healthy human brain using a variety of non-invasive neurophysiological measures which have relevant mechanistic underpinnings. This investigation also controlled for the influence of a sweet placebo and time-locked measurements to individually measured peak blood glucose latencies. It was hypothesized that glucose would increase corticospinal excitability and SICI. The primary observation was that glucose did not change SICI, SAI, LAI or corticospinal excitability. Furthermore, corticospinal excitability and blood pressure increased over time when data were averaged across all treatments.

2. Materials and Methods

2.1. Participants

Healthy, young (n = 18, 22.8 ± 2.4 years), right-handed, male non-smokers were recruited from the McMaster University student population. Participants passed a screening for TMS contraindications [18] and were identified as right-handed using a modified handedness questionnaire [19]. Inactive individuals were excluded using the International Physical Activity Questionnaire (IPAQ, <600 MET-minutes/week) to reduce the risk of influence from prediabetic impairment of glucose metabolism, which is inversely correlated with physical activity level [20]. This study was approved by the Hamilton Integrated Research Ethics Board (HiREB) and conformed to the declaration of Helsinki.

2.2. Electromyography

Motor-evoked potentials were recorded via surface electromyography (EMG) over the first dorsal interosseous (FDI) muscle. FDI was chosen because MEPs from this muscle have demonstrated good inter- and intra-session reliability [21]. Adhesive electrodes (9 mm diameter Ag-AgCl) were placed over the FDI muscle belly and the metacarpal head of the index finger. The EMG signal was amplified 1000x and sampled at 5 kHz with low and high pass signal filters of 2.5 kHz and 20 Hz, respectively. EMG data was recorded using an analog-to-digital interface (Power 1401; Cambridge Electronics Design, Cambridge, UK) in combination with Signal/CED analysis software (Signal version 6.02; Cambridge Electronics Design).

2.3. Transcranial Magnetic Stimulation

Participants sat upright in the testing chair with their palms resting supine and elbows at an approximate 45° angle. Single and paired-pulse TMS was delivered with a custom 50 mm figure-of-eight coil, connected to a Magstim Bistim stimulator (Magstim, Whitland, UK). The coil was held over the left primary motor cortex (M1) and the optimal stimulation location or "motor hotspot" was targeted using Brainsight neuro-navigation software (Rogue Research, Canada). The motor hotspot for FDI muscle was determined by delivering pulses at 50% of the maximum stimulator output (%MSO) over the approximate location of M1 while adjusting coil placement until the TMS pulses reliably evoked large MEPs in the FDI muscle. The angle of the coil relative to the midsagittal plane was maintained at 45° to induce posterior-to-anterior current in cortical tissue.

2.4. Resting Motor Threshold

Resting motor threshold (RMT) was defined as the stimulus intensity (%MSO) that evokes an MEP (i.e., peak-to-peak amplitude >50 μV) 50% of the time. This value was determined using TMS_MTAT_2.0 freeware (http://clinicalresearcher.org/software.htm). The starting stimulus intensity was set to 37% MSO. Twenty TMS pulses were then delivered over M1, adjusting the intensity after each pulse, as determined by the MTAT software based on the MEP occurrence (or lack thereof) in the previous trial [22].

2.5. MEP Recruitment Curve

Corticospinal excitability was measured using single-pulse TMS to obtain MEP recruitment curves. Eight TMS pulses were delivered at 90%, 100%, 110%, 120%, 130%, 140%, 150%, 160%, 170%, 180%, 190%, and 200% of RMT in a randomized order, with an inter-stimulus interval of 4 s. The MEP amplitude was plotted against stimulus intensity and data were fit with a Boltzmann sigmoidal curve. The regression line was segmented 1000 times and the area under the recruitment curve (AURC) was quantified by trapezoidal integration.

2.6. Short Interval Intracortical Inhibition

Short-interval intracortical inhibition (SICI) was measured using a paired-pulse TMS protocol. The suprathreshold test stimulus (TS) was adjusted to the intensity that evoked an MEP with a size of approximately 1 mV in peak–peak amplitude. The conditioning stimulus (CS) preceded the TS by 2 ms, with a stimulus intensity of 80% of RMT. Twelve unconditioned (MEPTS) and 12 conditioned (MEPCS-TS) pulses were delivered in a randomized order, with a 5 s inter-trial interval. The magnitude of SICI was quantified using the ratio of conditioned to unconditioned MEP size (MEPCS-TS/ MEPTS).

2.7. Afferent Inhibition

Electroencephalography (EEG) electrodes were positioned over C3′ (located 2 cm posterior to C3) and referenced to Fz (International 10–20 system). A bar electrode was positioned over the median nerve at the wrist (cathode proximal) to deliver square wave electrical pulses (0.2 s pulse width) using a constant current stimulator (DS7AH; Digitimer, Welwyn Garden City, UK). Nerve stimulation was delivered at the minimum intensity that evoked a visible twitch in the abductor pollicis brevis (APB) muscle. Time-locked averaging of five hundred stimuli delivered at 3 Hz was used to determine the latency of the N20 peak of the somatosensory-evoked potential (SEP).

To acquire afferent inhibition, the TMS intensity was set to evoke a MEP of ~1 mV in the right FDI muscle. Electrical stimulation was delivered to the median nerve at the wrist at the minimum intensity that evoked a visible twitch in the APB muscle. The average intensity of nerve stimulation was 11.1 ± 3.8 mA. For SAI, the interstimulus interval (ISI) between peripheral nerve stimulation and TMS was 4 ms longer than the N20 latency (i.e., N20 + 4 ms). An ISI of 200 ms was used to acquire LAI. Twelve unconditioned stimuli (MEPTS) were randomly presented among 36 conditioned stimuli (nerve stimulation followed by TMS, twelve stimuli per ISI), with a 5 s inter-trial interval. The magnitude of SAI/LAI was expressed as the ratio of the conditioned to the unconditioned MEP amplitude.

2.8. Blood Glucose and Blood Pressure

Capillary blood glucose measurements were performed via the glucose oxidase method using a hand-held diabetes monitoring device (Abbott MediSense FreeStyle Precision Neo Blood Glucose and Ketone Monitoring System, Abbott). Since previous research has indicated that blood pressure may be elevated by ingestion of a large glucose bolus [23,24], mean arterial blood pressure was measured using an automated blood pressure monitor (OMRON Blood Pressure Monitor, OMRON Healthcare). The mean arterial pressure (MAP) was calculated from the systolic (SBP) and diastolic blood pressure (DBP) as indicated below:

$$\text{MAP} = (2\text{DBP} + \text{SBP})/3 \quad (1)$$

2.9. Experimental Design

This study implemented a double-blinded, three-way crossover design in which fasted participants were assessed before and after ingestion of water, a sucralose-flavored placebo or a 75 g glucose bolus. All solutions were 300 mL. Prior to the first experimental testing session, participants completed a preliminary testing session. A schematic of the study schedule for each participant is shown in Figure 1.

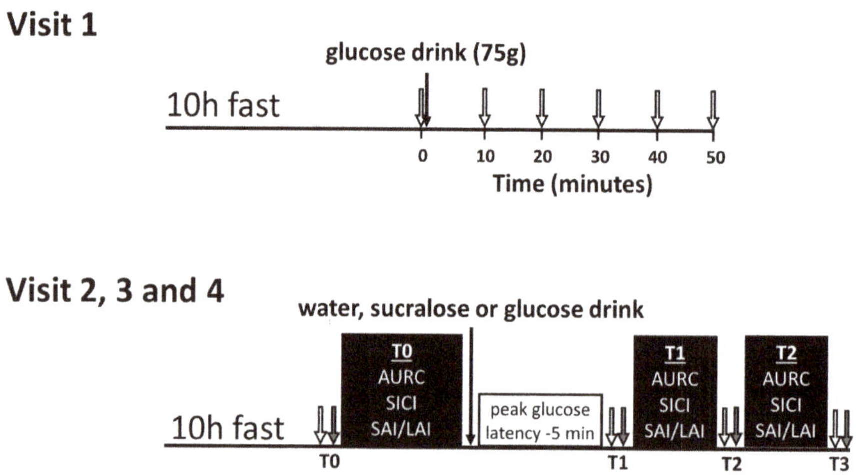

Figure 1. Timeline for the preliminary visit (visit 1) and the three experimental visits (visits 2, 3 and 4). For visit 1, baseline blood glucose was measured (0 min) and then participants drank 75 g of glucose. Next, blood glucose was measured every 10 min for 50 min or until peak was observed (open arrows). For visits 2, 3 and 4, the transcranial magnetic stimulation (TMS) testing bouts (black boxes) at T0 and T1 were separated by a rest period equal to the glucose latency minus 5 min. Mean arterial pressure (grey arrows) and blood glucose (open arrows) were measured before, after and in between the TMS testing bouts and are labeled T0, T1, T2 and T3. TMS—transcranial magnetic stimulation; AURC—area under the recruitment curve; SICI—short-interval intracortical inhibition; SAI/LAI—short-latency afferent inhibition/long-latency afferent inhibition.

Visit 1 was used to assess a time-course for glucose metabolism, allowing the subsequent TMS measures on visits 2, 3, and 4 to be individualized. Participants arrived in the lab having fasted for a minimum of 10 h, and then ingested a 75 g glucose bolus in 300 mL of solution. Finger-prick blood samples were collected and analyzed at 10 min intervals, as indicated in Figure 1 (top). The latency at which peak blood glucose occurred was used to ensure that TMS tests are conducted during a period of high-circulating glucose for each individual.

Visits 2, 3 and 4 were scheduled at least 48 h apart. On the day of each visit, participants arrived in the lab having fasted for 10 h and then ingested a 300 mL solution containing either plain water, sucralose-sweetened placebo (5 g/300 mL Splenda® solution), or a 75 g oral glucose tolerance test bolus. TMS measures were acquired before ingestion (T0), 5 min before each participant's peak blood glucose latency (~30 min after drink ingestion) as determined in Visit 1 (T1), and ~1 h after ingestion, corresponding to the approximated peak of glucose levels in the cerebrospinal fluid (T2) occurring ~30 min after plasma glucose [25]. SAI, LAI, SICI and MEP recruitment curves were measured in a pseudorandomized order which was determined using an online Latin square generator (https://hamsterandwheel.com/grids/index2d.php). Capillary blood glucose and blood pressure were measured before and after each of the post-drink bouts, as denoted by the labels T1, T2 and T3 (see Figure 1, bottom).

The McMaster University Medical Centre (MUMC) research pharmacy provided a randomized treatment schedule. All treatment solutions were provided in uniform, shrouded bottles, with a letter code corresponding to the order of delivery. MUMC pharmacy held the drink randomization (i.e., drink identity) key until collection was complete to ensure that the experimenters were blind to the identity of the drink. Blood glucose and subjective ratings of sweetness were recorded by an unblinded researcher who did not otherwise take part in data collection or analysis. The sucralose-sweetened placebo was taste-matched with the 75 g glucose solution by MUMC pharmacy. Participants were explicitly asked not to comment on the taste of the drink to the researchers and it was made clear that this was very important to the integrity of the study. The participants were blind to the identity of the drink to the degree that they could not distinguish between the sucralose placebo and the glucose solutions (water was not masked with any taste).

2.10. Statistical Analyses

Trials were discarded if the EMG activity was >100 μV in the 100 ms preceding the stimulation artefacts, similar to previous work [26]. Normality was assessed with the Shapiro–Wilks test and a square-root or log transformation was applied in cases where data was not normally distributed.

First, to confirm that inhibition was observed for measures of SICI, LAI and SAI, two-way ANOVAs with the factors PATTERN (two levels: unconditioned MEP and conditioned MEP) and TIME (three levels: T0, T1, T2) were performed. Next, one-way ANOVAs using the within-subjects factor of TREATMENT (three levels: glucose, sucralose, water) were used to confirm that T0 data were not different between visits. Next, outlier analysis was performed using SPSS. Four outliers were removed from the SICI data and one was removed from the SAI data. Data were subsequently analyzed using repeated-measure ANOVAs with factors TREATMENT (three levels: glucose, sucralose, water) and TIME (three levels: T0, T1, T2). Post-hoc testing was performed with Bonferroni-corrected two-tailed paired t-tests. A Conover's ANOVA [27] was performed in lieu of a parametric ANOVA in cases where the data was not normally distributed (even after attempted transformations), with the Wilcoxon-signed rank test used for post-hoc testing. Supplementary measures of capillary blood glucose and MAP were assessed using two-way repeated measures ANOVA with four levels of TIME (T0, T1, T2, T3).

Wilcoxon signed-rank tests were used to assess the hypotheses that glucose would strengthen SICI and increase AURC, as well as all indicated post-hoc comparisons. Effect sizes were calculated using Cohen's d for paired t-tests and r for Wilcoxon's tests. Significance for all statistical tests was set to alpha <0.05. Data from TMS, blood glucose and blood pressure measurements are included in the supplementary materials.

3. Results

3.1. Blood Glucose

The peak plasma glucose concentration after ingestion of the glucose bolus on Visit 1 was 9.5 ± 1.0 mmol/L. This represents a 2-fold increase from the fasting glucose level of 4.8 ± 0.4 mmol/L, with a rise of 4.7 ± 0.9 mmol/L. The majority of participants had peak glucose latencies of 40 min

($n = 10$), followed by six peaking at 30 min, and one participant peaking at 20 min and 50 min each. The observed peak glucose latency spread of 30 min emphasizes the importance of the individualized approach tested herein.

As expected, glucose levels for the experimental visits exhibited a significant effect of TREATMENT (F (2,17) = 108.261; $p < 0.001$), TIME (F (3,17) = 31.037; $p < 0.001$) and TREATMENT × TIME (F (6,17) = 29.722; $p < 0.001$). In the glucose delivery condition, plasma levels measured at T1 were 3.4 ± 1.9 mmol/L higher than fasting and remained elevated by 2.1 ± 1.2 mmol/L at T3. These data confirm that glucose levels were substantially increased during both post-drink TMS testing bouts in the glucose delivery condition. Notably, there was a slight increase in glucose level (<1 mmol/L) at T1 after the sucralose-sweetened placebo and a slight reduction at T1 following water. However, blood glucose returned to fasting levels at T2 and T3 in both cases. These data are displayed in Figure 2a and p-values and effect sizes for significant changes are reported in Table 1.

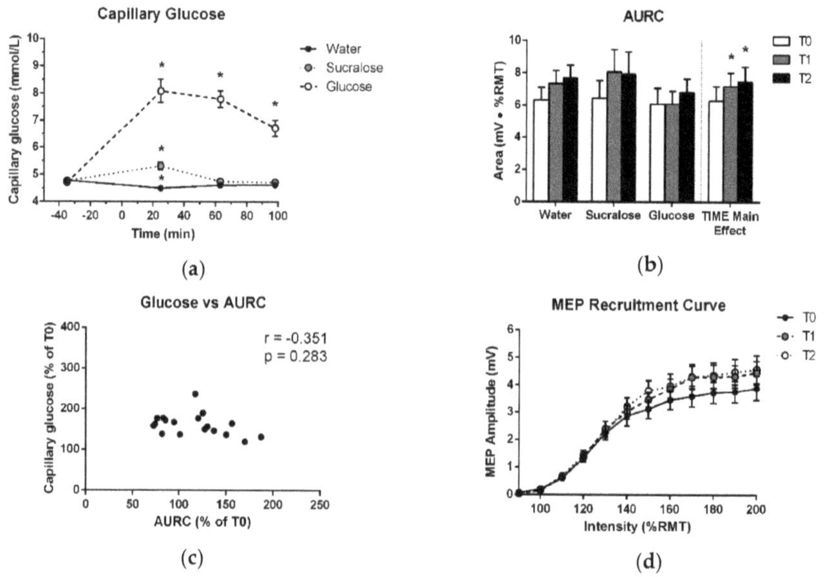

Figure 2. (**a**) Mean and standard error of capillary blood glucose levels over the course of each testing session. Measurements are plotted at each timepoint before and after ingestion of glucose, sucralose or water, which were consumed at 0 min. (**b**) Means and standard errors of AURC data entered into the ANOVA and treatment-averaged data, showing the significant effect of TIME. (**c**) Average relative increase in plasma glucose across T1, T2 and T3, plotted against the average relative increase in AURC across T1 and T2. (**d**) Treatment-averaged motor-evoked potentials (MEP) responses with mean and standard error at each TMS intensity of the recruitment curve for T0, T1 and T2. * indicates a significant difference from T0 ($p < 0.05$).

Table 1. Test statistics, effect sizes and p-values for all significant differences.

Measure	Timepoint	Test Statistic	Effect Size	p-Value
GLU_{glu}	T1–T0	7.781	1.834	<0.001
	T2–T0	9.370	2.209	<0.001
	T3–T0	6.963	1.641	<0.001
GLU_{water}	T1–T0	−3.044	−0.717	0.043
GLU_{suc}	T1–T0	7.434	1.752	<0.001
MAP_{all}	T1–T0	7.817	1.842	<0.001
	T2–T0	5.123	1.208	<0.001
	T3–T0	6.922	1.632	<0.001
$AURC_{all}$ *	T1–T0	2.940	0.490	0.003
	T2–T0	2.983	0.497	0.003

$AURC_{all}$: area under the recruitment curve averaged across all treatments, GLU: glucose data for each of the three drinks, MAP_{all}: mean arterial pressure averaged across all treatments * indicates that effect test statistics and effect size are derived from Wilcoxon's test (i.e., z and r, respectively).

3.2. MEP Recruitment Curve

As shown in Figure 2b, there was no increase in AURC after glucose at T1 or T2 (p = 0.110). There was only a significant effect of TIME, which was independent of treatment (Table 2, Figure 2b,d). No other effects in the ANOVA and no association between capillary blood glucose levels and the change in AURC at T1 or T2 were found (Table 3, Figure 2c). Further, no difference was detected between T0 data for the three treatments (Table 4).

Table 2. Results of two-way ANOVAs with factors TREATMENT and TIME, and the grand coefficient of variation (CV) of each TMS measurement.

Measure	Grand CV (%)	Factor	df	F	p-Value
AURC ¶	60.8	TREATMENT	2,32	1.555	0.226
		TIME	2,32	10.429	0.001
		TREATMENT × TIME	4,64	2.290	0.069
SICI #	80.8	TREATMENT	2,30	0.162	0.851
		TIME	2,30	2.640	0.088
		TREATMENT × TIME	4,60	1.712	0.200
SAI	44.6	TREATMENT	2,30	0.059	0.943
		TIME	2,30	0.118	0.889
		TREATMENT × TIME	4,60	0.268	0.897
LAI *	60.8	TREATMENT	2,32	2.663	0.085
		TIME	2,32	2.813	0.075
		TREATMENT × TIME	4,64	0.591	0.679

AURC: area under the recruitment curve, SICI: short-interval intracortical inhibition, SAI: short-latency afferent inhibition, LAI: long-latency afferent inhibition. * indicates data was square root transformed, # indicates log transformation, ¶ indicates data was ranked.

Table 3. Results from correlations between area under the recruitment curve (AURC) and capillary glucose (GLU) or mean arterial pressure (MAP).

Measure/Timepoint A	Measure/Timepoint B	Correlation Coefficient	p-Value
AURC $T1_{glu}$	GLU $T1_{glu}$	−0.432 *	0.261
AURC $T1_{glu}$	GLU $T2_{glu}$	−0.346	0.502
AURC $T1_{all}$	MAP $T1_{all}$	−0.127	0.978
AURC $T1_{all}$	MAP $T2_{all}$	0.078	0.996
AURC $T2_{glu}$	GLU $T2_{glu}$	−0.366	0.440
AURC $T2_{glu}$	GLU $T3_{glu}$	−0.492 *	0.144
AURC $T2_{all}$	MAP $T2_{all}$	0.017	1.000
AURC $T2_{all}$	MAP $T3_{all}$	−0.167	0.941
Ave AURC $T1_{glu}$ & $T2_{glu}$	Ave GLU $T1_{glu}$, $T2_{glu}$ & $T3_{glu}$	−0.351	0.283
Ave AURC $T1_{all}$ & $T2_{all}$	Ave MAP $T1_{all}$, $T2_{all}$ & $T3_{all}$	−0.089	0.924

Correlations were tested between relative change in AURC and GLU for the adjacent timepoint on the glucose visit only ($T1_{glu}$, $T2_{glu}$ and $T3_{glu}$). Mean arterial pressure correlations were calculated for treatment-averaged data at each timepoint ($T1_{all}$, $T2_{all}$ and $T3_{all}$). Correlations between the average change across all post-drink measures were also examined (bottom). All correlations were carried out using Pearson's r unless data were not normally distributed, in which case Spearman's rho (indicated by *) was used. Bonferroni correction was applied to all correlations.

Table 4. Results from preliminary ANOVAs confirming the presence of inhibition and no differences between T0 data for each measure.

Measure.	Factor	*df*	F	p-Value
AURC	TREATMENT (T0)	2,34	0.162	0.851
SICI	TREATMENT (T0)	2,32	0.206	0.732
Water	PATTERN	1,16	14.902	0.001
	TIME	2,32	0.463	0.560
	PATTERN × TIME	2,32	1.348	0.269
Sucralose	PATTERN	1,15	40.673	<0.001
	TIME	2,30	1.381	0.267
	PATTERN × TIME	2,30	3.372	0.069
Glucose	PATTERN	1,17	28.103	<0.001
	TIME	2,34	4.191	0.024
	PATTERN × TIME	2,34	1.384	0.264
SAI	TREATMENT (T0)	2,34	0.325	0.725
Water	PATTERN	1,16	49.518	<0.001
	TIME	2,32	0.002	0.998
	PATTERN × TIME	2,32	0.664	0.552
Sucralose	PATTERN	1,17	52.058	<0.001
	TIME	2,34	0.892	0.419
	PATTERN × TIME	2,34	0.345	0.710
Glucose	PATTERN	1,17	24.709	<0.001
	TIME	2,34	0.477	0.625
	PATTERN × TIME	2,34	0.100	0.905
LAI	TREATMENT (T0)	2,34	0.458	0.637
Water	PATTERN	1,16	16.100	0.001
	TIME	2,32	0.214	0.809
	PATTERN × TIME	2,32	2.256	0.121
Sucralose	PATTERN	1,17	28.422	<0.001
	TIME	2,34	0.858	0.414
	PATTERN × TIME	2,34	0.017	0.983
Glucose	PATTERN	1,17	8.101	0.011
	TIME	2,34	2.978	0.064
	PATTERN × TIME	2,34	1.024	0.370

AURC: area under the recruitment curve, SICI: short-interval intracortical inhibition, SAI: short-latency afferent inhibition, LAI: long-latency afferent inhibition.

3.3. SICI

The presence of SICI was confirmed by two-way ANOVAs which showed the main effects of PATTERN for each treatment (Table 4), indicating that inhibition was observed at all timepoints. There was no increase in SICI following glucose (Figure 3a) and there was no effect of TREATMENT, TIME or TREATMENT × TIME observed in the ANOVA (Table 2).

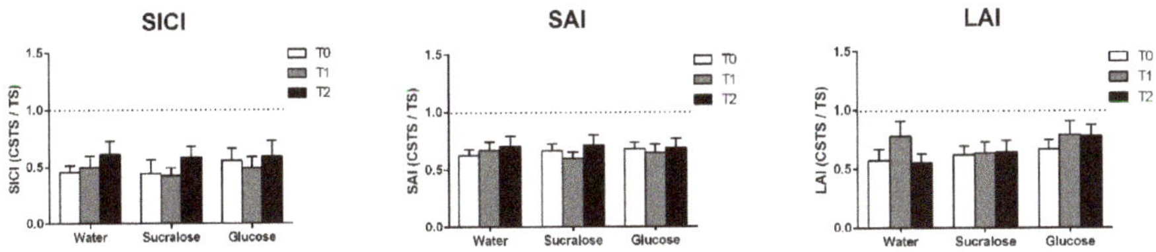

(a) (b) (c)

Figure 3. Means and standard errors of (**a**) SICI, (**b**) SAI and (**c**) LAI data. All data are expressed as the ratio of the conditioned response (CSTS) to the unconditioned response (TS) such that the degree to which the ratio falls below 1.0 (i.e., the dotted line) reflects the magnitude of inhibition observed.

3.4. Afferent Inhibition

For SAI and LAI, two-way ANOVAs confirmed the presence of significant inhibition (Table 4). There was no effect of TREATMENT, TIME or TREATMENT × TIME (Table 2, Figures 2c and 3b).

3.5. Mean Arterial Pressure

Figure 4 shows a main effect of TIME ($F_{(2,12} = 15.119$; $p < 0.001$), with significant increases in mean arterial pressure at all post-drink pressure which were observed across all treatments. Post-hoc analysis of treatment-averaged MAP data indicates that this increase from baseline (T0) was evident at T1, T2 and T3 ($p < 0.001$) (Figure 4a). There were also no correlations between the increase in mean arterial pressure across treatments and changes in AURC at T1 or T2 (Table 3).

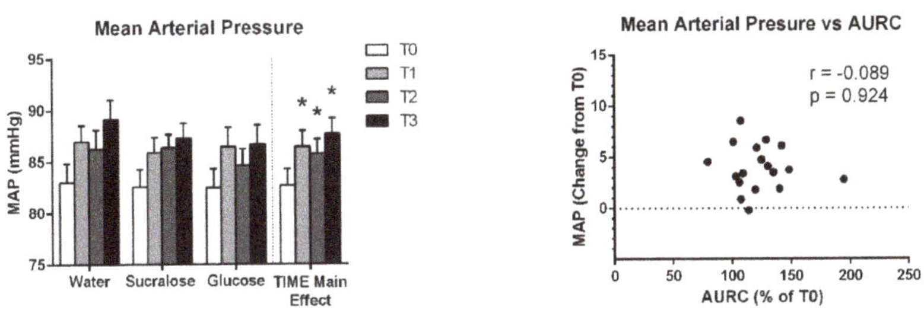

(a) (b)

Figure 4. (**a**) Means and standard errors of mean arterial pressure for each treatment and for the treatment-averaged data showing the effect of TIME. (**b**) Average increase in mean arterial pressure across T1, T2 and T3 plotted against the average relative increase in AURC across the T1 and T2 TMS testing bouts. * indicates a significant difference from baseline ($p < 0.05$).

4. Discussion

The primary finding of this study is that glucose did not lead to an increase in corticospinal excitability. Further, there was no change in SICI, SAI or LAI. However, we did observe that all treatments contributed to an increase in corticospinal excitability and mean arterial pressure. This increase in excitability was not related to the increase in mean arterial blood pressure. We will discuss below the lack of glucose effects on TMS measures and the possible role of hydration and prolonged brain stimulation on changes in mean arterial pressure.

The increase in AURC across all conditions contrasts with previous research reporting an increase in corticospinal excitability following glucose ingestion and not after a no-calorie placebo [10]. Notably, Specterman and colleagues [10] tested a small sample size ($n = 4$) compared to the present study. They also measured corticospinal excitability by quantifying the average size of MEPs evoked at 110% of RMT, while the present study did so across the multiple intensities of a recruitment curve. Therefore, it is possible that the contrasting results reflect a combination of differences in the study population and measurement protocol.

The finding that glucose did not alter SICI is in line with the results reported by Badawy and colleagues [9], who found that SICI was not significantly different after 12 h of fasting versus 2 h after a meal. It should be noted that the authors did report significantly greater LICI after feeding than after fasting. These findings suggest that glucose has a different effect on LICI versus SICI. Indeed, each measure is thought to reflect the activity of different neurotransmitter receptors, with $GABA_A$ receptors mediating SICI and $GABA_B$ receptors involved in LICI [28,29].

Intake of glucose, sucralose or water did not change measures of SAI or LAI. SAI and LAI are impaired in multiple neurodegenerative conditions, including Alzheimer's and Parkinson's disease [30], and have potential clinical utility as diagnostic tools or biomarkers of sensorimotor function. Therefore, it is important to establish whether factors of daily living influence the acquisition of these measures. The present study suggests that neither elevated glucose levels or hydration change the magnitude of SAI or LAI. The results also suggest that the weakening of SAI/LAI in these neurodegenerative populations [12,13] does not reflect the impairment in glucose metabolism.

Increased AURC at T1 and T2 coincided with a treatment-nonspecific rise in mean arterial pressure, suggesting influence from factors intrinsic to the testing protocol. These factors may include hydration, discomfort with the testing setup (i.e., a white-coat effect) or the delivery of a high number of TMS pulses over M1. Since participants were not asked to dehydrate themselves overnight, and the 300 mL drink is a relatively low volume of fluid, it seems unlikely that hydration was a major factor with respect to mean arterial pressure. It is more likely that changes in emotional state due to fatigue or level of interest, or the repetitive stimulation itself could have contributed to the observed TIME effects. The sensorimotor cortex has functional connections with brainstem structures involved in the regulation of vasomotor tone, which are modulated by cortical stimulation [31]. Transcranial direct-current stimulation of the sensorimotor area has been shown to acutely suppress blood pressure and adrenocorticotropic hormone levels [32]. If TMS changes blood pressure similarly to direct-current stimulation, it is possible that an association between changes in mean arterial pressure and AURC was masked by acute effects of TMS delivered at T1 and T2. In addition, Binkofski et al. [32] described longer-latency changes (1–2 h) in cerebral energy metabolism and glucose uptake following brain stimulation, suggesting that baseline TMS could have masked relationships between glucose and AURC at T1 and T2. This may provide an explanation for the contrasting results with Specterman et al. [10], who delivered only 45 TMS pulses at baseline and 15 per timepoint. They observed a positive association between glucose level and MEP size, while we delivered 180 pulses at T0, T1 and T2 and did not observe the same relationship.

In the present study, the timing of T1 was based on the latency of peak glucose levels obtained on Visit 1. The purpose of this protocol was to maximize the opportunity to observe a change in corticospinal excitability following glucose ingestion. However, the day-to-day variability in glucose metabolism was not assessed. Therefore, while we observed a significant rise in glucose levels following ingestion of the solution in the glucose session, the latency of peak glucose may have varied from day-to-day.

Future Considerations

It is important for future research to attempt to replicate previously observed effects of glucose on TMS measurements, using testing protocols which consider potential confounding factors such as the number of TMS pulses, session duration and hydration. The effect of TMS pulse load on changes in glucose levels and sympathetic tone should also be investigated to facilitate the interpretation and design of future TMS research. While glucose did not change our TMS measures, other dietary factors such as caffeine consumption [10,33], prolonged fasting [34] and ketogenic diets [35] merit further investigation as these may be important and easily modifiable factors in TMS research.

5. Conclusions

The present study found no explicit effect of glucose on corticospinal excitability, intracortical inhibition or afferent inhibition, but corticospinal excitability and mean arterial pressure increased across all treatments over the course of the experiment. These non-treatment-specific increases suggest that TMS measurements could be sensitive to various confounding factors related to repeated magnetic stimulation of the cortex, hydration, or fatigue. Further investigation of the influence of diet and acute carbohydrate consumption is warranted. However, studies should first directly examine the impact of the aforementioned confounding factors so that they can be effectively taken into account during the design and interpretation of research on this topic. Findings from such studies will work to reduce the likelihood that confounding effects arising from changing brain metabolism or autonomic modulation complicate the interpretation of TMS data.

Supplementary Materials: The following are available online at http://www.mdpi.com/2076-3425/9/12/339/s1, Table S1: AURC, Table S2: SICI MEP size and SICI ratio, Table S3: SAI MEP size and SAI ratio, Table S4: LAI MEP size and LAI ratio, Table S5: Glucose, Table S6: Mean Arterial Pressure.

Author Contributions: Conceptualization, S.L.T., C.V.T. and A.J.N.; Data curation, S.L.T. and A.J.N.; Formal analysis, S.L.T. and C.V.T.; Funding acquisition, A.J.N.; Investigation, S.L.T., C.V.T., M.B.L., C.N. and R.R.; Methodology, S.L.T., C.V.T. and A.J.N.; Project administration, S.L.T. and A.J.N.; Resources, A.J.N.; Software, A.J.N.; Supervision, S.L.T. and A.J.N.; Validation, A.J.N.; Visualization, S.L.T., C.V.T. and A.J.N.; Writing—original draft, S.L.T.; Writing—review and editing, S.L.T., C.V.T., M.B.L., C.N., R.R. and A.J.N.

Funding: This research was funded in part by a Natural Sciences and Engineering Research Council of Canada grant (NSERC RGPIN-2015-06309) to AJN.

Acknowledgments: The authors would like to thank Gita Sobhi and Megan Jutting of the McMaster University Research Pharmacy, who provided the treatment drinks and randomization schedule for this study. We would also extend our gratitude to Diana Harasym, Patrick Dans, and Jenin El-Sayes, all of whom assisted with data collection.

Conflicts of Interest: The authors declare no conflict of interest. The funders had no role in the design of the study; in the collection, analyses, or interpretation of data; in the writing of the manuscript, or in the decision to publish the results.

References

1. Ashrafi, G.; Ryan, T.A. Glucose metabolism in nerve terminals. *Curr. Opin. Neurobiol.* **2017**, *45*, 156–161. [CrossRef] [PubMed]
2. Dienel, G.A. Fueling and Imaging Brain Activation. *ASN Neuro* **2012**, *4*, AN20120021. [CrossRef] [PubMed]
3. Dienel, G.A. Brain Glucose Metabolism: Integration of Energetics with Function. *Physiol. Rev.* **2018**, *99*, 949–1045. [CrossRef] [PubMed]
4. Mergenthaler, P.; Lindauer, U.; Dienel, G.A.; Meisel, A. Sugar for the brain: The role of glucose in physiological and pathological brain function. *Trends Neurosci.* **2013**, *36*, 587–597. [CrossRef]
5. Simeone, T.A.; Simeone, K.A.; Rho, J.M. Ketone Bodies as Anti-Seizure Agents. *Neurochem. Res.* **2017**, *42*, 2011–2018. [CrossRef]
6. Lu, C.L.; Chang, Y.H.; Sun, Y.; Li, C.Y. A population-based study of epilepsy incidence in association with type 2 diabetes and severe hypoglycaemia. *Diabetes Res. Clin. Pract.* **2018**, *140*, 97–106. [CrossRef]

7. Desilles, J.P.; Meseguer, E.; Labreuche, J.; Lapergue, B.; Sirimarco, G.; Gonzalez-Valcarcel, J.; Lavallée, P.; Cabrejo, L.; Guidoux, C.; Klein, I.; et al. Diabetes mellitus, admission glucose, and outcomes after stroke thrombolysis: A registry and systematic review. *Stroke* **2013**, *44*, 1915–1923. [CrossRef]
8. Andersen, H.; Nielsen, S.; Nielsen, J.F. Motor cortical excitability remains unaffected of short-term hyperglycemia in Type 1 diabetic patients. *J. Diabetes Complicat.* **2006**, *20*, 51–55. [CrossRef]
9. Badawy, R.A.B.; Vogrin, S.J.; Lai, A.; Cook, M.J. Cortical excitability changes correlate with fluctuations in glucose levels in patients with epilepsy. *Epilepsy Behav.* **2013**, *27*, 455–460. [CrossRef]
10. Specterman, M.; Bhuiya, A.; Kuppuswamy, A.; Strutton, P.H.; Catley, M.; Davey, N.J. The effect of an energy drink containing glucose and caffeine on human corticospinal excitability. *Physiol. Behav.* **2005**, *83*, 723–728. [CrossRef]
11. Turco, C.V.; El-Sayes, J.; Savoie, M.J.; Fassett, H.J.; Locke, M.B.; Nelson, A.J. Short- and long-latency afferent inhibition; uses, mechanisms and influencing factors. *Brain Stimul.* **2018**, *11*, 59–74. [CrossRef] [PubMed]
12. Borghammer, P.; Hansen, S.B.; Eggers, C.; Chakravarty, M.; Vang, K.; Aanerud, J.; Hilker, R.; Heiss, W.D.; Rodell, A.; Munk, O.L.; et al. Glucose metabolism in small subcortical structures in Parkinson's disease. *Acta Neurol. Scand.* **2012**, *125*, 303–310. [CrossRef] [PubMed]
13. Kapogiannis, D.; Mattson, M.P. Disrupted energy metabolism and neuronal circuit dysfunction in cognitive impairment and Alzheimer's disease. *Lancet Neurol.* **2011**, *10*, 187–198. [CrossRef]
14. Stagg, C.J.; Bestmann, S.; Constantinescu, A.O.; Moreno Moreno, L.; Allman, C.; Mekle, R.; Woolrich, M.; Near, J.; Johansen-Berg, H.; Rothwell, J.C. Relationship between physiological measures of excitability and levels of glutamate and GABA in the human motor cortex. *J. Physiol.* **2011**, *589*, 5845–5855. [CrossRef]
15. Ikemoto, A.; Bole, D.G.; Ueda, T. Glycolysis and glutamate accumulation into synaptic vesicles: Role of glyceraldehyde phosphate dehydrogenase and 3-phosphoglycerate kinase. *J. Biol. Chem.* **2003**, *278*, 5929–5940. [CrossRef]
16. Pellerin, L.; Magistretti, P.J. Glutamate uptake into astrocytes stimulates aerobic glycolysis: A mechanism coupling neuronal activity to glucose utilization. *Proc. Natl. Acad. Sci. USA* **1994**, *91*, 10625–10629. [CrossRef]
17. Roberts, B.L.; Zhu, M.; Zhao, H.; Dillon, C.; Appleyard, S.M. High glucose increases action potential firing of catecholamine neurons in the nucleus of the solitary tract by increasing spontaneous glutamate inputs. *Am. J. Physiol. Regul. Integr. Comp. Physiol.* **2017**, *313*, R229–R239. [CrossRef]
18. Rossi, S.; Hallett, M.; Rossini, P.M.; Pascual-Leone, A.; Avanzini, G.; Bestmann, S.; Berardelli, A.; Brewer, C.; Canli, T.; Cantello, R.; et al. Safety, ethical considerations, and application guidelines for the use of transcranial magnetic stimulation in clinical practice and research. *Clin. Neurophysiol.* **2009**, *120*, 2008–2039. [CrossRef]
19. Oldfield, R.C. The assessment and analysis of handedness: The Edinburgh inventory. *Neuropsychologia* **1971**, *9*, 97–113. [CrossRef]
20. Hu, X.; Yu, W.; Yang, L.; Pan, W.; Si, Q.; Chen, X.; Li, Q.; Gu, X. Inverse association between physical activity and blood glucose is independent of sex, menopause status and first-degree family history of diabetes. *J. Diabetes Investig.* **2019**, 1–8. [CrossRef]
21. Vaseghi, B.; Zoghi, M.; Jaberzadeh, S. Inter-pulse interval affects the size of single-pulse TMS-induced motor evoked potentials: A reliability study. *Basic Clin. Neurosci.* **2015**, *6*, 44–51. [PubMed]
22. Ah Sen, C.B.; Fassett, H.J.; El-Sayes, J.; Turco, C.V.; Hameer, M.M.; Nelson, A.J. Active and resting motor threshold are efficiently obtained with adaptive threshold hunting. *PLoS ONE* **2017**, *12*, e0186007. [CrossRef] [PubMed]
23. Rebello, T.; Hodges, R.E.; Smith, J.L. Short-term effects of various sugars on antinatriuresis and blood pressure changes in normotensive young men. *Am. J. Clin. Nutr.* **1983**, *38*, 84–94. [CrossRef] [PubMed]
24. Synowski, S.J.; Kop, W.J.; Warwick, Z.S.; Waldstein, S.R. Effects of glucose ingestion on autonomic and cardiovascular measures during rest and mental challenge. *J. Psychosom. Res.* **2013**, *74*, 149–154. [CrossRef] [PubMed]
25. Shestov, A.A.; Emir, U.E.; Kumar, A.; Henry, P.G.; Seaquist, E.R.; Öz, G. Simultaneous measurement of glucose transport and utilization in the human brain. *Am. J. Physiol. Endocrinol. Metab.* **2011**, *301*, E1040–E1049. [CrossRef] [PubMed]
26. Turco, C.V.; Pesevski, A.; McNicholas, P.D.; Beaulieu, L.-D.; Nelson, A.J. Reliability of transcranial magnetic stimulation measures of afferent inhibition. *Brain Res.* **2019**, *1723*, 146394. [CrossRef]
27. Conover, W.J.; Iman, R.L. Analysis of Covariance Using the Rank Transformation. *Biometrics* **1982**, *38*, 715. [CrossRef]

28. Di Lazzaro, V.; Pilato, F.; Dileone, M.; Profice, P.; Ranieri, F.; Ricci, V.; Bria, P.; Tonali, P.A.; Ziemann, U. Segregating two inhibitory circuits in human motor cortex at the level of GABAA receptor subtypes: A TMS study. *Clin. Neurophysiol.* **2007**, *118*, 2207–2214. [CrossRef]
29. Müller-Dahlhaus, J.F.M.; Liu, Y.; Ziemann, U. Inhibitory circuits and the nature of their interactions in the human motor cortex—A pharmacological TMS study. *J. Physiol.* **2008**, *586*, 495–514. [CrossRef]
30. Turco, C.V.; Locke, M.B.; El-sayes, J.; Tommerdahl, M.; Nelson, A.J. Exploring Behavioral Correlates of Afferent Inhibition. *Brain Sci.* **2018**, *8*, 64. [CrossRef]
31. Sequeira, H.; Viltart, O.; Ba-M'Hamed, S.; Poulain, P. Cortical control of somato-cardiovascular integration: Neuroanatomical studies. *Brain Res. Bull.* **2000**, *53*, 87–93. [CrossRef]
32. Binkofski, F.; Loebig, M.; Jauch-Chara, K.; Bergmann, S.; Melchert, U.H.; Scholand-Engler, H.G.; Schweiger, U.; Pellerin, L.; Oltmanns, K.M. Brain energy consumption induced by electrical stimulation promotes systemic glucose uptake. *Biol. Psychiatry* **2011**, *70*, 690–695. [CrossRef] [PubMed]
33. De Carvalho, M.; Marcelino, E.; De Mendonça, A. Electrophysiological studies in healthy subjects involving caffeine. *J. Alzheimer's Dis.* **2010**, *20*, 9–13. [CrossRef] [PubMed]
34. Mattson, M.P.; Longo, V.D.; Harvie, M. Impact of intermittent fasting on health and disease processes. *Ageing Res. Rev.* **2017**, *39*, 46–58. [CrossRef] [PubMed]
35. Newman, J.C.; Verdin, E. Ketone bodies as signaling metabolites. *Trends Endocrinol. Metab.* **2014**, *25*, 42–52. [CrossRef]

brain sciences

MDPI

Article

Exercise-Induced Fatigue in One Leg Does Not Impair the Neuromuscular Performance in the Contralateral Leg but Improves the Excitability of the Ipsilateral Corticospinal Pathway

Saied Jalal Aboodarda [1], Cindy Xin Yu Zhang [1], Ruva Sharara [1], Madeleine Cline [1] and Guillaume Y Millet [1,2,*]

1 Faculty of Kinesiology, University of Calgary, Calgary, AB T2N 1N4, Canada; saiedjalal.aboodarda@ucalgary.ca (S.J.A.); xinyu.zhang2@ucalgary.ca (C.X.Y.Z.); ruvarashe.sharara@ucalgary.ca (R.S.); madeleine.cline@ucalgary.ca (M.C.)

2 Inter-University Laboratory of Human Movement Biology, University of Lyon, UJM-Saint-Etienne, EA 7424, F-42023 Saint-Etienne, France

* Correspondence: guillaume.millet@univ-st-etienne.fr; Tel.: +33-477-421-894

Received: 5 September 2019; Accepted: 20 September 2019; Published: 25 September 2019

Abstract: To investigate the influence of pre-induced fatigue in one leg on neuromuscular performance and corticospinal responses of the contralateral homologous muscles, three experiments were conducted with different exercise protocols; A ($n = 12$): a 60 s rest vs. time-matched sustained left leg knee extension maximum voluntary contraction (MVC), B ($n = 12$): a 60 s rest vs. time-matched left leg MVC immediately followed by 60 s right leg MVC, and C ($n = 9$): a similar protocol to experiment B, but with blood flow occluded in the left leg while the right leg was performing the 60 s MVC. The neuromuscular assessment included 5 s knee extensions at 100%, 75%, and 50% of MVC. At each force level, transcranial magnetic and peripheral nerve stimuli were elicited to investigate the influence of different protocols on the right (tested) knee extensors' maximal force output, voluntary activation, corticospinal excitability, and inhibition. The pre-induced fatigue in the left leg did not alter the performance nor the neuromuscular responses recorded from the right leg in the three experiments (all $p > 0.3$). However, enhanced corticospinal pathway excitability was evident in the tested knee extensors ($p = 0.002$). These results suggest that the pre-induced fatigue and muscle ischemia in one leg did not compromise the central and peripheral components of the neuromuscular function in the tested contralateral leg.

Keywords: cross-over fatigue; isometric contraction; force; voluntary activation; transcranial magnetic stimulation

1. Introduction

Non-local muscle fatigue (NLMF) refers to a condition in which exercise-induced fatigue in a muscle group transiently impairs neuromuscular function in another limb [1]. In the absence of change in the contractile machinery of the non-exercised muscles (i.e., the index of peripheral fatigue), any attenuation in the neuromuscular function of these muscles is attributed to central mechanisms [2–4]. Presently, these central modulations are not well understood. However, it has been postulated that the fatiguing exercise-induced activation of group III and IV metabo- and mechano-receptors could compromise the motor control and execution of the non-fatigued muscles [5,6].

Transcranial magnetic stimulation (TMS) of the motor cortex in conjunction with electrical stimulation of the peripheral nerve (PNS) are non-invasive techniques that could be employed to investigate the central and peripheral mechanisms of fatigue [7]. Although several investigations have explored the influence of lower limb exercise-induced fatigue on responsiveness of the corticospinal pathway innervating upper limb muscles (heterologous muscle groups) [3,5,8], currently, there is little empirical evidence pertaining to the influence of a fatiguing exercise performed by a lower limb muscle group on central voluntary activation and corticospinal responses of the contralateral homologous muscles.

A conventional research paradigm used to study the concept of NLMF includes a fatiguing exercise in one limb followed by the neuromuscular function assessment of the contralateral (non-fatigued) limb quantified during a single (4–6 s) maximum voluntary contraction (MVC) [9–13]. Given that in this approach the post-fatigue MVC is very brief, it has been postulated that participants might be able to overcome the sense of the pre-induced fatigue and maintain the MVC force output and central voluntary activation (VA) in the tested muscle group [4,10]. In an alternative paradigm, some investigators have recently used long-term sustained or intermittent MVCs performed by the tested muscle groups following the already fatigued contralateral limb [6,14,15]. The pre-induced fatigue in this approach has been shown to promote an augmented sense of fatigue and consequently, to attenuate the cycling performance in the tested contralateral limb [6]. However, there is little documented research that has studied this approach using sustained maximal isometric contractions. Halperin et al. [15] demonstrated a significant decline in the force output and VA (using the twitch interpolated technique) recorded during 12 (5 s) knee extension (KE) MVCs following a fatiguing exercise in the contralateral leg. On the contrary, Kennedy et al. [9] did not find any impairment in the MVC force, and VA recorded during 8 (2–3 s) KE MVCs. The controversy regarding these paradoxical findings was further fuelled by the fact that Kennedy et al. [9] maintained muscle ischemia in the fatigued leg to further activate the group III and IV muscle afferents, yet they did not find any change in MVC force and VA of the contralateral tested muscles. These findings argue against the idea that activation of group III and IV muscle afferents in one lower limb muscle could compromise the isometric performance in the contralateral limb. In addition to these controversies, none of the above-mentioned studies have explored the voluntary cortical activation (using TMS) nor the excitability and inhibition of the corticospinal pathway innervating the tested muscle groups following fatigue induced in the contralateral leg.

Thus, the purpose of the present study was threefold. In experiment A, we investigated whether fatiguing exercise in one leg (e.g., left) could alter the neuromuscular performance, corticospinal excitability, and inhibition of the tested (right) knee extensors recorded during brief maximal contraction performed immediately after the pre-induced fatigue. In experiment B, we explored whether pre-induced fatigue (in the left leg) combined with the subsequent sustained contraction (in the right leg) could alter the muscle performance and corticospinal responses in the tested (right) leg. In these two experiments, the group III and IV muscle afferents originating from the left knee extensors would stop firing after the termination of the sustained MVC in the leg [16]. Therefore, we hypothesized that the pre-induced left leg fatigue would not alter the neuromuscular performance recorded in the tested contralateral leg. In order to extend the input from group III and IV muscle afferents, we chose to maintain muscle ischemia in the fatigued (left) leg [16,17]. Thus, in experiment C, we investigated whether the prolonged activation of afferent receptors could modulate the neuromuscular performance and corticospinal responses in the tested (right) knee extensors. We hypothesized that maintaining blood ischemia in the fatigued leg could modulate the central motor drive and, consequently, the neuromuscular responses in the tested muscles.

2. Materials and Methods

2.1. Participants

Twelve healthy, recreationally active participants (six females: 22 ± 3 years, 168 ± 7 cm, 65 ± 1 kg; and six males: 25 ± 1 years, 175 ± 4 cm, 74 ± 9 kg) provided informed consent to participate in experiments A and B. Five participants from these two experiments and four additional participants volunteered to complete experiment C (three females: 26 ± 3 years, 169 ± 9 cm, 66 ± 9 kg; and six males: 30 ± 4 years, 173 ± 6 cm, 76 ± 8 kg). Participants did not have any history of neurological, cardiovascular, or musculoskeletal injuries and were determined as right-leg dominant based on the preferred leg used to kick a ball [18]. They were instructed to refrain from rigorous physical activity, as well as ingesting caffeine and alcohol at least 24 hours prior to each testing session. Participants also completed the TMS safety checklist [19] and Physical Activity Readiness Questionnaire for Everyone PAR-Q+ form [20]. The procedures were conducted in accordance with the declaration of Helsinki and approved by the Health Research Ethics Authority of the University of Calgary (REB17-0147).

2.2. Experimental Setup and Procedures

The experimental protocol was performed on an isometric chair (Kin–Com, Chattecx Corporation, TN, USA) while the hips and knees were fixed at 90°. To ensure that the upper body did not contribute to the KE force, straps along the trunk and waist were secured, and participants crossed arms to shoulders during contractions. Left and right legs were each secured inside a padded ankle cuff attached to a strain gauge (LC101-2K; Omegadyne, Sunbury, OH, USA) with a non-extensible strap. Force data was sampled at 2000 Hz, digitally converted using the Power Lab acquisition hardware and Lab Chart software (ADInstruments, Bella Vista, Australia), and monitored on a computer. The electromyography (EMG) was recorded from the left and right knee-extensors, and both TMS and PNS were employed during the neuromuscular evaluations [21].

Bipolar self-adhesive Ag/AgCl surface EMG electrodes (Kendall MediTrace, MA) were positioned on the muscle belly of the right and left vastus lateralis (VL), rectus femoris (RF), and biceps femoris (BF) [22]. A reference electrode was placed on the patella of the right leg. Before the placement of electrodes, the area of skin was shaved and abraded to remove dead skin with sandpaper and cleaned with an isopropyl alcohol swab to decrease skin resistance. An inter-electrode impedance of <5 kΩ was obtained prior to recording to ensure an adequate signal-to-noise ratio. All EMG signals were digitized at a sampling rate of 2000 Hz by the PowerLab system (16/30-ML880/P, ADInstruments) and amplified with an octal bio-amplifier (ML138, ADInstruments). EMG signals were bandpass filtered (5–500 Hz) and all data were analyzed offline using the Labchart 8 software (ADInstruments).

Single-pulse TMS was manually delivered to the left motor cortex via a 110 mm concave double-cone coil (maximum output of 1.4 T) connected to a magnetic stimulator (Magstim 200^2 Company Ltd., Whitland, UK). The optimal location of the coil was determined on the scalp for every centimeter from 2 cm anterior and 1 cm to the left of the vertex (i.e., the midpoint from nasal–inion and tragi). Single pulses of TMS at 50% of maximum stimulator output (MSO) were delivered during brief (2–3 s) knee extensions at 20% of MVC and the location which evoked the highest twitch force and motor evoked potential (MEP) amplitudes for VL and RF (without an increase in BF MEP amplitude) was marked on a latex swim cap worn by participants. The optimal stimulator intensity was determined by delivering four stimuli at each intensity between 20 to 80% of the MSO (with 10% increments, in random order) and the intensity that elicited the highest twitch force, as well as VL and RF MEP amplitudes, was used for the rest of the session. The size of superimposed twitch (SIT), VL, or RF amplitude showed a plateau up to 80% of MSO for all participants, thus further increase in TMS intensities (90% or 100% MSO) was not necessary. The average stimulating intensity was 61 ± 8% (range: 40%–80%) of the maximal stimulator output for the six testing sessions.

Square-wave percutaneous electrical stimuli were delivered to the femoral nerve using a constant current stimulator (DS7A, Digitimer, Welwyn Garden City, UK). The cathode electrode (Kendall MediTrace, MA, USA) was secured on top of the inguinal triangle, and the anode (50 × 90 mm electrode, Duractick Plus, CA, USA) was placed in between greater trochanter and suprailiac projections. The stimuli intensity (pulse duration: 1000 μs, maximal voltage: 400 V) was increased in incremental steps until a maximal twitch force and compound muscle action potential (i.e., VL and RF Mmax) were observed. The supramaximal stimulation (i.e., 130% of stimulator output that achieved maximal twitch and Mmax responses) was held constant throughout the protocol. The average stimulating intensity was 126 ± 52 mA (range: 65–185 mA) for the six testing sessions.

Experiment A (n = 12). Neurophysiological responses recorded immediately after a contralateral fatiguing exercise. In two randomly selected testing sessions, the fatigued (left) leg underwent either 60 s sustained KE MVC (Fatigue; Ftg-0) or 60 s rest (Rest-0) (Figure 1A). Before and immediately after these two protocols, the tested (right) leg performed a neuromuscular evaluation including a 5 s KE MVC superimposed with a PNS delivered on the force plateau and another PNS evoked 2 s after the MVC on the relaxed muscle (i.e., potentiated twitch, Pt). Three seconds after the Pt, subjects performed another 5 s MVC followed by contractions at 75% and 50% of MVC. Single-pulse TMS and PNS were delivered at 100%, 75%, and 50% of MVC with ~2 s interstimulus intervals, during which participants would reobtain the target force following each TMS interference to the voluntary force output. The resting interval between 100%, 75%, and 50% MVC was ~5 s. Two sets of neuromuscular evaluations were performed before and one set was performed immediately after each intervention. Additionally, immediately before and after each left leg protocol (either rest or 60 s sustained MVC), a PNS was evoked to the right leg at rest to account for any potential influence of contralateral protocols on contractile property and excitability of the tested muscles.

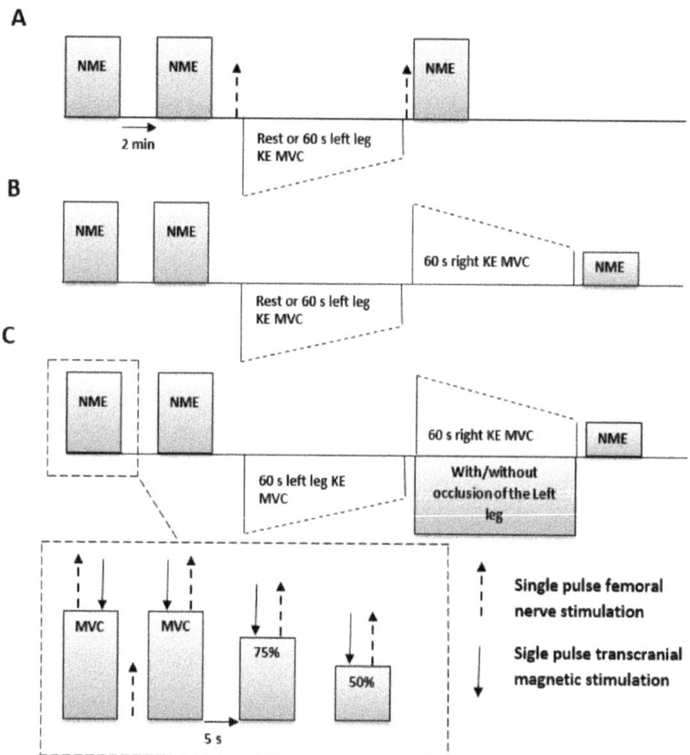

Figure 1. Details of experimental protocols. In experiment A (panel **A**) (n = 12), a 60 s rest or 60 s sustained left leg knee extension (KE) maximum voluntary contraction (MVC) were followed by a neuromuscular evaluation (NME) including 5 s sustained right leg KEs at 100%, 75%, and 50% of MVC. In experiment B (panel **B**) (n = 12), a 60 s rest or 60 s left leg MVC was immediately followed by a 60 s right leg MVC. In experiment C (panel **C**) (n = 9), a similar protocol as experiment B was performed while blood flow was occluded in the left leg while the right leg was performing a 60 s MVC.

Experiment B (n = 12). Neurophysiological responses recorded after a sustained MVC superimposed upon the pre-induced contralateral fatigue. In two randomly selected testing sessions (Figure 1B), participants undertook 60 s rest (Rest-60) or 60 s left leg sustained MVC (Ftg-60), followed immediately by 60 s sustained right leg MVC. Neuromuscular evaluations, as described in experiment A, were performed before and immediately after each experimental protocol.

Experiment C (n = 9). Influence of maintaining muscle ischemia in the fatigued contralateral leg on the neurophysiological responses of the tested muscles. In two randomly selected testing sessions, participants underwent the Ftg-60 protocol (see experiment B) with the exception of the "cuff session (Ftg-60-Occl), where blood flow to the left (fatigued) leg was occluded 2 s prior to the end of 60 s MVC (Figure 1C). A standard single-bladder adult thigh cuff connected to an automatic rapid inflation system (Hokanson E20 AG101, Bellevue, WA, USA) was located at the distal part of the thigh. The cuff was inflated to 300 mmHg using compressed air within 2 s.

2.3. Data Analysis

Performance. A force-time integral including the sum of the data points, multiplied by the sample interval during the 60 s sustained MVCs (performed by the tested leg) was calculated in experiments B and C.

Neuromuscular function. The maximal force output and the corresponding root mean square EMG [rmsEMG of the VL and RF muscles] was quantified during the first and last 2 s intervals of the sustained 60 s left and right leg MVCs, as well as over 500 ms prior to each TMS evoked during brief right leg MVCs. The VL and RF rmsEMG were normalized to the amplitude of the corresponding muscle compound action potential (Mmax) recorded during the same contraction to calculate the rmsEMG·Mmax^{-1} ratio (rmsEMG$_{100}$).

The central voluntary activation (VA$_{PNS}$) was measured using the twitch interpolation technique [23] with the following formula, where Fb is the voluntary force output just before the PNS, D is the difference between Fb and the maximum SIT force evoked by the stimulus, F_{MAX} is the highest force output during MVC, and FPt is the size of Pt force evoked at rest [24].

$$\text{VA}~(\%) = 100 - \text{D} \times (\text{Fb}/\text{F}_{\text{MAX}})/\text{FPt} \times 100 \quad (1)$$

The voluntary cortical activation (VA$_{TMS}$) was assessed using the amplitude of SITs evoked by TMS during 5 s contractions at 100%, 75%, and 50% of MVC [25,26]. The y-intercept of the linear regression ($r^2 > 0.9$) between the SITs was computed to identify the estimated resting twitch (ERT). VA$_{TMS}$ was then calculated using the following equation:

$$\text{VA}_{\text{TMS}} = [1 - (\text{SIT}/\text{ERT})] \times 100 \quad (2)$$

In order to explore the excitability of the corticospinal pathway, the areas under the MEP and the Mmax signals were measured for both VL and RF muscles. The onset of MEP and Mmax were defined as the point at which the voltage trace became tangential to baseline in either the positive or negative direction. The responses to VL and RF MEP were normalized to the subsequent Mmax recorded during the same contraction to calculate MEP·Mmax^{-1} ratios (MEP$_{100}$, MEP$_{75}$, MEP$_{50}$). The duration (ms) of the silent period (i.e., indicative of corticospinal inhibition) was assessed for VL and RF MEPs as the interval from the stimulus artifact to the return of the continuous EMG by visual inspection during 100%, 75%, and 50% of MVC contractions (SP$_{100}$, SP$_{75}$, SP$_{50}$). The end of the SP for two participants in experiments A and B was not distinguishable, so the values were removed from the data pool.

2.4. Statistical Analysis

Statistical analyses were computed using SPSS software (version 23.0; SPSS, Inc., Chicago, IL, USA). Shapiro–Wilk and Mauchly tests were used to ensure the assumptions of normality and sphericity for all dependent variables, respectively. All variables were normally distributed. Greenhouse–Geisser correction factor was applied when the assumption of sphericity was violated. Unless otherwise notified, sphericity assumed data were reported. For each experiment separately, a paired sample *t*-test was used to compare baseline measures and then a two-way repeated measure analysis of variances (ANOVA) was run to explore the effect of 2times (baseline vs. post-intervention values) × 2 conditions for all outcome variables. When ANOVAs showed significant main effects of times or conditions, Bonferroni post hoc test was used to compare values. When an interaction effect was observed, paired *t*-tests with Holm–Bonferroni corrections were applied. In one case where baseline measures revealed a significant difference between conditions (i.e., MEP_{100} in experiment A), the post-intervention values were normalized to the baseline (to account for the day to day variations), and then the paired *t*-test was used to compare the relative values between different conditions. Paired *t*-test was used to compare the force-time integral values (calculated during the 60 s MVCs performed by the tested leg) between the two conditions in each experiment. The effect size was calculated for repeated measure ANOVAs by converting partial eta-squared to Cohen's d and for paired comparisons by using the mean ± SD values [27]. According to Cohen [27], the magnitude of effect size was classified as small ($0.2 \leq d < 0.5$), medium ($0.5 \leq d < 0.8$), and large ($d \geq 0.8$). The mean ± SD of variables are presented in the text, Figures 1–5, and the Table.

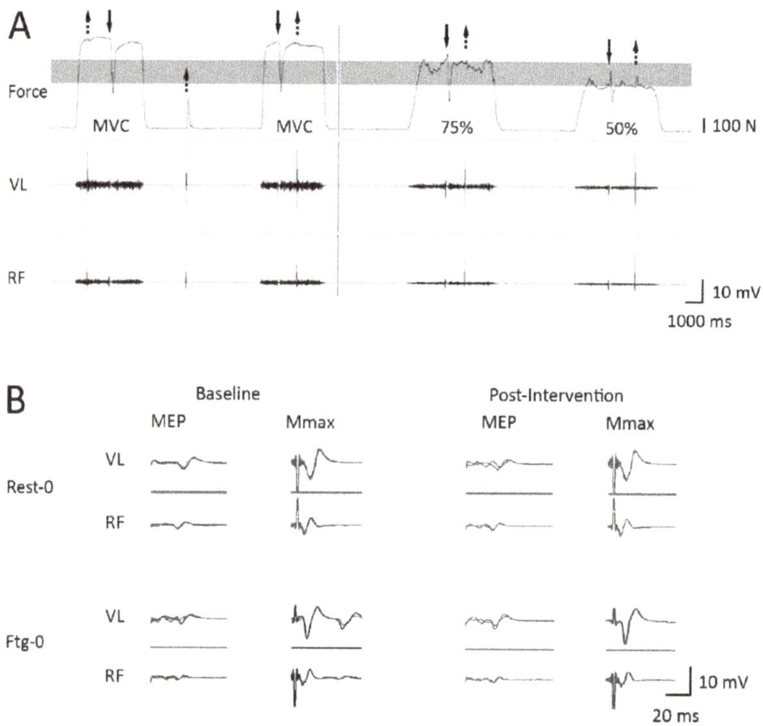

Figure 2. Representative traces from a single subject for the force output and VL and RF EMG signals at 100%, 75%, and 50% of MVC (panel **A**) and the MEPs and Mmaxs recorded from VL and RF at 100% of MVC at baseline and post-intervention levels following a 60 s rest (Rest-0) or 60 s knee extension MVC with the contralateral limb (Ftg-0) (panel **B**). The top and the bottom line of the shaded grey box (in panel **A**) represents 75% and 50% of MVC, respectively. In panel **B**, four MEPs and four Mmaxs recorded at baseline and two MEPs and two Mmaxs recorded at the post-intervention level are superimposed (see Figure 1 for more details).

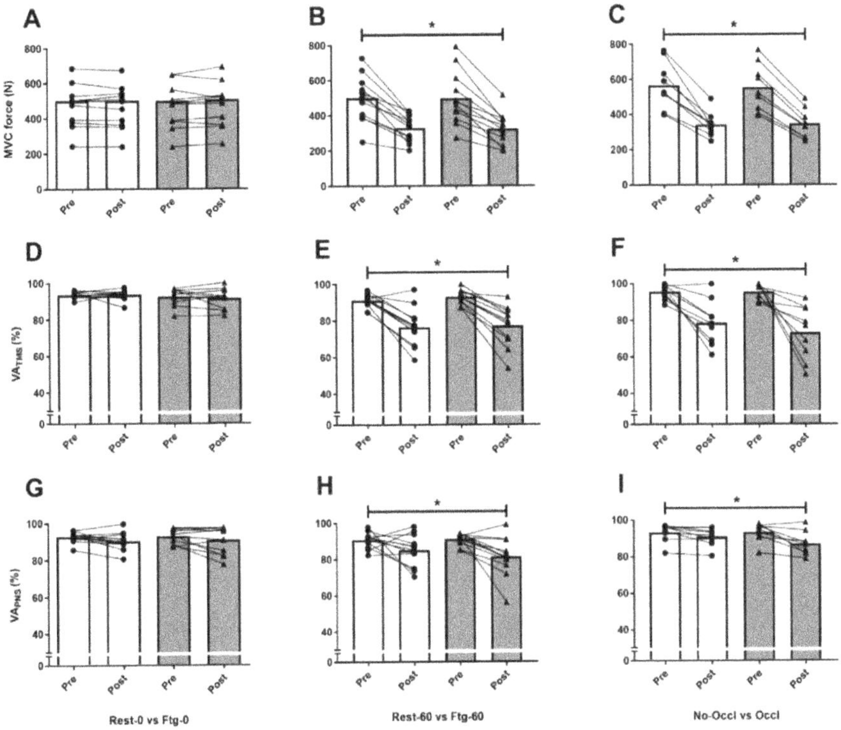

Figure 3. Mean and SD of knee extension (right leg) maximum voluntary contraction (MVC), voluntary activation (VA) using transcranial magnetic stimulation (VA_{TMS}), and peripheral nerve stimulation (VA_{PNS}) values normalized to the baseline. In experiments A (panels **A**,**D**,**G**), a 60 s rest (Rest-0) vs. a 60 s left leg MVC (Ftg-0) were compared. In experiment B (panels **B**,**E**,**H**), a 60 s rest (Rest-60) or a 60 s left leg MVC (Ftg-60) followed immediately by a 60 s right leg MVC were compared. In experiment C (panels **C**,**F**,**I**), Ftg-60 was performed at the absence or presence of a blood flow occlusion in the left leg (Ftg-60-Occl). * Significantly different from baseline (time effect).

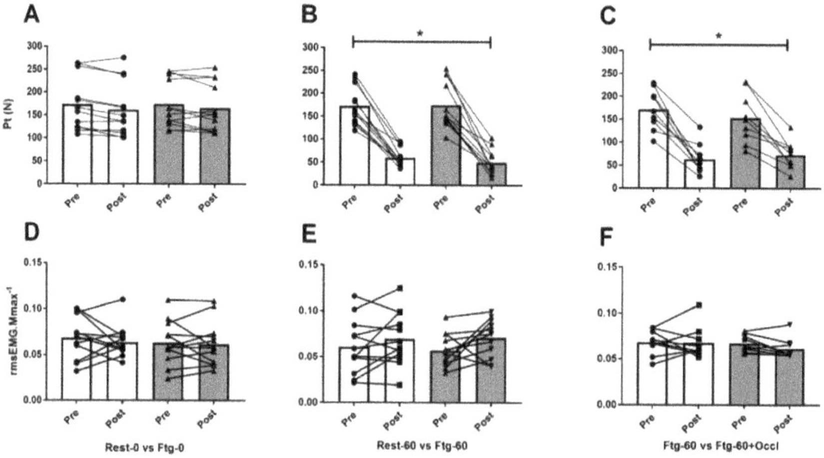

Figure 4. Mean and SD of potentiated twitch (Pt) and rectus femoris (RF) root mean square EMG (rmsEMG) values normalized to the baseline. In experiments A (panels **A**,**D**), a 60 s rest (Rest-0) or a 60 s left leg MVC (Ftg-0) were compared. In experiment B (panels **B**,**E**), a 60 s rest (Rest-60) or a 60 s left leg MVC (Ftg-60) followed immediately by a 60 s right leg MVC were compared. In experiment C (panels **C**,**F**), Ftg-60 was performed at the absence or presence of a blood flow occlusion in the left leg (Ftg-60-Occl). * Significantly different from baseline (time effect).

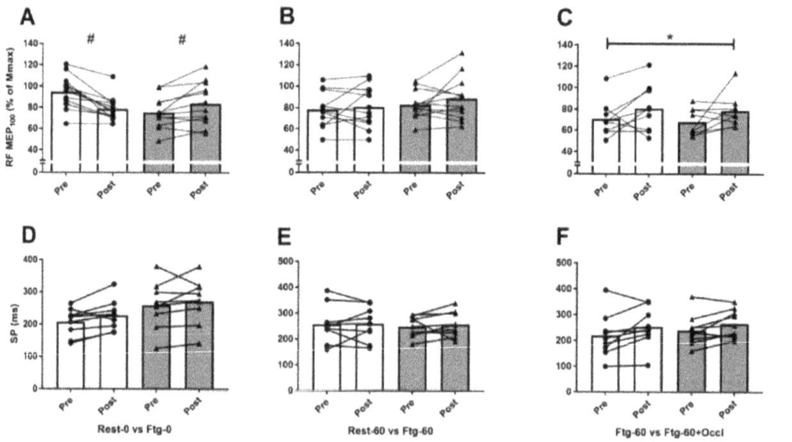

Figure 5. Mean and SD of motor evoked potential normalized to the subsequent muscle compound action potential (MEP·Mmax^{-1} [MEP]) and the duration of silent period (SP) recorded from the rectus femoris (RF) during MVC. The values are presented as a percentage of baseline. In experiments A (panels **A**,**D**), a 60 s rest (Rest-0) or a 60 s left leg MVC (Ftg-0) were compared. In experiment B (panels **B**,**E**), a 60s rest (Rest-60) or a 60 s left leg MVC (Ftg-60) followed immediately by a 60 s right leg MVC were compared. In experiment C (panels **C**,**F**), Ftg-60 was performed in the absence or presence of a blood flow occlusion in the left leg (Ftg-60-Occl). * Significantly different from baseline (time effect), # significantly different from baseline (interaction effect).

3. Results

Force changes during left (fatigued) leg sustained MVCs. The decline in the maximal force output recorded during the 60 s left KE MVC in three experiments was: −55% (mean ± SD: 471 ± 137 to 209 ± 67 N) following Ftg-0 in experiment A, −54% (477 ± 133 to 218 ± 54 N) following Ftg-60 in experiment

B, and −51% (529 ± 139 to 260 ± 35 N), and −56% (571 ± 170 to 251 ± 93 N) after Ftg-60 and Ftg-60-Occl in experiment C, respectively.

3.1. Experiment A

The pre-induced fatigue in the contralateral (left) leg did not alter MVC, $rmsEMG_{100}$, Pt, Mmax, VA_{PNS}, and VA_{TMS} recorded from the right (tested) leg in the Ftg-0 compared to Rest-0 condition (Figures 3A,D,G, and 4A,D). The resting twitch and corresponding Mmax evoked before and after 60 s left leg MVC (Figures 1A and 2B) also did not show any difference between the two conditions (data not shown). However, a significant interaction effect of time × group was observed for the corticospinal excitability ($MEP \cdot Mmax^{-1}$ ratio) recorded from VL ($F_{1,11} = 6.29$, $p = 0.031$, d = 1.58), and RF ($F_{1,11} = 13.33$, $p = 0.004$, d = 0.84) at 100% MVC (MEP_{100}). While, higher values were observed for VL ($p = 0.032$, d = 0.39) and RF ($p = 0.020$, d = 0.43) following the contralateral leg fatigue (in Ftg-0), a decline was observed for the resting condition (in Rest-0) (Figure 5A).

3.2. Experiment B

As expected, the MVC force ($F_{1,11} = 50.9$, $p < 0.001$, d = 4.29), Pt ($F_{1,11} = 109.3$, $p < 0.001$, d = 6.32), VA_{PNS} ($F_{1,11} = 7.21$, $p = 0.023$, d = 1.69) and VA_{TMS} ($F_{1,11} = 27.7$, $p = 0.001$, d = 3.71) significantly declined (all time effects) following 60 s right leg MVC. However, the pre-induced fatigue in the left leg did not result in any significant difference in the neuromuscular performance and corticospinal excitability and inhibition measures between Rest-60 and Ftg-60 conditions (Figures 3B,E,H, 4B,E and 5B,E).

3.3. Experiment C

The VA_{TMS} values from one participant were removed from the data pool because the linear regression line exhibited $r^2 < 0.9$. Similar to experiment B, the MVC force ($F_{1,8} = 84.3$, $p < 0.001$, d = 6.13), Pt ($F_{1,8} = 47.4$, $p < 0.001$, d = 4.87), VA_{PNS} ($F_{1,8} = 16.8$, $p = 0.005$, d = 3.09) and VA_{TMS} ($F_{1,7} = 16.8$, $p = 0.003$, d = 2.90) significantly declined and MEP_{100} ($F_{1,8} = 29.9$, $p = 0.001$, d =1.49) increased (all time effects) following 60 s right leg MVC. However, the pre-induced fatigue and the subsequent blood occlusion in the left leg (while the right leg was performing the 60 s MVC) did not result in any significant difference in the neuromuscular performance and corticospinal excitability and inhibition between Ftg-60 and Ftg-60-Occl conditions (Figures 3C,F,I, 4C,F and 5C,F).

The left leg pre-induced fatigue in experiments B and C did not alter the force-time integral values recorded across 60 s sustained right leg MVCs between Rest-60 (19,737 ± 6847 N·s) vs. Ftg-60 (18,264 ± 7413 Ns) as well as the Ftg-60 (22,422 ± 3645 Ns) vs. Ftg-60-Occl (22,744 ± 4197 Ns). In addition, the corticospinal excitability (MEP) and inhibition (SP) recorded from VL and RF during contractions at 100%, 75%, and 50% of MVC did not demonstrate any significance in the three experiments unless otherwise stated above. Since VL and RF rmsEMG, MEP, and SP demonstrated similar patterns, only the results of RF were presented in Figures 3–5.

4. Discussion

The most important findings of the present study are that (i) in line with our first hypothesis, the pre-induced left leg fatigue did not alter the right knee extensors neuromuscular function, while (ii) the VL and RF corticospinal excitability (MEP_{100}) significantly increased following the 60 s contralateral MVC (in Ftg-0) compared to a time-matched rest (in Rest-0), and (iii) contrary to our second hypothesis, the pre-induced fatigue, in combination with the subsequent left leg blood occlusion did not alter the neuromuscular function or performance (i.e., force-time integral during 60 s MVC) nor corticospinal responses between Ftg-60-Occl vs. Ftg-60 condition. These results confirm that 60 s sustained MVC-induced fatigue and subsequent muscle ischemia in one lower limb muscle, activating the group III and IV muscle afferents, does not compromise the neuromuscular performance in the contralateral limb when performed in isometric conditions.

4.1. MVC Force and Central Motor Drive

Multiple studies have found that exercise-induced fatigue in one leg may impair the neuromuscular function in the contralateral leg (for review see [1]). However, the results of the current study, in line with several other investigations [9–13], have failed to show the NLMF effect in the lower limb muscles. This lack of observable NLMF is apparent in a research paradigm where the post-intervention neuromuscular function is assessed during a single, brief (4–6 s) MVC [6,11,28,29]. It was hypothesized that in this paradigm, despite a general sensation of fatigue created by the pre-induced fatiguing task, participants would be able to maintain the level of central motor commands to the contralateral tested muscles to prevent a deterioration in maximal performance. In line with this hypothesis, the indices of central fatigue, including $rmsEMG_{100}$ as well as the central neural drive (measured by VA_{PNS}) and neural drive associated with circuits at or above the cortical motor cells (measured by VA_{TMS}), did not change following Ftg-0 and Rest-0 conditions. These findings confirm that a 60 s MVC in one leg does not result in any central constraint to the performance of the tested contralateral muscles when the post-fatigue assessment includes a brief MVC [4,30].

In an alternative research paradigm, Halperin et al. [4,15] demonstrated that MVC force and VA_{PNS} were compromised when 12 × 5 s MVCs (as opposed to a single MVC) were performed following the fatiguing contralateral limb. Contrary to their observation, however, Kennedy et al. [9] did not find any impairment in the MVC force and VA_{PNS} recorded during 8 × 3 s MVCs, despite the fact that these investigators maintained muscle ischemia in the fatigued leg to further activate the group III and IV muscle afferents. The results of our study (in experiments B and C) support those reported by Kennedy et al. [9] indicating that adding 60 s sustained MVC in combination with subsequent muscle ischemia to the contralateral leg does not alter the central ($rmsEMG_{100}$, VA_{PNS}, and VA_{TMS}) and peripheral (Pt and Mmax) indices of fatigue in the tested limb. Of note, in experiments B and C, while VA_{TMS} demonstrated 18 to 25% reduction (Table 1, Figure 3E,F), only 3% to 10% decrement was observed for VA_{PNS} (Table 1, Figure 3H,I). Although one may take these results to suggest that different sites within the central nervous system would play distinct roles in impaired maximal force output after the fatiguing task, limitations associated with the interpretation of the voluntary activation data [31] make it difficult to draw an explicit conclusion about the results. Nonetheless, pre-induced fatigue and muscle ischemia in the contralateral limb did not alter the contribution of cortical and motoneuronal circuits in the end-exercise central fatigue level.

Table 1. Group data (mean and SD) for variables recorded from the knee extensor muscles during 100%, 75%, and 50% of MVC in the three experiments.

Variables		Experiment A		Experiment B		Experiment C	
		Rest-0	Ftg-0	Rest-60	Ftg-60	Ftg-60	Ftg-60-Occl
MVC force (N)	Baseline	501.7 ± 159.9	499.9 ± 158.7	507.1 ± 148.2	495.3 ± 150.3	555.0 ± 128.8	573.9 ± 144.1
	Post-test	501.1 ± 170.7	506.3 ± 158.6	334.8 ± 88.9	322.2 ± 89.9	349.9 ± 71.1	363.6 ± 96.7
rmsEMG·Mmax^{-1} ratio (rmsEMG$_{100}$)	Baseline	0.074 ± 0.031	0.070 ± 0.037	0.067 ± 0.034	0.066 ± 0.026	0.063 ± 0.030	0.068 ± 0.026
	Post-test	0.069 ± 0.028	0.069 ± 0.044	0.066 ± 0.026	0.078 ± 0.029	0.066 ± 0.031	0.072 ± 0.038
PT (N)	Baseline	173.2 ± 54.4	173.7 ± 51.7	171.8 ± 41.7	174.5 ± 50.4	171.5 ± 44.3	153.4 ± 54.5
	Post-test	161.4 ± 59.1	165.6 ± 52.9	59.9 ± 20.2	49.8 ± 26.1	63.5 ± 33.2	72.8 ± 31.1
VA$_{PNS}$ (%)	Baseline	92.7 ± 2.9	93.1 ± 3.8	90.7 ± 4.5	91.1 ± 3.2	93.1 ± 5.1	93.2 ± 5.2
	Post-test	90.6 ± 4.9	90.7 ± 7.5	85.1 ± 9.1	81.2 ± 11.2	90.6 ± 5.2	86.7 ± 6.7
VA$_{TMS}$ (%)	Baseline	93.6 ± 1.9	92.6 ± 5.2	91.1 ± 4.2	93.1 ± 4.4	94.8 ± 5.5	85.4 ± 4.5
	Post-test	93.4 ± 3.1	91.2 ± 6.1	73.3 ± 18.3	76.6 ± 12.6	78.4 ± 12.2	70.1 ± 17.1
RF MEP·Mmax^{-1} (MEP$_{100}$)	Baseline	0.93 ± 0.18	0.80 ± 0.21	0.77 ± 0.18	0.83 ± 0.14	0.70 ± 0.16	0.67 ± 0.12
	Post-test	0.82 ± 0.18 #	0.90 ± 0.25 #	0.79 ± 0.21	0.88 ± 0.21	0.80 ± 0.22	0.78 ± 0.22
RF MEP·Mmax^{-1} (MEP$_{75}$)	Baseline	0.94 ± 0.19	0.84 ± 0.23	0.87 ± 0.20	0.92 ± 0.15	0.82 ± 0.15	0.81 ± 0.15
	Post-test	0.87 ± 0.22	0.82 ± 0.20	0.83 ± 0.22	0.89 ± 0.11	0.92 ± 0.24	0.84 ± 0.33
RF MEP·Mmax^{-1} (MEP$_{50}$)	Baseline	0.96 ± 0.15	0.90 ± 0.19	0.89 ± 0.18	0.97 ± 0.13	0.89 ± 0.14	0.81 ± 09
	Post-test	0.95 ± 0.16	0.88 ± 0.19	0.87 ± 0.14	0.88 ± 0.22	0.97 ± 0.34	0.79 ± 0.28
RF SP$_{100}$ (ms)	Baseline	206.7 ± 45.2	258.5 ± 78.3	257.5 ± 78.6	249.3 ± 41.6	219.1 ± 84.6	240.6 ± 60.5
	Post-test	227.3 ± 50.0	270.9 ± 74.7	260.1 ± 69.7	259.3 ± 52.0	253.4 ± 75.3	267.4 ± 53.7
RF SP$_{75}$ (ms)	Baseline	211.4 ± 48.3	256.8 ± 59.6	267.7 ± 64.0	252.0 ± 54.3	214.1 ± 77.7	208.5 ± 96.8
	Post-test	226.1 ± 67.3	270.3 ± 77.0	254.1 ± 70.3	254.9 ± 56.5	231.6 ± 67.5	250.2 ± 55.9
RF SP$_{50}$ (ms)	Baseline	231.6 ± 40.9	276.4 ± 61.4	270.0 ± 69.1	273.1 ± 45.9	207.3 ± 71.8	213.1 ± 47.4
	Post-test	248.9 ± 63.2	269.8 ± 61.2	256.9 ± 61.2	261.5 ± 55.6	248.5 ± 74.6	253.2 ± 77.3
BF MEP$_{100}$ (mV.s)	Baseline	0.009 ± 0.005	0.010 ± 0.008	0.007 ± 0.004	0.009 ± 0.005	0.008 ± 0.005	0.008 ± 0.004
	Post-test	0.007 ± 0.006	0.009 ± 0.006	0.007 ± 0.005	0.007 ± 0.005	0.008 ± 0.005	0.006 ± 0.005
BF MEP$_{75}$ (mV.s)	Baseline	0.007 ± 0.005	0.009 ± 0.008	0.007 ± 0.004	0.009 ± 0.006	0.007 ± 0.005	0.008 ± 0.005
	Post-test	0.007 ± 0.006	0.009 ± 0.007	0.005 ± 0.005	0.007 ± 0.007	0.006 ± 0.004	0.006 ± 0.005
BF MEP$_{50}$ (mV.s)	Baseline	0.007 ± 0.006	0.009 ± 0.008	0.007 ± 0.006	0.007 ± 0.003	0.007 ± 0.006	0.006 ± 0.004
	Post-test	0.006 ± 0.006	0.008 ± 0.007	0.004 ± 0.002	0.005 ± 0.003	0.005 ± 0.003	0.004 ± 0.002

Note: The shaded boxes indicate that there was a significant time effect between baseline and post-test measures; # means significantly different from baseline (interaction effect).

As opposed to previous experiments that used intermittent brief MVCs as the fatiguing task for the tested leg [9,15], we used 60 s sustained MVC to generate high intramuscular pressure and decreased oxygen delivery due to muscle ischemia induced by sustained MVC [32]. The 60 s MVC was chosen to match the total duration of intermittent MVCs (12 × 5 s MVCs) used by Halperin et al. [15]. However, the left leg pre-induced fatigue in combination with the blood ischemia did not alter the force-time integral values recorded during the 60 s right leg sustained MVCs. These results suggest that the activation of group III and IV muscle afferents induced by a 60 s sustained MVC in one leg does not modulate the motor performance in the contralateral leg during either sustained or brief MVCs.

A point of deliberation in the present study may be that the magnitude of the pre-induced fatigue generated by the left leg was not enough to constitute an NLMF. Indeed, two 100 s sustained KE MVCs used by Halperin et al. [15] resulted in ~70% force drop in the fatigue limb whereas one 60 s MVC resulted in ~55% reduction in MVC in our study. However, the neuromuscular performance was not altered in the study by Kennedy et al. [9] despite a 77% decline in MVC force. We also minimized the time delay between the fatiguing tasks and the post-intervention neuromuscular assessments to avoid recovery of central and peripheral fatigue, yet no evidence of NLMF was found. This time delay in previous studies was approximately 15–60 s [9,15]. It is also worth clarifying that the failure in observing NLMF phenomenon in the current study might be associated with the size of muscle mass involved (i.e., knee extensors in one leg) and the mode of exercise (i.e., sustained MVC) performed during the fatiguing task. Prior investigations that used bilateral dynamic knee extensions [8], leg cycling [5], and arm cranking exercises [32] found NLMF in the rested/remote muscle groups. The accelerated neuromuscular fatigue within the tested muscle groups in these studies have been attributed to central inhibitory factors mediated by afferent feedback from pre-fatigued muscles. Nonetheless, further studies using a longer duration of sustained MVC and larger muscle mass during whole-body dynamic exercises are required to explore the influence of group III and IV muscle afferents on the NLMF phenomenon.

4.2. Corticospinal Excitability and Inhibition

The corticospinal pathway excitability recorded from RF and VL muscles demonstrated relatively greater values (MEP_{100}) following 60 s contralateral MVC (in Ftg-0) compared to a time-matched rest (in Rest-0). Although the current study showed a small effect size, this so-called "cross-over facilitation" of the corticospinal pathway had previously been shown in the upper limb musculature [33] and was attributed to (i) transfer of excitatory signals from one hemisphere to another via the callosal commissure [34,35], (ii) increase in the excitability of motoneurons innervating the tested muscles via release of monoaminergic neuromodulators [36], and (iii) activation of additional brain regions including ipsilateral and contralateral prefrontal and sensorimotor areas as a compensatory mechanism to prevent reduction in maximal force output [37]. With the single-pulse TMS used in the present study, it is indeed difficult to confirm the contribution of the above-mentioned mechanisms; particularly because the MEP_{75} and MEP_{50} recorded from both VL and RF did not show an increase following contralateral fatiguing contraction. However, in support of these notions contributing to cross-over facilitation of MEP recorded during MVC (MEP_{100}), Tanaka and Watanabe [37] have suggested that additional brain regions are involved in a task if participants choose to mobilize all of their mental effort to execute a "maximal" rather than a "submaximal" contraction.

The failure in observing cross-over facilitation of MEP_{100} following Ftg-60 (experiment B) and Ftg-60-Occl (experiment C) is not clear; however, this could be attributed to the 60 s right leg MVC. More specifically, the corticospinal excitability (i.e., MEP_{100}) demonstrated a significant increase following the 60 s sustained right leg MVC (time effect) (Figure 5B,C). This MEP facilitation was observed regardless of the preceding conditions undertook by the left leg (i.e., 60 s MVC vs. 60 s MVC + occlusion in experiment C). Thus, our data suggest that the 60 s sustained MVC performed by the right (tested) knee extensors facilitated a strong excitatory effect to the corticospinal system while the contralateral fatiguing contraction did not enhance this facilitation any further.

Our data in all three experiments demonstrated no statistical difference in the duration of SP compared within or between conditions. This finding indicates that the pre-induced fatigue and the subsequent blood ischemia in one leg do not alter the rate of corticospinal inhibition in the contralateral homologous muscles recorded during 100%, 75%, and 50% of MVC contractions.

4.3. Limitations

Methodological considerations of the present study include: (i) The sequence of contractions in the neuromuscular assessment (i.e., 100%, 75%, and 50% of MVC) was kept consistent throughout the experiment whereas an early recruitment of high threshold motoneurons during the MVC could have offset the influence of contralateral contraction on MEP_{75} and MEP_{50}, (ii) the size of MEP monitors the excitability of the pyramidal tract neurons as well as the spinal motoneurons and spinal interneurons [38]; thus, further research is required to elucidate the influence of pre-induced fatiguing exercises on the cortical and spinal excitability responses, (iii) some may argue that the antagonist muscle excitability (M-wave amplitude) and contractile property (peak twitch) might have affected the slope of the regression line used to calculate the estimated resting twitch in VA_{TMS} measurement, however, the hamstring Mmax and peak twitch were not measured due to complexity of the sciatic nerve electrical stimulation; therefore, caution should be taken in interpretation of the VA_{TMS} data, as it is difficult to determine the contribution of hamstring coactivation on TMS-evoked knee extensors SITs, (iv) pre-induced fatiguing exercise in one limb may modulate central and peripheral homodynamic associates such as cardiorespiratory responses, neurotransmitters, neuromodulators, hormonal factors as well as autonomic responses, however, these parameters were not directly measured in the current study, thus they were not discussed, and finally, (v) more investigation should be directed towards the measurement of intracortical facilitation and inhibition using the paired-pulse TMS paradigm.

5. Conclusions

In conclusion, the results of the present study suggest that the pre-induced fatigue evoked by the 60 s sustained KE MVC, with or without subsequent blood ischemia, in the fatigued leg does not compromise the performance and neuromuscular function of the tested contralateral leg. However, the pre-induced fatigue may enhance the responsiveness of the ipsilateral corticospinal pathway innervating the tested limb, provided that the neuromuscular evaluation in the tested muscles is performed immediately after the pre-induced fatigue.

Author Contributions: Conceptualization, S.J.A., C.X.Y.Z., M.C. and G.Y.M.; Data curation, S.J.A., C.X.Y.Z., R.S., M.C. and G.Y.M.; Formal analysis, S.J.A., C.X.Y.Z., R.S., M.C. and G.Y.M.; Funding acquisition, M.C. and G.Y.M.; Investigation, S.J.A., C.X.Y.Z., R.S., M.C. and G.Y.M.; Methodology, S.J.A. and G.Y.M.; Project administration, S.J.A., C.X.Y.Z. and M.C.; Resources, G.Y.M.; Supervision, S.J.A., R.S. and G.Y.M.; Validation, S.J.A.; Visualization, S.J.A.; Writing—original draft, S.J.A. and G.Y.M.; Writing—review & editing, S.J.A., C.X.Y.Z., R.S., M.C. and G.Y.M.

Funding: This research was partially funded by the Eyes High Postdoctoral Scholars, Program for Undergraduate Research Experience (PURE) and Markin Undergraduate Student Research Program (USRP) in Health & Wellness awards, University of Calgary, Alberta, Canada.

Acknowledgments: We would like to thank Elaheh Mousavian for working on raw data figure (Figure 2).

Conflicts of Interest: The authors declare no conflict of interest.

References

1. Halperin, I.; Chapman, D.W.; Behm, D.G. Non-local muscle fatigue: Effects and possible mechanisms. *Eur. J. Appl. Physiol.* **2015**, *115*, 2031–2048. [CrossRef] [PubMed]
2. Martin, P.G.; Rattey, J. Central fatigue explains sex differences in muscle fatigue and contralateral cross-over effects of maximal contractions. *Pflugers Arch.* **2007**, *454*, 957–969. [CrossRef] [PubMed]
3. Šambaher, N.; Aboodarda, S.J.; Behm, D.G. Bilateral Knee Extensor Fatigue Modulates Force and Responsiveness of the Corticospinal Pathway in the Non-fatigued, Dominant Elbow Flexors. *Front. Hum. Neurosci.* **2016**, *10*, 18. [CrossRef] [PubMed]

4. Halperin, I.; Aboodarda, S.J.; Behm, D.G. Knee extension fatigue attenuates repeated force production of the elbow flexors. *Eur. J. Sport Sci.* **2014**, *14*, 823–829. [CrossRef] [PubMed]
5. Sidhu, S.K.; Weavil, J.C.; Venturelli, M.; Garten, R.S.; Rossman, M.J.; Richardson, R.S.; Gmelch, B.S.; Morgan, D.E.; Amann, M. Spinal μ-opioid receptor-sensitive lower limb muscle afferents determine corticospinal responsiveness and promote central fatigue in upper limb muscle. *J. Physiol.* **2014**, *592*, 5011–5024. [CrossRef] [PubMed]
6. Amann, M.; Venturelli, M.; Ives, S.J.; McDaniel, J.; Layec, G.; Rossman, M.J.; Richardson, R.S. Peripheral fatigue limits endurance exercise via a sensory feedback-mediated reduction in spinal motoneuronal output. *J. Appl. Physiol. (1985)* **2013**, *115*, 355–364. [CrossRef] [PubMed]
7. Gandevia, S.C. Spinal and supraspinal factors in human muscle fatigue. *Physiol. Rev.* **2001**, *81*, 1725–1789. [CrossRef] [PubMed]
8. Aboodarda, S.J.; Šambaher, N.; Millet, G.Y.; Behm, D.G. Knee extensors neuromuscular fatigue changes the corticospinal pathway excitability in biceps brachii muscle. *Neuroscience* **2017**, *340*, 477–486. [CrossRef] [PubMed]
9. Kennedy, D.S.; Fitzpatrick, S.C.; Gandevia, S.C.; Taylor, J.L. Fatigue-related firing of muscle nociceptors reduces voluntary activation of ipsilateral but not contralateral lower limb muscles. *J. Appl. Physiol. (1985)* **2015**, *118*, 408–418. [CrossRef] [PubMed]
10. Prieske, O.; Aboodarda, S.J.; Benitez Sierra, J.A.; Behm, D.G.; Granacher, U. Slower but not faster unilateral fatiguing knee extensions alter contralateral limb performance without impairment of maximal torque output. *Eur. J. Appl. Physiol.* **2017**, *117*, 323–334. [CrossRef] [PubMed]
11. Paillard, T.; Chaubet, V.; Maitre, J.; Dumitrescu, M.; Borel, L. Disturbance of contralateral unipedal postural control after stimulated and voluntary contractions of the ipsilateral limb. *Neurosci. Res.* **2010**, *68*, 301–306. [CrossRef] [PubMed]
12. Regueme, S.C.; Barthèlemy, J.; Nicol, C. Exhaustive stretch-shortening cycle exercise: No contralateral effects on muscle activity in maximal motor performances. *Scand. J. Med. Sci. Sports* **2007**, *17*, 547–555. [CrossRef] [PubMed]
13. Aboodarda, S.J.; Copithorne, D.B.; Power, K.E.; Drinkwater, E.; Behm, D.G. Elbow flexor fatigue modulates central excitability of the knee extensors. *Appl. Physiol. Nutr. Metab.* **2015**, *40*, 924–930. [CrossRef] [PubMed]
14. Johnson, M.A.; Mills, D.E.; Brown, P.I.; Sharpe, G.R. Prior upper body exercise reduces cycling work capacity but not critical power. *Med. Sci. Sports Exerc.* **2014**, *46*, 802–808. [CrossRef] [PubMed]
15. Halperin, I.; Copithorne, D.; Behm, D.G. Unilateral isometric muscle fatigue decreases force production and activation of contralateral knee extensors but not elbow flexors. *Appl. Physiol. Nutr. Metab.* **2014**, *39*, 1338–1344. [CrossRef] [PubMed]
16. Kaufman, M.P.; Rybicki, K.J.; Waldrop, T.G.; Ordway, G.A. Effect of ischemia on responses of group III and IV afferents to contraction. *J. Appl. Physiol. Respir. Environ. Exerc. Physiol.* **1984**, *57*, 644–650. [CrossRef] [PubMed]
17. Hayward, L.; Wesselmann, U.; Rymer, W.Z. Effects of muscle fatigue on mechanically sensitive afferents of slow conduction velocity in the cat triceps surae. *J. Neurophysiol.* **1991**, *65*, 360–370. [CrossRef] [PubMed]
18. van Melick, N.; Meddeler, B.M.; Hoogeboom, T.J.; Nijhuis-van der Sanden, M.W.G.; van Cingel, R.E.H. How to determine leg dominance: The agreement between self-reported and observed performance in healthy adults. *PLoS ONE* **2017**, *12*, e0189876. [CrossRef] [PubMed]
19. Rossi, S.; Hallett, M.; Rossini, P.M.; Pascual-Leone, A. Screening questionnaire before TMS: An update. *Clin. Neurophysiol.* **2011**, *122*, 1686. [CrossRef] [PubMed]
20. Warburton, D.; Jamnik, V.; Bredin, S.; Gledhill, N. The Physical Activity Readiness Questionnaire for Everyone (PAR-Q+): English North America Version. *Health Fit. J. Canada* **2011**, *4*, 18–20.
21. Vernillo, G.; Temesi, J.; Martin, M.; Millet, G.Y. Mechanisms of Fatigue and Recovery in Upper versus Lower Limbs in Men. *Med. Sci. Sports Exerc.* **2018**, *50*, 334–343. [CrossRef] [PubMed]
22. Hermens, H.J.; Freriks, B.; Disselhorst-Klug, C.; Rau, G. Development of recommendations for SEMG sensors and sensor placement procedures. *J. Electromyogr. Kinesiol.* **2000**, *10*, 361–374. [CrossRef]
23. Behm, D.G.; St-Pierre, D.M.; Perez, D. Muscle inactivation: Assessment of interpolated twitch technique. *J. Appl. Physiol.* **1996**, *81*, 2267–2273. [CrossRef] [PubMed]
24. Strojnik, V.; Komi, P.V. Neuromuscular fatigue after maximal stretch-shortening cycle exercise. *J. Appl. Physiol.* **1998**, *84*, 344–350. [CrossRef] [PubMed]

25. Todd, G.; Taylor, J.L.; Gandevia, S.C. Measurement of voluntary activation of fresh and fatigued human muscles using transcranial magnetic stimulation. *J. Physiol.* **2003**, *551*, 661–671. [CrossRef] [PubMed]
26. Todd, G.; Taylor, J.L.; Gandevia, S.C. Reproducible measurement of voluntary activation of human elbow flexors with motor cortical stimulation. *J. Appl. Physiol. (1985)* **2004**, *97*, 236–242. [CrossRef] [PubMed]
27. Cohen, J. *Statistical Power Analysis for the Behavioral Sciences*, 2nd ed.; L. Erlbaum Associates: Hillsdale, NJ, USA, 1988; pp. xxi, 567.
28. Arora, S.; Budden, S.; Byrne, J.M.; Behm, D.G. Effect of unilateral knee extensor fatigue on force and balance of the contralateral limb. *Eur. J. Appl. Physiol.* **2015**, *115*, 2177–2187. [CrossRef] [PubMed]
29. Grabiner, M.D.; Owings, T.M. Effects of eccentrically and concentrically induced unilateral fatigue on the involved and uninvolved limbs. *J. Electromyogr. Kinesiol.* **1999**, *9*, 185–189. [CrossRef]
30. Aboodarda, S.J.; Šambaher, N.; Behm, D.G. Unilateral elbow flexion fatigue modulates corticospinal responsiveness in non-fatigued contralateral biceps brachii. *Scand. J. Med. Sci. Sports* **2016**, *26*, 1301–1312. [CrossRef] [PubMed]
31. Todd, G.; Taylor, J.L.; Gandevia, S.C. Measurement of voluntary activation based on transcranial magnetic stimulation over the motor cortex. *J. Appl. Physiol. (1985)* **2016**, *121*, 678–686. [CrossRef] [PubMed]
32. Sadamoto, T.; Bonde-Petersen, F.; Suzuki, Y. Skeletal muscle tension, flow, pressure, and EMG during sustained isometric contractions in humans. *Eur. J. Appl. Physiol. Occup. Physiol.* **1983**, *51*, 395–408. [CrossRef] [PubMed]
33. Hortobágyi, T.; Taylor, J.L.; Petersen, N.T.; Russell, G.; Gandevia, S.C. Changes in segmental and motor cortical output with contralateral muscle contractions and altered sensory inputs in humans. *J. Neurophysiol.* **2003**, *90*, 2451–2459. [CrossRef] [PubMed]
34. Samii, A.; Wassermann, E.M.; Ikoma, K.; Mercuri, B.; Hallett, M. Characterization of postexercise facilitation and depression of motor evoked potentials to transcranial magnetic stimulation. *Neurology* **1996**, *46*, 1376–1382. [CrossRef] [PubMed]
35. Brasil-Neto, J.P.; Araújo, V.P.; Carneiro, C.R. Postexercise facilitation of motor evoked potentials elicited by ipsilateral voluntary contraction. *Muscle Nerve* **1999**, *22*, 1710–1712. [CrossRef]
36. Heckman, C.J.; Hyngstrom, A.S.; Johnson, M.D. Active properties of motoneurone dendrites: Diffuse descending neuromodulation, focused local inhibition. *J. Physiol.* **2008**, *586*, 1225–1231. [CrossRef] [PubMed]
37. Tanaka, M.; Watanabe, Y. Supraspinal regulation of physical fatigue. *Neurosci. Biobehav. Rev.* **2012**, *36*, 727–734. [CrossRef] [PubMed]
38. Taylor, J.L. Stimulation at the cervicomedullary junction in human subjects. *J. Electromyogr. Kinesiol.* **2006**, *16*, 215–223. [CrossRef]

brain sciences

MDPI

Article

Corticospinal-Evoked Responses from the Biceps Brachii during Arm Cycling across Multiple Power Outputs

Evan J. Lockyer [1,2], Katarina Hosel [1,3], Anna P. Nippard [1,2], Duane C. Button [1,2] and Kevin E. Power [1,2,*]

1 Human Neurophysiology Lab, School of Human Kinetics and Recreation, Memorial University of Newfoundland, St. John's, NL A1C 5S7, Canada
2 Faculty of Medicine, Memorial University of Newfoundland, St. John's, NL A1C 5S7, Canada
3 Department of Psychology, Memorial University of Newfoundland, St. John's, NL A1C 5S7, Canada
* Correspondence: kevin.power@mun.ca; Tel.: +1-709-864-7275

Received: 14 July 2019; Accepted: 15 August 2019; Published: 19 August 2019

Abstract: *Background*: We examined corticospinal and spinal excitability across multiple power outputs during arm cycling using a weak and strong stimulus intensity. *Methods*: We elicited motor evoked potentials (MEPs) and cervicomedullary motor evoked potentials (CMEPs) in the biceps brachii using magnetic stimulation over the motor cortex and electrical stimulation of corticospinal axons during arm cycling at six different power outputs (i.e., 25, 50, 100, 150, 200 and 250 W) and two stimulation intensities (i.e., weak vs. strong). *Results*: In general, biceps brachii MEP and CMEP amplitudes (normalized to maximal M-wave (M_{max})) followed a similar pattern of modulation with increases in cycling intensity at both stimulation strengths. Specifically, MEP and CMEP amplitudes increased up until ~150 W and ~100 W when the weak and strong stimulations were used, respectively. Further increases in cycling intensity revealed no changes on MEP or CMEP amplitudes for either stimulation strength. *Conclusions*: In general, MEPs and CMEPs changed in a similar manner, suggesting that increases and subsequent plateaus in overall excitability are likely mediated by spinal factors. Interestingly, however, MEP amplitudes were disproportionately larger than CMEP amplitudes as power output increased, despite being initially matched in amplitude, particularly with strong stimulation. This suggests that supraspinal excitability is enhanced to a larger degree than spinal excitability as the power output of arm cycling increases.

Keywords: MEP; CMEP; arm cranking; motoneurone; exercise

1. Introduction

The influence of muscle contraction intensity on the excitability of the corticospinal pathway in humans has been well-studied during isometric contractions. Most of this research has involved the use of non-invasive stimulation techniques to assess corticospinal and/or spinal excitability to muscles of the upper [1–3] and, to a lesser extent, the lower limb [4] across a wide range of isometric contraction intensities. In general, the findings from these studies indicate that motor evoked potentials (MEPs) elicited by transcranial magnetic stimulation (TMS) increase in size as the strength of isometric muscle contractions increase up until a peak, after which they plateau and subsequently decrease as contraction strength approaches maximal ((i.e., 100% maximal voluntary contraction (MVC)) [1–4]. This modulation in MEP is accompanied by a similar change in the cervicomedullary MEP (CMEP) elicited by transmastoid electrical stimulation (TMES) of corticospinal axons, suggesting that the change in corticospinal excitability is largely mediated by spinal factors [1,4].

Using a strong stimulus intensity (set to evoke responses equal to 65–80% the maximal compound muscle action potential (M_{max})), Martin et al. (2006) showed that MEP and CMEP areas increased linearly in size during weak isometric contractions (i.e., <50% MVC) of the biceps brachii as muscle contraction intensity increased, whereas during strong contractions (i.e., >50% MVC) MEP and CMEP areas plateaued at ~75% MVC, and subsequently decreased as the contraction intensity approached 100% MVC [1]. When a lower stimulus intensity (set to evoke responses equal to 30–50% M_{max}) was used, MEP and CMEP areas followed a similar pattern of modulation with contraction intensity, however, peak responses were not observed until ~90% MVC, after which MEP and CMEP areas decreased. Moreover, the decline in MEP and CMEP area with the lower stimulus intensity was less marked than that observed when the stronger stimulus intensity was used [1]. Thus, the intensity of stimulation is an important factor to consider in assessing corticospinal excitability given how it can influence the primary measurement(s), and the associated interpretation of the data.

Substantially less information, however, is available regarding the influence of muscle contraction intensity on the modulation of corticospinal excitability during rhythmic motor outputs such as those observed during cycling [5–7]. This is an important topic to consider given that rhythmic motor outputs, such as arm cycling, are partially generated by spinally located networks of interneurons referred to as central pattern generators [8,9], and that corticospinal excitability is modulated differently during rhythmic locomotor outputs than during isometric contractions, indicating task-specificity [5,10,11]. In two separate studies from our lab, we have investigated changes in corticospinal and spinal excitability as arm cycling intensity (i.e., power output) was increased [6,7]. However, changes in excitability were assessed across a small range of power outputs, and thus may not have observed potential changes in excitability that occurred at higher cycling intensities. Thus, it remains unknown whether a similar peak, plateau and subsequent decline in corticospinal and spinal excitability are observed with increasing arm cycling intensity, as observed in isometric contractions.

Accordingly, the purpose of the present study was to: (1) characterize the influence of muscle contraction intensity on changes in corticospinal and spinal excitability projecting to the biceps brachii over a wide range of arm cycling intensities, and (2) assess the influence of stimulation intensity on corticospinal and spinal outputs as cycling intensity increased. Specifically, we sought to examine the effects of using a weak and a strong stimulus intensity on corticospinal and spinal excitability as power output was increased during cycling. We hypothesized that: (1) using the weak stimulus, corticospinal and spinal excitability would increase similarly across all arm cycling power outputs, and (2) using the *strong* stimulus, corticospinal and spinal excitability would increase but experience a plateau, and subsequent decrease as cycling intensity increased towards the maximum power output examined.

2. Materials and Methods

2.1. Participants

This study consisted of a familiarization session and two experimental sessions; (1) a transcranial magnetic stimulation (TMS) session and (2) a transmastoid electrical stimulation (TMES) session (see *Protocol* below). A total of nine healthy, male volunteers (24.2 ± 5.9 years, 180.7 ± 7.8 cm, 82.2 ± 8.3 kg, 1 left-hand dominant) with no known neurological impairment participated in session one, and eight of those volunteers (1 left-hand dominant) returned on a separate day (>24 h) to complete session two. In accordance with the Tri-Council guidelines in Canada, all participants gave written, informed consent prior to participating in the study, and potential risks were fully disclosed. Prior to TMS, all participants were screened for contraindications to magnetic stimulation using a safety checklist [12]. To determine limb dominance, the Edinburgh handedness inventory [13] was used. This information was gathered because all evoked responses elicited by TMS and TMES (see *Stimulation Conditions* below) were taken from the dominant arm. Additionally, all participants filled out a Physical Activity Readiness Questionnaire for Everyone (PAR-Q+, Canadian Society for Exercise Physiology (CSEP)) to screen for any contraindications to physical activity. Participants also refrained from caffeine for 12 h and alcohol for 24 h prior to each experimental session. All procedures were performed in compliance with the Declaration of Helsinki and were approved by the Interdisciplinary Committee on Ethics in Human Research (ICEHR no. 20181196-HK) at Memorial University of Newfoundland.

2.2. Experimental Setup

Many of the experimental procedures and recording techniques herein are similar to those described previously [6,7,14]. All sessions were conducted with participants seated upright on an arm cycle ergometer (SCIFIT ergometer, model PRO2 Total Body, Tulsa, OK, USA). The seat height of the ergometer was adjusted so that participants' shoulders were approximately in line with the axis of rotation of the arm cranks, and the seat distance was manipulated to a position in which participants were at a comfortable distance (i.e., no reaching or trunk variation during cycling) from the hand pedals. The seat height and distance were recorded for each participant during the familiarization session and were used for the subsequent sessions. Arm cycling trials were performed in an asynchronous cranking pattern with the forearms fixed in a pronated position. Wrist braces were worn to limit the amount of wrist flexion and extension during cycling as a means to diminish the influence of short- and long-latency reflex connections that have been shown to exist between the wrist flexors and the biceps brachii (see Figure 1) [15].

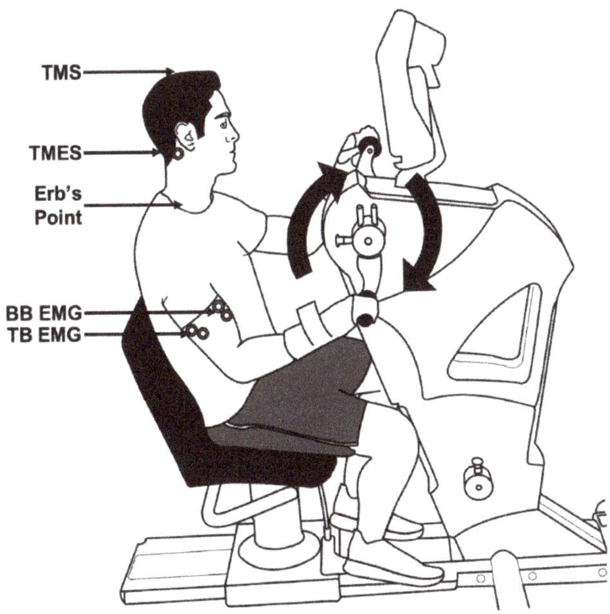

Figure 1. Experimental setup for arm cycling trials showing participant seated on the ergometer instrumented with surface EMG electrodes on the biceps and triceps brachii. Arrows point to the site of each stimulation technique. All arm cycling trials were conducted in the forward direction. Abbreviations: TMS, transcranial magnetic stimulation; TMES, transmastoid electrical stimulation; BB, biceps brachii; TB, triceps brachii; EMG, electromyography.

For this study, participants were required to cycle at 6 different power outputs: 25, 50, 100, 150, 200, and 250 Watts (W) all at a constant cadence of 60 revolutions per minute (rpm). These cycling conditions were repeated at two different stimulation intensities (see *Stimulation Conditions* below), for a total of 12 cycling trials.

2.3. Electromyography Recordings

Surface electromyography (EMG) was recorded from the biceps brachii of the dominant arm using pairs of disposable Ag-AgCl surface electrodes (MediTrace^{TM} 130 Foam Electrodes with conductive adhesive hydrogel, Covidien IIC, MA, USA). Electrodes were positioned approximately 2 cm apart (center to center) over the midline of the biceps brachii and on the lateral head of the triceps brachii in a bipolar configuration. A ground electrode was positioned on the lateral epicondyle of the dominant arm. To reduce the impedance for EMG recordings, the skin was thoroughly prepared by removing hair (via a handheld razor), abraded to remove dead skin cells (via abrasive paper), and cleaned using isopropyl alcohol swabs prior to electrode placement. The EMG signals were amplified (×300, CED 1902 amplifier; Cambridge Electronic Design Ltd., Cambridge, UK), and bandpass filtered using a 3-pole Butterworth filter with cut-off frequencies of 10–1000 Hz. All analog signals were digitized at a sampling rate of 5000 Hz and stored on a laboratory computer for off-line analysis (CED 1401 interface and Signal 5.11 software; Cambridge Electronic Design Ltd., Cambridge, UK).

2.4. Stimulation Conditions

Recordings were made of the motor responses in the biceps brachii to three different stimulation techniques: (1) brachial plexus stimulation at Erb's point, (2) magnetic stimulation of the motor cortex (i.e., TMS), and (3) electrical stimulation between the mastoids at the cervicomedullary junction (i.e., TMES). Motor responses were evoked during arm cycling at the 6 o'clock position, which corresponds to the mid-elbow flexion phase of arm cycling and when biceps brachii activity is relatively the largest (for a more detailed explanation of the phases of arm cycling see review by [16]). Stimulations were triggered automatically when the right hand passed a magnetic sensor on the ergometer, at either the 6 o'clock or 12 o'clock position for right-handed and left-handed participants, respectively. The intensities for all three stimulation techniques were set during arm cycling at a constant cadence of 60 rpm and power output of 25 W. For TMS and TMES, two different stimulation intensities were used: (1) a weak stimulation intensity (set to evoke responses equal to ~10% M_{max}), and (2) a strong stimulation intensity (set to evoke responses equal to ~40% M_{max}). These response amplitudes were chosen to provide insight into potential differences in excitability at different portions of the motoneurone pool as cycling intensity increased. All participants had prior experience with each of the stimulation procedures before participating.

2.5. Brachial Plexus Stimulation

For both sessions, single rectangular pulses (200-μs duration, 90–275 mA) were delivered via a DS7AH constant current stimulator (Digitimer Ltd., Welwyn Garden City, Hertfordshire, UK) to the brachial plexus at Erb's point to elicit maximal compound muscle action potentials (maximal M-wave (M_{max})) in the biceps brachii. The cathode was placed in the supraclavicular fossa and the anode over the acromion process. Stimulus intensity was initially set at 25 mA and was gradually increased until the size of the M-wave plateaued (i.e., M_{max}). At this point, the stimulation intensity was increased by 10% (supramaximal) to ensure that M_{max} was elicited throughout the remainder of the study.

2.6. TMS

TMS was delivered over the vertex of the motor cortex to elicit MEPs in the biceps brachii using a Magstim 200^2 magnetic stimulator (Magstim, Whitland, Dyfed, UK) and circular coil (13.5 cm outside diameter). The vertex was measured and marked on the participant's scalp with a felt-tip permanent marker. One investigator ensured proper and consistent coil placement directly over vertex throughout the experiment. The coil was held firmly against the participant's skull, parallel to the floor with the direction of current flow-oriented to preferentially activate either the left or right motor cortex, depending on hand dominance (i.e., "A" side up for right-handed participants, "B" side up for left-handed participants). Initially, TMS intensity was set at 25% of maximal stimulator output (MSO) and was increased until MEPs were observed in the biceps brachii equal in amplitude to ~10% M_{max}. Once found, a trial consisting of 8 TMS was performed to ensure that the average MEPs were ~10% M_{max}. This stimulation intensity was recorded as the weak stimulation intensity, and was then used for the remainder of the experiment. For the strong stimulation intensity, the same procedures were performed except the %MSO was increased until MEPs from the biceps brachii were equal in amplitude to ~40% M_{max}. Once again, a trial consisting of 8 TMS was performed to ensure that the intensity of TMS would evoke MEPs equal to ~40% M_{max}. Once determined, this intensity was recorded and then used as the strong intensity for the rest of the experiment.

2.7. TMES

TMES was delivered (200 µs pulse-width duration, DS7AH, Digitimer Ltd., Welwyn Garden City, Hertfordshire, UK) to the corticospinal axons at the cervicomedullary junction to elicit CMEPs in the dominant arm biceps brachii. Self-adhesive Ag-AgCl surface electrodes were placed on the skin at the grooves between the mastoid processes and the occipital bone, with the anode and cathode on the side corresponding to each participant's dominant and non-dominant arm, respectively. Similar to the procedures for setting the stimulation intensities for TMS (see *TMS* above), the intensity of electrical stimulation was gradually increased (initially from 25 mA) until the amplitudes of the CMEPs were equal in amplitude to ~10% M_{max} (for the weak stimulation intensity) and ~40% M_{max} (for the strong stimulation intensity). Trials of 8 CMEPs were evoked at each stimulation intensity and the average was calculated. These stimulation intensities were recorded, and were then used for the remainder of the experiment. The latency of responses was monitored carefully to ensure that stimulation did not activate the corticospinal axons at or near the ventral roots, which would be indicated by a reduction in latency by ~2 ms [17,18].

2.8. Protocol

Following familiarization, participants were randomly assigned to complete either session one (TMS) or session two (TMES) first. For both sessions, the procedures were identical with the exception of the stimulation type. Following EMG preparation and ergometer modifications, stimulation intensities were determined (see above). In both sessions, M_{max} was determined first followed by the setting of stimulation intensities for the weak and strong stimulations for either TMS (session one) or TMES (session two). Once stimulation intensities were determined, participants began the 12 cycling trials consisting of six power outputs (25, 50, 100, 150, 200, and 250 W) performed at a constant cadence of 60 rpm with either the weak or strong stimulation intensity (i.e., six cycling trials at each stimulation intensity). The order of the cycling trials was randomized for each participant. While cycling, as the dominant hand passed the 6 o'clock position, one M_{max} and either six MEPs or six CMEPs (depending on the session) were evoked in a randomized order. The time between stimulations was 5–6 s. The total length of each trial was approximately 30 s. To reduce the potential influence of fatigue, one-minute rest periods were given following completion of the lower power output trials (i.e., 25, 50, 100 W), and two-minute rest periods were given after the higher power output trials (i.e., 150, 200, 250 W). Additionally, half-way through the 12 trials (i.e., after trial six), a 5-min rest period was given before the remainder of the trials were completed.

2.9. Data Analysis

For analysis of M_{max}, MEP, and CMEP, the averaged peak-to-peak amplitudes from each cycling trial were measured from the biceps brachii of the dominant arm. Since M_{max} is thought to represent the maximal response of the motor system [4], averaged MEPs (n = 6) and CMEPs (n = 6) from each trial were normalized to the M_{max} within each cycling trial. Response latencies of all evoked responses were carefully monitored throughout all cycling trials as well. The latency for each response was classified as the duration from the stimulus artifact to the initial deflection in the voltage trace from baseline and was averaged across the total number of stimulation trials. Additionally, since the level of voluntary muscle contraction could potentially have an influence on changes in MEP and CMEP amplitudes, pre-stimulus EMG was measured from the rectified virtual channel created for the biceps and triceps brachii as the mean of a 50 ms window immediately prior to the stimulation artifact [14]. For two participants who completed CMEPs (n = 8), pre-stimulus EMG from the triceps brachii was not available due to a technical error during data collection. Therefore, the final sample size for CMEP pre-stimulus EMG data from the triceps brachii was n = 6.

2.10. Statistical Analysis

Group data are presented as means ± SD in the text and means ± SE in the figures (with n in the legends). All statistics were performed using IBM's SPSS Statistics (IBM SPSS Statistics for Windows, Version 23.0. Armonk, NY, USA: IBM Corp.). Mauchly's test was employed to assess the assumption of sphericity for repeated measures analysis. In cases where sphericity was violated, the appropriate correction was applied (i.e., Greenhouse Geisser or Huynh-Feldt) and the degrees of freedom were adjusted. Separate two-way repeated-measures ANOVAs were used to assess the effects of stimulation intensity and cycling intensity (and any interaction) on the M_{max}, MEP, and CMEP amplitudes (both normalized to M_{max}), the average pre-stimulus EMG, and the MEP/CMEP ratios. Post hoc pairwise comparisons were made between means using the Bonferroni correction. Additionally, because one of our aims was to examine the effects of cycling intensity on corticospinal excitability measures within each stimulation intensity (weak and strong), separate one-way repeated-measures ANOVAs were conducted for both the weak and strong stimulus on M_{max}, MEP, and CMEP amplitudes (normalized to M_{max}), pre-stimulus EMG, and MEP/CMEP ratios as cycling intensity increased. If a main effect was identified, post hoc pairwise comparisons were made between means using the Bonferroni correction. Independent samples t-tests were conducted to compare whether MEPs and CMEPs (normalized to M_{max}) at both stimulation intensities were matched appropriately. To compare between MEP and CMEP amplitudes (normalized to M_{max}) at each power output, independent sample t-tests were used with a Bonferroni correction. Paired samples t-tests were conducted on MEP/CMEP ratios between stimulation strengths (weak vs. strong) at each power output. All statistics were performed on group data and statistical significance was set at $p < 0.05$.

3. Results

Evoked responses (i.e., M_{max}, MEPs, and CMEPs) were recorded from the dominant arm biceps brachii at two different stimulation intensities while participants performed arm cycling bouts over a range of contraction strengths. MEPs and CMEPs (normalized to M_{max}) were evoked on separate days but were initially matched to equal 10% (weak stimulus) and 40% (strong stimulus) of the M_{max} on each day. MEPs and CMEPs were not significantly different when either the weak or the strong stimulation intensity were examined ($p > 0.05$ for both conditions), suggesting that the responses were indeed matched initially between days.

3.1. Biceps Brachii Evoked Responses

3.1.1. MEP Amplitude

Figure 2 (top panel) and Figure 3A show representative and grouped data, respectively for MEP amplitudes from the biceps brachii during arm cycling across the various contraction intensities. Figure 2 shows evoked potential traces from one participant during arm cycling with the weak stimulation intensity. In this example, the amplitudes of the MEPs show a progressive and generally consistent increase from the lowest (25 W) to the highest (250 W) arm cycling/muscle contraction intensity. Results from the two-way ANOVA on MEP amplitudes showed a significant main effect for both stimulation intensity (strong > weak, $F_{5,40} = 96.81$, $p < 0.001$) and cycling intensity ($F_{1,8} = 65.30$, $p < 0.001$). Bonferroni post hoc tests revealed that MEP amplitudes at 25 W and 50 W were not different from one another ($p = 0.187$) but were significantly smaller than MEP amplitudes evoked during the 100, 150, 200, and 250 W trials ($p < 0.05$ for all comparisons). Additionally, there was a significant interaction between the intensity of stimulation and the intensity of cycling on MEP amplitudes ($F_{5,40} = 65.30$, $p < 0.001$). Further analysis, through use of one-way ANOVAs for each stimulation intensity, showed a significant main effect for cycling intensity on MEP amplitudes at both the weak ($F_{5,40} = 55.61$, $p < 0.001$) and strong ($F_{5,40} = 41.28$, $p < 0.001$) stimulation conditions. Using the weak stimulation, Bonferroni post hoc tests revealed that MEP amplitudes increased as cycling intensity increased up until 200 W (200 W > 150 W >100 W > 50 W > 25 W, $p < 0.05$ for all comparisons) after which MEPs plateaued ($p > 0.05$). Using the strong stimulation, MEP amplitudes similarly increased with cycling intensity, however, a peak was observed at 100 W (100 W > 50 W > 25 W, $p < 0.05$ for all comparisons), at a lower power output than that observed using the weaker stimulation condition (i.e., 200 W). Beyond 100 W, there were no further increases in MEP amplitudes ($p > 0.05$).

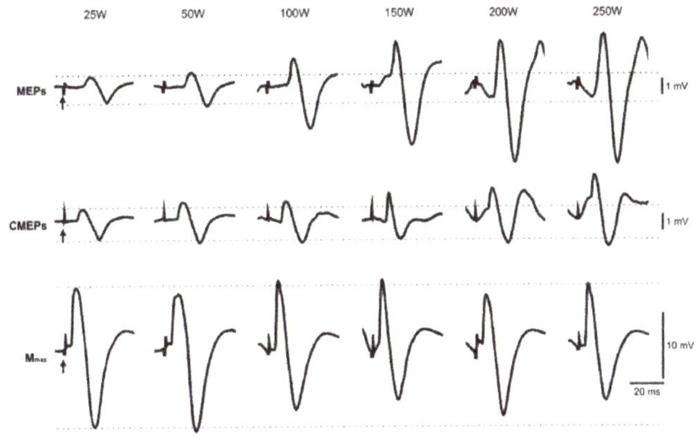

Figure 2. Raw traces for MEPs (top row), CMEPs (middle row), and M_{max} (bottom row) from the biceps brachii of a single participant ($n = 1$) across arm cycling power outputs using the weak stimulation intensity. Each MEP and CMEP waveform represent the average of six evoked potentials. Arrows indicate the stimulus artifact, and dashed lines portray the initial amplitudes of evoked potentials with the *weak* stimulation (~10% M_{max}). In this example, MEP and CMEP amplitudes show a general progressive increase as power output increases towards 250 W, while M_{max} gradually decreases.

3.1.2. Biceps Brachii Pre-stimulus EMG

Figure 3C shows group data for biceps brachii pre-stimulus EMG prior to MEPs during arm cycling. Results from the two-way ANOVA showed that mean biceps brachii pre-stimulus EMG in the 50 ms preceding an MEP was not different between the weak and strong stimulation intensity ($F_{1,8} = 1.42$, $p = 0.267$). Therefore, the average pre-stimulus EMG was pooled between the weak and strong stimulation conditions, which are represented in Figure 3C. There was a significant main effect on biceps brachii pre-stimulus EMG for cycling intensity ($F_{1.76,14.12} = 29.33$, $p < 0.001$), but, there was no interaction between stimulation intensity and cycling intensity ($F_{1.96,27.35} = 1.96$, $p = 0.137$). To further examine changes in pre-stimulus EMG with cycling intensity, one-way ANOVAs were performed. Pre-stimulus EMG increased as cycling intensity increased up until 200 W (Figure 3C, $p < 0.05$), and no differences were observed between the 200 W and 250 W conditions ($p = 1.00$).

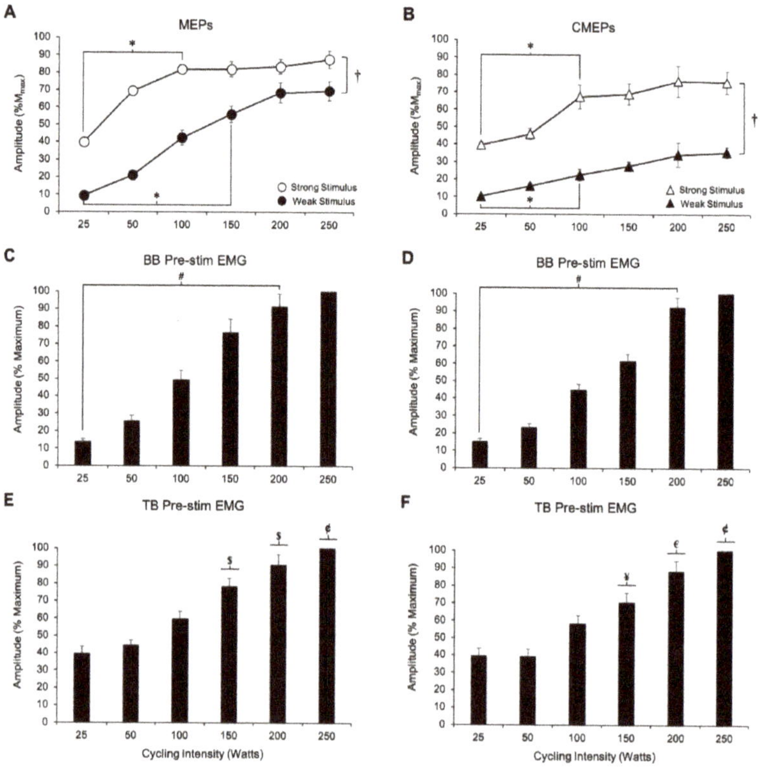

Figure 3. (**A**,**B**) Normalized grouped data (means ± SE) of the peak-to-peak amplitudes for MEPs (**A**) and CMEPs (**B**) obtained from the biceps brachii at each power output examined. MEPs and CMEPs were normalized to M_{max} at each corresponding cycling intensity. In both A and B, filled data points represent when the weak stimulus was used, while unfilled points represent data from the strong stimulus. For clarity, circles were used for MEPs, while triangles were used for CMEPs. In some cases, data points are bigger than SE bars. * Significant difference between illustrated data points. † Significant main effect for stimulation strength ($p < 0.05$). (**C**,**D**) Pre-stimulus EMG (means ± SE) from the biceps brachii which has been pooled and averaged between both stimulation intensities for the TMS session (**C**) and TMES session (**D**), respectively. # Significant difference between all data points. (**E**,**F**) Pre-stimulus EMG (means ± SE) from the triceps brachii which has been pooled and averaged between both stimulation intensities for the TMS session (**E**) and TMES session (**F**), respectively. $ denotes significant difference from all previous power outputs. ¥ denotes significant difference from the 25 W condition. € denotes significant difference from the 25, 50, and 100 W conditions. ¢ denotes significant difference from the 25, 50, 100, and 150 W conditions.

3.1.3. Triceps Brachii Pre-Stimulus EMG

Figure 3E shows group data for triceps brachii pre-stimulus EMG prior to MEPs. Similar to the biceps, results from the ANOVA showed no effect of stimulation intensity on triceps brachii EMG activity prior to a MEP (Figure 3E, $F_{1,8} = 0.100$, $p = 0.760$), but there was a significant main effect of cycling intensity ($F_{1.62,12.94} = 19.32$, $p < 0.001$). Also, there was no significant interaction between cycling intensity and stimulation intensity ($F_{5,40} = 0.803$, $p = 0.554$). To further examine the effect of cycling intensity on triceps brachii pre-stimulus EMG, one-way ANOVAs were performed using the pooled data. Results from these tests indicated that as cycling intensity increased, triceps brachii pre-stimulus EMG values were only significantly different at 150 W and 200 W. Specifically, triceps brachii pre-stimulus EMG was larger at 150 W than at 100 W ($p = 0.006$) and was larger at 200 W than 150 W ($p = 0.044$).

3.1.4. CMEP Amplitude

Figure 2 (middle panel) and Figure 3B show representative and grouped data, respectively for CMEP amplitudes during the arm cycling bouts. Figure 2 portrays data from one participant from the weak stimulation intensity condition. Similar to the MEP amplitudes, in this example, CMEP amplitudes increase in a relatively consistent and progressive manner. The results from the two-way ANOVA on CMEP amplitudes showed significant main effects for both stimulation intensity (strong > weak, $F_{1,7} = 91.50$, $p < 0.001$) and cycling intensity ($F_{3.81,26.65} = 20.16$, $p < 0.001$), however, there was no significant interaction between the two factors ($F_{5,35} = 1.34$, $p = 0.271$). For cycling intensity, Bonferroni post hoc analysis revealed that CMEPs at 25 and 50 W are smaller than those at all other cycling intensities (i.e., 100, 150, 200, and 250 W) ($p < 0.05$ for all comparisons). To decipher specific effects of cycling intensity within each stimulation condition, separate one-way ANOVAs for the weak and strong stimulation conditions were performed on CMEP amplitudes. The results from the one-way ANOVAs showed a significant main effect for cycling intensity on CMEP amplitudes at both the weak ($F_{5,35} = 21.11$, $p < 0.001$) and strong ($F_{5,35} = 9.95$, $p < 0.001$) stimulation conditions. For the weak stimulation condition, Bonferroni post hoc analyses revealed that CMEP amplitudes increased up until 150 W (150 W > 100 W > 50 W > 25 W; $p < 0.05$ for all comparisons), after which CMEP amplitudes did not change ($p > 0.05$). When the strong stimulation intensity was used, post hoc analyses revealed that CMEP amplitudes increased up until 100 W (100 W > 50 W > 25 W, $p < 0.05$ for all comparisons), after which CMEPs plateaued ($p > 0.05$).

3.1.5. Biceps Brachii Pre-Stimulus EMG

Figure 3D shows group data for biceps brachii pre-stimulus EMG prior to CMEPs during arm cycling. Results from the two-way ANOVA showed that mean biceps brachii pre-stimulus EMG in the 50 ms preceding CMEPs was not influenced by stimulation intensity ($F_{1,7} = 0.02$, $p = 0.906$), thus the data was pooled between the weak and strong stimulation conditions as shown in Figure 3D. There was a significant main effect on biceps brachii pre-stimulus EMG for cycling intensity ($F_{1.49,10.41} = 43.08$, $p < 0.001$), but, there was no interaction between stimulation intensity and cycling intensity ($F_{5,35} = 1.22$, $p = 0.320$). To further examine changes in pre-stimulus EMG with cycling intensity, one-way ANOVAs were performed using the pooled data. Similar to MEPs, pre-stimulus EMG for CMEPs increased as cycling intensity increased up until 200 W (Figure 3D, $p < 0.05$), and there was no difference between the 200 W and 250 W conditions ($p = 0.885$).

3.1.6. Triceps Brachii Pre-Stimulus EMG

Figure 3F shows group data for triceps brachii pre-stimulus EMG prior to CMEPs. Similar to above, results from the two-way ANOVA showed no effect of stimulation intensity ($F_{1,5} = 0.761$, $p = 0.423$) and thus, the data was pooled between the week and strong stimulation intensities (Figure 3F). There was, however, a significant main effect of cycling intensity ($F_{1.31,6.55} = 14.04$, $p = 0.006$) on triceps brachii pre-stimulus EMG, but no significant interaction ($F_{5,25} = 0.961$, $p = 0.460$). To further examine the effect of cycling intensity on triceps brachii pre-stimulus EMG, one-way ANOVAs were performed using the pooled data. Results from these tests indicated that triceps brachii pre-stimulus EMG values for CMEPs were only increased at 150 W, 200 W and 250 W compared to the 25 W condition ($p < 0.05$ for all comparisons). However, triceps brachii pre-stimulus EMG was not significantly different with increased cycling intensity from 150 W to 250 W ($p > 0.05$ for all comparisons).

3.1.7. MEP/CMEP Ratios

Although MEPs and CMEPs were evoked on separate days, the responses were initially matched in amplitude to approximately 10% or 40% M_{max} for the weak and strong stimulation conditions, respectively ($p > 0.05$ for both stimulation conditions). Thus, MEP amplitudes were expressed relative to CMEP amplitudes and multiplied by 100% to obtain MEP/CMEP percentages for each participant (Figure 4). This was done in an attempt to isolate whether changes in overall excitability could be attributed to changes in supraspinal and/or spinal excitability. Values greater than 100% indicate that MEP amplitudes are larger than CMEP amplitudes, suggesting that supraspinal excitability may be increased. Similarly, values less than 100% indicate that MEP amplitudes are less than CMEP amplitudes, suggesting that changes in spinal excitability are important factors in maintaining excitability of the corticospinal pathway. Results from the two-way ANOVA revealed a significant main effect for stimulation intensity (weak > strong, $F_{1,7} = 6.94$, $p = 0.034$) and cycling intensity ($F_{5,35} = 9.71$, $p < 0.001$). Bonferroni post hoc tests revealed that MEP/CMEP at 25 W and 50 W were not different from one another ($p = 0.413$) but were significantly smaller than MEP/CMEP at 100, 150, 200, and 250 W trials ($p < 0.05$ for all comparisons). As well, there was a significant interaction effect ($F_{5,35} = 8.18$, $p < 0.001$) between stimulation intensity and cycling intensity on MEP/CMEP ratios. To examine changes in MEP/CMEP with increased power output, one-way ANOVAs were conducted within each stimulation intensity. Results from the one-way ANOVAs showed a significant main effect for cycling intensity on MEP/CMEP ratios at both the weak ($F_{5,35} = 9.44$, $p < 0.001$) and strong ($F_{5,35} = 4.60$, $p = 0.003$) stimulation conditions. When the weak stimulation was used, Bonferroni post hoc analysis revealed that MEP/CMEP were only significantly larger than that at 25 W at 150 W ($p = 0.037$), and 200 W ($p = 0.05$). When the strong stimulation intensity was used, MEP/CMEP were significantly larger at 50 W than at 25 W ($p = 0.026$) but were not different for any other comparison. To compare changes in MEP/CMEP between the weak and strong stimulation intensities, paired samples *t*-tests were performed at each power output. Thus, a total of six comparisons were made. The *t*-tests revealed that the MEP/CMEP ratios were not significantly different at 25 W ($t_{(7)} = 1.22$, $p = 0.261$) or 50 W ($t_{(7)} = 0.52$, $p = 0.622$) when either the weak or strong stimulus was used. However, MEP/CMEP ratios were significantly larger at 100 W ($t_{(7)} = 2.51$, $p = 0.041$), 150 W ($t_{(7)} = 3.24$, $p = 0.014$), 200 W ($t_{(7)} = 3.03$, $p = 0.019$), and 250 W ($t_{(7)} = 2.41$, $p = 0.047$) when the weak stimulation was used.

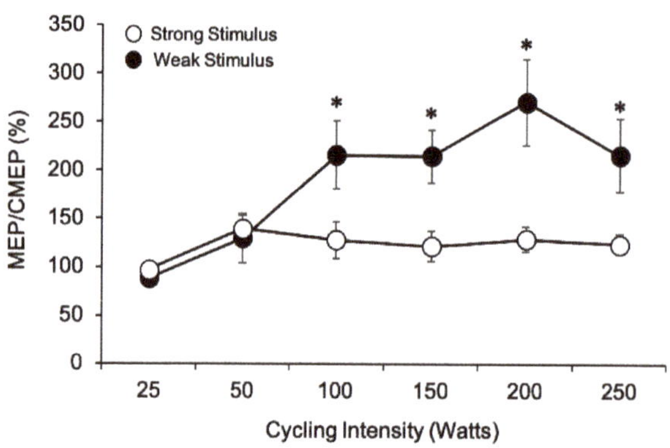

Figure 4. Comparison of MEP/CMEP ratios for the weak (filled circles) and strong (unfilled circles) stimulation intensities as power output increased from 25 W to 250 W. * represents significant difference between stimulation intensities at each given power output ($p < 0.05$). In some cases, SE bars were smaller than the symbols for the data points.

3.1.8. M_{max} Amplitude

For both the TMS and TMES sessions, the results from the two-way ANOVA revealed similar effects on biceps brachii M_{max} amplitudes. For both sessions, there was no effect of stimulation intensity (TMS: $F_{1,8} = 0.093$, $p = 0.769$, TMES: $F_{1,7} = 1.06$, $p = 0.337$), but there was a significant main effect for cycling intensity (TMS: $F_{5,40} = 15.66$, $p < 0.001$; TMES: $F_{1,7} = 8.89$, $p < 0.001$) on M_{max} amplitudes (Figure 5). As cycling intensity increased M_{max} amplitudes decreased (Figure 5A,B). Additionally, there was no interaction observed between factors on either day (TMS: $F_{5,40} = 0.836$, $p = 0.532$, TMES: $F_{5,35} = 0.430$, $p = 0.825$). Since there was no effect of stimulation intensity on M_{max} values, the averages from each stimulation condition (weak and strong) were pooled across the cycling intensities for each session (as shown in Figure 5). For cycling intensity, Bonferroni post hoc analyses indicated that M_{max} values decreased for the TMS and TMES session as cycling intensity increased from 25 to 250 W.

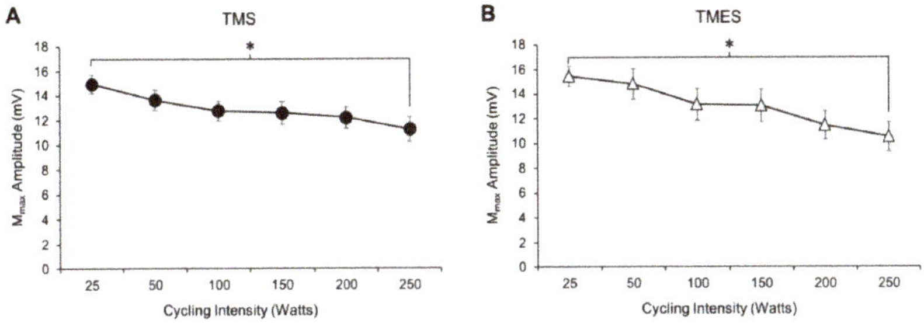

Figure 5. Changes in M_{max} amplitudes with increasing power output pooled between stimulation intensities for the TMS (**A**) and TMES (**B**) session. * denotes significant main effect of power output on M_{max} amplitude. M_{max} decreased by approximately 24.9 and 31.7% as power output increased from 25 to 250 W for the TMS and TMES sessions, respectively.

4. Discussion

This study shows that the amplitudes of TMS-evoked MEPs and TMES-evoked CMEPs increase with power output and plateau, but do not decrease in amplitude as has been previously shown by others during intense tonic contractions [1,4]. MEP amplitudes were much larger than CMEP amplitudes as power output increased regardless of stimulation strength, despite being initially matched in amplitude (Figures 3A,B and 4). This finding suggests that supraspinal factors mediate the change in overall corticospinal excitability observed during arm cycling as intensity increases. Importantly, stimulus strength had a substantial effect on MEP and CMEP amplitudes as cycling power output increased. Responses evoked by the weak stimulation (10% M_{max}) increased up to approximately 200 W for MEPs (Figures 3A and 4) and 150 W for CMEPs (Figures 3B and 4), whereas with the strong stimulation (40% M_{max}), responses reached a peak at 100 W for both MEPs and CMEPs and did not change afterward. Thus, the MEP/CMEP ratio used as a measure of supraspinal excitability was influenced by stimulation strength, which would lead to different conclusions on mechanisms of enhanced corticospinal excitability during arm cycling as power output increases.

4.1. Modulation of Corticospinal and Spinal Excitability with Cycling Intensity

Past research involving isometric contractions has shown that biceps brachii MEPs and CMEPs increase up until a peak at ~75–90% MVC [1–3], a finding which has been attributed to the motor unit firing and recruitment characteristics of the biceps brachii during progressively stronger isometric contractions [19,20]. Following the peak, there is a subsequent decline in responses as contraction intensity approaches 100% MVC [1] which is thought to reflect the inability for some motoneurones to fire in response to artificial excitatory input at strong contraction strengths, given the high degree of voluntary input to the motoneurone pool and the associated changes in their intrinsic properties [1]. In the present study, we did not observe a decline in corticospinal excitability as arm cycling intensity increased to the maximum intensity employed. Instead, we observed a plateauing of responses for both MEPs and CMEPs at intensities below 250 W, which were differentially influenced by stimulus strength (Figure 3A,B). Our results, however, do coincide with findings from the only other study to examine corticospinal excitability changes during a locomotor-like output over a wide range of contraction intensities [5]. In that study, MEPs and CMEPs from the knee extensors during leg cycling increased in amplitude up to 300 W, after which there was a plateauing, but no decline as cycling intensity increased to 400 W [5]. Taken together, these studies suggest task-dependent changes in corticospinal and spinal excitability may be present, a finding we have previously reported [10,16,21].

In the current study, MEP and CMEP amplitudes increased at the lower, but not higher power outputs (Figure 3A,B), suggesting that the increase in overall corticospinal excitability at the low intensities (i.e., 25 to 100 W) is partially generated by increased spinal excitability. These finding are partially supported by biceps brachii pre-stimulus EMG values which increase for both stimulation types (Figure 3C,D) at the low cycling intensities, but are not significantly different between the highest cycling intensities (200 and 250 W). While this may explain the enhanced spinal excitability at the low power outputs, it does not explain why we observed a plateau in CMEP amplitudes beyond 150 W for the weak stimulus and 100 W for the strong stimulus in the present study, since EMG was still increasing beyond these power outputs. It is noted, however, that Weavil and colleagues showed increased EMG and workloads without changes in MEP and CMEP amplitudes. During isometric contractions, the biceps brachii is capable of recruiting additional motor units during contractions up to and beyond 90% MVC [19,20], which help to explain why CMEPs continue to increase beyond 90% MVC [1]. Corticospinal excitability to the biceps brachii is also task- [10,16] and forearm position-dependent [21] which is an important consideration when a comparison to tonic contractions is made. However, the lack of increase in CMEP amplitudes beyond 150 W and 100 W during arm cycling in the current study, while MEPs and background EMG are still increasing is unlikely to be explained by reaching the maximum motor unit recruitment of the biceps, given that these cycling intensities are not maximal, at least relative to a sprint test [7]. It is possible, however, that motoneurone recruitment strategies during a rhythmic motor output such as arm cycling may be different from those observed during isometric contractions (Power et al., 2018), and therefore could cause motoneurones to be maximally recruited sooner than 90% of maximal cycling power. Work in adult decerebrate cats and rats, for example, demonstrated that spinal motoneurones are characterized by changes in their electrical properties during locomotor outputs that would act to enhance their recruitment and firing [22–24]. These same changes in motoneurone excitability do not occur during tonic motor output [23].

4.2. Modulation of Supraspinal Excitability with Cycling Intensity

In the current study, MEP/CMEP ratios increased with power output, in particular when the weak stimulation intensity was used (Figure 4) suggesting that supraspinal excitability was enhanced to a larger degree than spinal excitability. It is plausible that changes in the excitability of interneuronal circuits and/or interhemispheric connections may be involved. During tonic contractions, short-interval intracortical inhibition (SICI) is reduced as muscle contraction intensity increases [25–27], a finding that is thought to downregulate the activity of the inhibitory neurons which project onto corticospinal cells involved in producing the movement. We recently showed that SICI was present during arm cycling, albeit not different than a tonic contraction [28]. Thus, it is possible that reductions in SICI during arm cycling as power output increases may underlay increases in MEP amplitudes as has been shown during tonic contractions.

Another potential mechanism involves cortical spread from the non-dominant to the dominant motor cortex as we have previously hypothesized [6,7,14]. Since arm cycling is a bilateral motor output it is possible that cortical excitation arising from the active, non-dominant motor cortex could facilitate excitability in the dominant motor cortex, which could reduce the input required to induce an MEP by a given TMS pulse. However, when the strong stimulation intensity was used, the changes in MEP/CMEP ratios were less marked and did not increase as cycling intensity increased suggesting a ceiling effect in the MEP amplitudes had been reached.

4.3. Differences between Stimulation Intensities

This study highlights the importance of stimulation intensity selection for experimental design during locomotor outputs. Notably, MEPs continued to increase with cycling intensity up until approximately 200 W when elicited with weak stimulation intensity (10% M_{max}), while they plateaued at approximately 100 W under strong (40% M_{max}) stimulation. This led us to conclude that supraspinal excitability increases with increased power output, an effect only observed when weak stimulus intensity was used. In contrast, using the strong stimulation intensity leads one to believe, perhaps falsely, that spinal factors were driving the change in overall corticospinal excitability as a function of power output, a conclusion also reached by Weavil and colleagues (2015) who used a strong stimulation intensity (MEPs and CMEPs were ~50% M_{max}). The use of a weak stimulation intensity yielded a more precise measure of corticospinal excitability in this specific study as MEPs were less susceptible to ceiling effects than at the strong stimulation.

4.4. Methodological Considerations

An important methodological consideration in interpreting the current data is that we did not make the power outputs relative to each individual as we have recently done in two separate studies during arm cycling [6,7]. In Spence et al. (2016) we used 5 and 15% of peak power output determined by a sprint test (modified Wingate) while in Lockyer et al. (2018) we used 20, 40, and 60% of peak power output determined via a standard incremental aerobic test (20 W increases every two minutes) [29]. These methods were not without limitations, however. The former used a sprint test to prescribe aerobic cycling intensity at 60 RPM and the latter incremental test resulted in most of the participants reaching a similar peak power output of ~120 W. In the present study we used absolute power outputs as has been used by others [5,30] and all participants were able to cycle well above the aerobic test maximum power output of 120 W obtained in our prior work. We were thus able to have participants cycle at supramaximal intensities, albeit we did not quantify exertion levels. Additionally, the sample size of (n = 9) for MEPs and (n = 8) for CMEPs was not determined by a power analysis and therefore, it is unclear whether a larger sample size would have influenced the present results.

5. Conclusions

The present study describes the influence of stimulation strength over a wide range of cycling intensities on corticospinal and spinal excitability during arm cycling. We have demonstrated that corticospinal excitability to the biceps brachii is increased with cycling intensity during low power outputs, a finding that is partially mediated by spinal factors. As cycling intensity increases, however, it appears as though supraspinal factors may play more of a role in modulating overall corticospinal excitability. Additionally, this study highlights the importance of stimulation intensity selection to assess corticospinal excitability during motor output. It is concluded that the use of a weaker stimulation intensity provides a more precise measure of corticospinal excitability during locomotor outputs at high intensities as they are less susceptible to potential ceiling effects.

Author Contributions: All authors participated in data collection, analysis, interpretation and manuscript preparation. All authors have approved the submitted version of this manuscript.

Funding: This research was funded by the Natural Sciences and Engineering Research Council of Canada, grant number (NSERC-#RGPIN-2015-05765).

Acknowledgments: This study was supported by PGS-D NSERC funding to E.J.L. as well as NSERC Discovery Grant to K.E.P. We would like to thank Thamir Alkanani for technical support and the participants for volunteering their time.

Conflicts of Interest: The authors declare no conflict of interest.

References

1. Martin, P.G.; Gandevia, S.C.; Taylor, J.L. Output of human motoneuron pools to corticospinal inputs during voluntary contractions. *J. Neurophysiol.* **2006**, *95*, 3512–3518. [CrossRef] [PubMed]
2. Todd, G.; Taylor, J.L.; Gandevia, S.C. Measurement of voluntary activation of fresh and fatigued human muscles using transcranial magnetic stimulation. *J. Physiol.* **2003**, *551*, 661–671. [CrossRef] [PubMed]

3. Taylor, J.L.; Allen, G.M.; Butler, J.E.; Gandevia, S.C. Effect of contraction strength on responses in biceps brachii and adductor pollicis to transcranial magnetic stimulation. *Exp. Brain Res.* **1997**, *117*, 472–478. [CrossRef] [PubMed]
4. Oya, T.; Hoffman, B.W.; Cresswell, A.G. Corticospinal-evoked responses in lower limb muscles during voluntary contractions at varying strengths. *J. Appl. Physiol.* **2008**, *105*, 1527–1532. [CrossRef] [PubMed]
5. Weavil, J.C.; Sidhu, S.K.; Mangum, T.S.; Richardson, R.S.; Amann, M. Intensity-dependent alterations in the excitability of cortical and spinal projections to the knee extensors during isometric and locomotor exercise. *Am. J. Physiol. Regul. Integr. Comp. Physiol.* **2015**, *308*, R998–R1007. [CrossRef]
6. Lockyer, E.J.; Benson, R.J.; Hynes, A.P.; Alcock, L.R.; Spence, A.J.; Button, D.C.; Power, K.E. Intensity matters: effects of cadence and power output on corticospinal excitability during arm cycling are phase- and muscle-dependent. *J. Neurophysiol.* **2018**, *120*, 2908–2921. [CrossRef] [PubMed]
7. Spence, A.J.; Alcock, L.R.; Lockyer, E.J.; Button, D.C.; Power, K.E. Phase- and Workload-Dependent Changes in Corticospinal Excitability to the Biceps and Triceps Brachii during Arm Cycling. *Brain Sci.* **2016**, *6*, 60. [CrossRef]
8. Zehr, E.P.; Carroll, T.J.; Chua, R.; Collins, D.F.; Frigon, A.; Haridas, C.; Hundza, S.R.; Thompson, A.K. Possible contributions of CPG activity to the control of rhythmic human arm movement. *Can. J. Physiol. Pharmacol.* **2004**, *82*, 556–568. [CrossRef]
9. Zehr, E.P.; Collins, D.F.; Frigon, A.; Hoogenboom, N. Neural control of rhythmic human arm movement: Phase dependence and task modulation of hoffmann reflexes in forearm muscles. *J. Neurophysiol.* **2003**, *89*, 12–21. [CrossRef]
10. Forman, D.; Raj, A.; Button, D.C.; Power, K.E. Corticospinal excitability of the biceps brachii is higher during arm cycling than an intensity-matched tonic contraction. *J. Neurophysiol.* **2014**, *112*, 1142–1151. [CrossRef]
11. Forman, D.A.; Monks, M.; Power, K.E. Corticospinal excitability, assessed through stimulus response curves, is phase-, task-, and muscle-dependent during arm cycling. *Neurosci. Lett.* **2018**, *692*, 100–106. [CrossRef]
12. Rossi, S.; Hallett, M.; Rossini, P.M.; Pascual-Leone, A. Safety of TMS Consensus Group. Safety, ethical considerations, and application guidelines for the use of transcranial magnetic stimulation in clinical practice and research. *Clin. Neurophysiol.* **2009**, *120*, 2008–2039. [CrossRef]
13. Veale, J.F. Edinburgh Handedness Inventory—Short Form: A revised version based on confirmatory factor analysis. *Laterality* **2014**, *19*, 164–177. [CrossRef]
14. Forman, D.A.; Philpott, D.T.; Button, D.C.; Power, K.E. Cadence-dependent changes in corticospinal excitability of the biceps brachii during arm cycling. *J. Neurophysiol.* **2015**, *114*, 2285–2294. [CrossRef]
15. Manning, C.D.; Bawa, P. Heteronymous reflex connections in human upper limb muscles in response to stretch of forearm muscles. *J. Neurophysiol.* **2011**, *106*, 1489–1499. [CrossRef]
16. Power, K.E.; Lockyer, E.J.; Forman, D.A.; Button, D.C. Modulation of motoneurone excitability during rhythmic motor outputs. *Appl. Physiol. Nutr. Metab.* **2018**, *43*, 1176–1185. [CrossRef]
17. Taylor, J.L.; Gandevia, S.C. Noninvasive stimulation of the human corticospinal tract. *J. Appl. Physiol.* **2004**, *96*, 1496–1503. [CrossRef]
18. Taylor, J.L. Stimulation at the cervicomedullary junction in human subjects. *J. Electromyogr. Kinesiol.* **2006**, *16*, 215–223. [CrossRef]
19. De Luca, C.J.; LeFever, R.S.; McCue, M.P.; Xenakis, A.P. Behaviour of human motor units in different muscles during linearly varying contractions. *J. Physiol.* **1982**, *329*, 113–128. [CrossRef]
20. Kukulka, C.G.; Clamann, H.P. Comparison of the recruitment and discharge properties of motor units in human brachial biceps and adductor pollicis during isometric contractions. *Brain Res.* **1981**, *219*, 45–55. [CrossRef]
21. Forman, D.A.; Richards, M.; Forman, G.N.; Holmes, M.W.; Power, K.E. Changes in Corticospinal and Spinal Excitability to the Biceps Brachii with a Neutral vs. Pronated Handgrip Position Differ between Arm Cycling and Tonic Elbow Flexion. *Front. Hum. Neurosci.* **2016**, *10*, 543. [CrossRef]
22. MacDonell, C.W.; Power, K.E.; Chopek, J.W.; Gardiner, K.R.; Gardiner, P.F. Extensor motoneurone properties are altered immediately before and during fictive locomotion in the adult decerebrate rat. *J. Physiol.* **2015**, *593*, 2327–2342. [CrossRef]
23. Power, K.E.; McCrea, D.A.; Fedirchuk, B. Intraspinally mediated state-dependent enhancement of motoneurone excitability during fictive scratch in the adult decerebrate cat. *J. Physiol.* **2010**, *588*, 2839–2857. [CrossRef]

24. Krawitz, S.; Fedirchuk, B.; Dai, Y.; Jordan, L.M.; McCrea, D.A. State-dependent hyperpolarization of voltage threshold enhances motoneurone excitability during fictive locomotion in the cat. *J. Physiol.* **2001**, *532*, 271–281. [CrossRef]
25. Ridding, M.C.; Taylor, J.L.; Rothwell, J.C. The effect of voluntary contraction on cortico-cortical inhibition in human motor cortex. *J. Physiol.* **1995**, *487*, 541–548. [CrossRef]
26. Kujirai, T.; Caramia, M.D.; Rothwell, J.C.; Day, B.L.; Thompson, P.D.; Ferbert, A.; Wroe, S.; Asselman, P.; Marsden, C.D. Corticocortical inhibition in human motor cortex. *J. Physiol.* **1993**, *471*, 501–519. [CrossRef]
27. Ortu, E.; Deriu, F.; Suppa, A.; Tolu, E.; Rothwell, J.C. Effects of volitional contraction on intracortical inhibition and facilitation in the human motor cortex. *J. Physiol.* **2008**, *586*, 5147–5159. [CrossRef]
28. Alcock, L.R.; Spence, A.J.; Lockyer, E.J.; Button, D.C.; Power, K.E. Short-interval intracortical inhibition to the biceps brachii is present during arm cycling but is not different than a position- and intensity-matched tonic contraction. *Exp. Brain Res.* **2019**, *237*, 2145–2154. [CrossRef]
29. Price, M.J.; Collins, L.; Smith, P.M.; Goss-Sampson, M. The effects of cadence and power output upon physiological and biomechanical responses to incremental arm-crank ergometry. *Appl. Physiol. Nutr. Metab.* **2007**, *32*, 686–692. [CrossRef]
30. Christensen, L.O.; Johannsen, P.; Sinkjaer, T.; Petersen, N.; Pyndt, H.S.; Nielsen, J.B. Cerebral activation during bicycle movements in man. *Exp. Brain Res.* **2000**, *135*, 66–72. [CrossRef]

brain sciences

MDPI

Communication

Modulating Observation-Execution-Related Motor Cortex Activity by Cathodal Transcranial Direct Current Stimulation

Fengxue Qi [1,2,3,*], Michael A. Nitsche [2,4] and Volker R. Zschorlich [1,5,6]

1 Department of Movement Science, Faculty of Philosophy, University of Rostock, 18057 Rostock, Germany; volker.zschorlich@uni-rostock.de
2 Department of Psychology and Neurosciences, Leibniz Research Centre for Working Environment and Human Factors, 44139 Dortmund, Germany; nitsche@ifado.de
3 Department of Sport Training, Sport Coaching College, Beijing Sport University, Beijing 100084, China
4 Department of Neurology, University Medical Hospital Bergmannsheil, 44789 Bochum, Germany
5 Faculty of Medicine, University of Rostock, 18055 Rostock, Germany
6 Department Ageing of Individuals and Society, Faculty of Interdisciplinary Research, University of Rostock, 18051 Rostock, Germany
* Correspondence: fengxue.qi@hotmail.com or fengxue.qi@uni-rostock.de; Tel.: +49-231-108-4320

Received: 10 April 2019; Accepted: 23 May 2019; Published: 26 May 2019

Abstract: The aim of this randomized sham-controlled study was to examine the impact of cathodal transcranial direct current stimulation (ctDCS) of the primary motor cortex (M1) during movement observation on subsequent execution-related motor cortex activity. Thirty healthy participants received sham or real ctDCS (1 mA) over the left M1 for 10 minutes, respectively. The participants observed a video showing repeated button pressing tasks of the right hand during the sham or real ctDCS, followed by performance of these tasks by the right hand. Motor-evoked potentials (MEP) were recorded from the resting right first dorsal interosseous muscle before movement observation during the sham or real ctDCS, immediately after observation of actions, and after subsequent movement execution. The results of the ANOVA showed a significant main effect on the group ($F_{1,28} = 4.60$, $p = 0.041$) and a significant interaction between time and the group ($F_{2,56} = 5.34$, $p = 0.008$). As revealed by respective post hoc tests, ctDCS induced a significant reduction of MEP amplitudes in connection with movement observation ($p = 0.026$, Cohen's $d = 0.861$) and after subsequent movement execution ($p = 0.018$, Cohen's $d = 0.914$) in comparison with the sham stimulation. It is concluded that ctDCS during movement observation was effective in terms of modulating motor cortex excitability. Moreover, it subsequently influenced execution-related motor cortex activity. This indicates a possible application for rehabilitative treatment in syndromes with pathologically enhanced cortical activity.

Keywords: movement observation; movement execution; transcranial direct current stimulation; motor cortex activity

1. Introduction

Movement observation or execution-related cortical networks are activated when individuals are conducting an action or observe the identical movement performed by another individual. These networks include the primary somatosensory cortex, primary motor cortex (M1), ventral premotor cortex and inferior frontal gyrus [1,2]. Alpha and beta frequency oscillations desynchronize within these networks during action observation and execution [3–5]. These alpha (8–13 Hz) and beta (14–30 Hz) oscillations in the electroencephalogram (EEG) are referred to as mu and beta rhythms and reflect mirror neuron activity [5–7].

Modulation of the activity of respective networks via non-invasive brain stimulation might alter respective task-related physiological processes and ultimately affect performance. Transcranial electrical stimulation is suitable to modulate cortical excitability and activity by a weak electrical current, which modulates synaptic activity and/or neuronal resting membrane potentials [8,9]. Cathodal transcranial direct current stimulation (ctDCS) results in long-term depression-like effects [8,10]. The effects of transcranial direct current stimulation (tDCS) on motor cortex activity are dependent on the state of the stimulated cortical region before and/or during the application [9]. With respect to treatment of brain diseases associated with pathological enhancement of cortical activity or excitability via cathodal tDCS, some studies have reported that ctDCS applied over an epileptogenic focus reduces epileptic EEG abnormalities [11,12] and seizure frequency in epilepsy [13]. Clinical studies moreover showed that cathodal tDCS over the contralesional M1 combined with motor training rebalances bi-hemispheric activity and facilitates motor performance in stroke patients [14,15] and in children with cerebral palsy [16,17]. Furuya et al. describe that ctDCS applied over the affected motor cortex in musicians cramp, combined with anodal tDCS over the contralateral motor cortex and bimanual finger movements, reduces dystonic symptoms in these patients, most probably due to a reduction of pathologically enhanced activity of the affected M1 [18].

In this study, we contrasted the effects of ctDCS and sham stimulation application to M1 during movement observation on subsequent execution-related motor cortex activity. It was expected that cathodal tDCS would reduce movement observation-related motor cortex activity, and that it subsequently would reduce execution-related motor cortex activity, as compared to a sham stimulation.

2. Materials and Methods

2.1. Participants

Thirty healthy adults (mean age, 25.93 ± 4.75 years; 12 females) participated in this study and were randomized to receive sham or real cathodal tDCS. They gave written informed consent. The ethics committee of the University Medicine Rostock approved this study and the study meets the standards of the Declaration of Helsinki. None of the participants had pregnancy, family history of epilepsy, the presence of neurological, psychiatric or musculoskeletal disorders, central nervous system acting medications, major medical diseases, cardiac pacemakers, skin lesions near the region of stimulation, or metal implants. They had normal or corrected-to-normal vision. The Oldfield's Edinburgh Handedness Inventory [19] was used to evaluate handedness, and all participants were right-handed. Each stimulation condition was performed in 15 participants.

2.2. Monitoring of Motor Cortex Excitability

A single-pulse transcranial magnetic stimulation (TMS) over the left M1 representation of the relaxed right resting first dorsal interosseus (FDI) muscle induced motor-evoked potentials (MEP) which were recorded to monitor motor cortex excitability changes. TMS with a biphasic pulse was conducted by a MagPro R100 magnetic stimulator (Medtronic, Skovlunde, Denmark) with a D-B80 coil. In order to minimize head to coil movement, a chin-forehead rest was used to stabilize the head of the participants. The magnetic coil was held tangentially to the skull over the left M1. A mechanical arm (Manfrotto Feltre, Italy) fixed the handle pointing backwards and laterally 45° from the midline. The optimal magnetic coil position (hotspot) was determined by a moderate suprathreshold stimulation intensity to constantly elicit the largest MEP in the right resting FDI muscle. The intensity of TMS was determined as the percentage of maximal stimulator output (% MSO) which elicited peak-to-peak MEP amplitudes of approximately 1 mV (SI1mV) at baseline before the intervention. This intensity was kept constant for the remaining experiment. Twenty MEPs were obtained from the right FDI muscle for each time bin using a pair of Ag-AgCl cup electrodes with a surface area of 3 mm^2 (GE Medical Systems, Milwaukee, USA) in a belly-tendon montage. The ground electrode was positioned over the right lateral biceps brachii muscle. The electromyographic (EMG) signals were amplified (input resistance of 10 GΩ, bandwidth of 1–1000 Hz, Biovision, Wehrheim, Germany) with an amplification rate of 1000. DIAdem software was used to process EMG signals.

2.3. tDCS

Cathodal tDCS (1 mA) was delivered for 10 minutes (5 s of ramp-up and ramp-down) by a battery-driven electrical stimulator (BrainSTIM, EMS, Bologna, Italy) and applied through a pair of 25 cm^2 surface saline-soaked sponge electrodes. The cathode electrode was placed over the defined left M1 'hotspot' and the anode electrode was positioned over the right supra-orbital area. The direct current was switched on for 30 s and then turned off during sham stimulation.

2.4. Movement Observation and Execution

The participants kept their hands in a relaxed position. They comfortably seated in front of a computer screen (24-inch) to watch one movement observation video, which displayed the right hand pressing buttons. The video was 10 minutes long and incorporated 20 short clips. Twenty-second long clips in natural speed were presented 10 times. At half of the natural speed, 40 s long clips were presented 10 times. At the beginning, the 20 s long clip was displayed, followed by a 40 s long clip. Low-speed movements were included because these result in a more prominent modulation of motor cortex activity [20]. The clips showed that whenever a black spot changed into a red spot in a regular sequence, a human hand reached for the appropriate round button immediately, pressed it with the index finger only, and quickly got back to the original position afterwards. Participants were instructed to concentrate on the performing finger and the button press task. They were instructed to count the number of button presses, because attention is an important mediator of performance, and performance-related physiological effects [21].

After movement observation, participants performed a movement execution task (160 s) which was identical to the observed video. A 24-inch computer screen and a custom-made button-box with 4 red round buttons were positioned on the table in front of the participant. The distance was 20 cm from the right hand to the button-box. The four buttons corresponded to the black spots shown on the computer screen. When a black spot changed into a red spot in a modeled response sequence on the computer screen (each 3 s), participants pressed the button with the right index finger immediately, and quickly moved back to the initial hand position afterwards. The left hand always remained in the resting position.

2.5. Experimental Procedures

The participants received either ctDCS or a sham stimulation, while watching the movement observation video. Movement execution was conducted after the termination of action observation and tDCS. MEPs were recorded before (T0) and after (T1) movement observation combined with the sham or real ctDCS, and then after movement execution (T2). For an overview of the experimental procedure, refer to Figure 1.

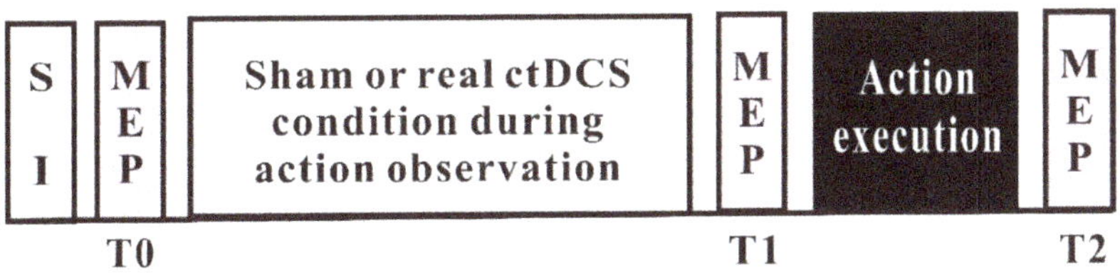

Figure 1. Timeline of the experimental protocol.

2.6. Statistical Analysis

SPSS (version 22.0; IBM, Armonk, NY, USA) and Prism (Version 8; GraphPad Software Inc., San Diego, CA, USA) were used to perform statistical analyses. Normal distribution of the data was confirmed by Kolmogorov-Smirnov tests. Independent-samples t-tests were used to examine differences of demographic and physiological data between experimental conditions respectively, including age, SI1mV, and baseline measurements of MEP amplitudes. Gender distribution differences between groups were examined via a χ^2 test.

For MEP analysis, first, the mean values of the peak-to-peak amplitudes of 20 MEPs for each time bin were calculated individually. If EMG activity larger than 50 μV was present in the 300 ms time-window before TMS, the respective MEP was excluded. A repeated measures ANOVA was performed to analyze MEP amplitude data with the between-subject factors tDCS (sham and real ctDCS condition), and Time course. The assumption of sphericity was examined by Mauchly's test and the Greenhouse-Geisser correction was applied if necessary. Fisher's post hoc tests were performed to determine differences between groups. Cohen's d was used to calculate effect sizes. A significance level of $p < 0.05$ was used for all statistical tests.

3. Results

All participants tolerated ctDCS well. There were not significant differences with respect to age, gender, SI1mV, and baseline MEP amplitudes between groups (all values of $p \geq 0.136$; Table 1). The results revealed a significant main effect for the group ($F_{1,28} = 4.60$, $p = 0.041$) and interaction between time and the group ($F_{2,56} = 5.34$, $p = 0.008$), but the main effect of time was not significant ($F_{2,56} = 2.76$, $p = 0.072$). The post hoc tests showed no significant differences between all time points in the sham stimulation condition. MEP amplitudes were however significantly decreased following ctDCS in comparison with the sham stimulation at time point T1 (after movement observation) ($p = 0.026$, Cohen's $d = 0.861$) and T2 (after movement execution) ($p = 0.018$, Cohen's $d = 0.914$, Figure 2A). In comparison to the respective baseline values, there were significant differences of MEP amplitudes from T0 to T1 ($p = 0.001$, Cohen's $d = 1.156$) and T0 to T2 ($p = 0.005$, Cohen's $d = 0.772$) in the real ctDCS group. In the real ctDCS group, 13 of 15 (86.67 %) participants at T1, and 11 of 15 (73.33 %) participants at T2 had reduced MEP amplitudes in contrast to the baseline (T0) (see Figure 2C).

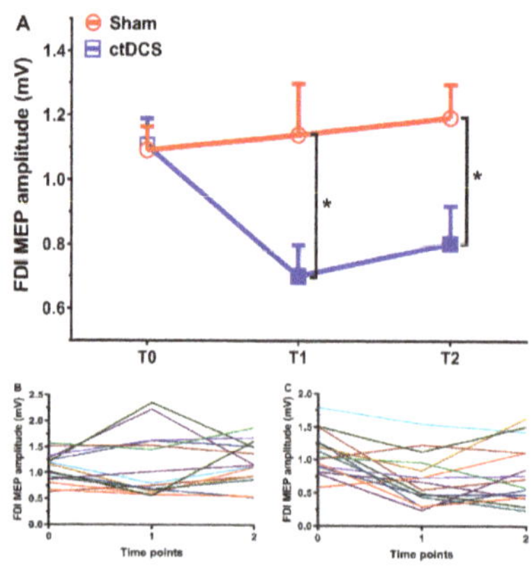

Figure 2. Results of the study. For time points T0, T1, and T2, averaged MEP amplitudes of the first dorsal interosseus (FDI) muscle for all participants (**A**) and individual MEP amplitude in the sham (**B**) and cathodal transcranial direct current stimulation (ctDCS) (**C**) groups are depicted. Filled symbols indicate significant differences of MEP amplitudes after movement observation and execution in comparison to baseline values. Error bars represent standard error of means. * denotes significant differences between groups at $p < 0.05$.

Table 1. Demographic information, SI1mV, and baseline measurements of motor-evoked potentials (MEP) amplitudes (mV) of the intervention groups.

Group	*n*	Gender (F/M)	Age (Years)	SI1mV	MEP (Baseline)
Sham	15	4/11	26.13 ± 3.98	44.13 ± 10.45	1.09 ± 0.28
Real	15	8/7	25.73 ± 5.55	45.67 ± 11.25	1.11 ± 0.32
Between group	-	$p = 0.136$	$p = 0.822$	$p = 0.702$	$p = 0.888$

SI1mV (% MSO) refers to the intensity of TMS required to elicit an average motor evoked potential (MEP) of approximately 1 mV. '-' refers to no data available. The mean values ± standard deviations are shown for the sham and real cathodal tDCS sessions.

4. Discussion

In the sham stimulation condition, there were not significant changes in MEP amplitudes at all time points. Thus, in accordance with another study with the same task, movement observation followed by movement execution did not promote MEP amplitude alterations when no effective stimulation was applied [22]. This result is also compatible with other studies in the field. Movement observation alone did not enhance contralateral motor cortex excitability immediately after movement observation [23,24]. Cortical activity returned to baseline levels within a few seconds [25], as shown by direct recordings of mirror neurons in monkeys [26,27] and indirect measures of neural activity in human motor regions [28].

Cathodal tDCS applied during movement observation induced a reduction of MEP amplitudes, and this effect remained after movement execution in comparison to the sham stimulation condition. Thus ctDCS had antagonistic effects on the modulation of motor cortex activity which was not abolished by movement observation or execution. This might have clinical implications, namely when symptoms are caused by task-related over-activity or lesions of respective cortical regions. Recent clinical studies have shown that ctDCS over the left sensorimotor cortex combined with action observation and electromyographic biofeedback training improved dystonic posture and writing movements of the right upper limb in a patient with writer's cramp [29]. Re-training of finger movements in musician's cramp [18] was improved by cathodal tDCS over the respective motor cortex. Cathodal tDCS combined with movement observation and/ or training thus might improve abnormal cortical excitability and enable a re-arrangement of motor representations. It was also reported that cathodal tDCS over the non-lesional M1 combined with movement execution facilitates motor performance in stroke patients [14,15] and in children with unilateral cerebral palsy [16,17]. The suggested mechanism is that ctDCS coupled with motor training downregulates activity of the non-lesioned hemisphere, while it upregulates activity of the lesioned hemisphere. The latter is via movement execution and reduced transcallosal inhibition, and thus re-balances bi-hemispheric activity and improves sensorimotor representations. In most motor performance studies, anodal tDCS enhanced cortical excitability and improved motor performance [30,31]. This is thought to be caused by improved induction of task-related LTP [32,33]. It was however also shown that cathodal tDCS can improve motor performance under certain conditions [34,35]. This could be due to a signal-to-noise-enhancing effect of the stimulation, which might be beneficial for noisy tasks [36], or when non-selective hyperactivity of a task-relevant area otherwise would reduce performance. This is assumed to be the case in occupation-related dystonia [18].

In the ctDCS group, 86.67 % of the participants at T1 and 73.33 % of the participants at T2 had reduced MEP amplitudes in comparison with the baseline. Given former descriptions of high variability of cathodal tDCS effects [37,38], which might be partially driven by relatively complex non-linear effects of this stimulation protocol [39,40], these effects are remarkably stable. One reason for this stability in the present study might be the known state-dependent effect of tDCS [9]. Brain states might be more variable under resting conditions, which is the context of most studies showing high variability. In the present study, brain states with respect to the target area might have been relatively stable because of task performance during stimulation.

The limitations of this study include the lack of electroencephalographic recordings and motor performance data (technical limitation of the custom-made button-box), which would have improved mechanistical understanding, and evaluation of behavioral consequences. In addition, this pilot study did not include a ctDCS condition alone, and thus information on how ctDCS interacted specifically with movement observation is not available. The aim of this study was not to induce effects of maximum size, but to deliver proof of principle data. Future studies might show if recently developed optimized ctDCS protocols deliver larger effects [40].

5. Conclusions

It was demonstrated that ctDCS during movement observation was effective in terms of modulating motor cortex activity. Moreover, it subsequently influenced execution-related motor cortex activity. This evidence might guide the implementation of ctDCS combined with task performance to enhance physical therapy in clinical practice in syndromes with pathologically enhanced cortical activity.

Author Contributions: F.Q. and M.A.N. contributed to the conception and design of the experiment. F.Q. and V.R.Z. collected the data. F.Q. drafted the paper. M.A.N. revised it critically for important intellectual content. All authors contributed to the analysis and interpretation of the data.

Funding: F.Q. is supported by the China Scholarship Council (201508080070). This publication was funded by the German Research Foundation (DFG, 325496636) and the Open Access Publishing Fund of the University of Rostock.

Conflicts of Interest: M.A.N. is a member of the scientific advisory board of Neuroelectrics.

References

1. Molenberghs, P.; Cunnington, R.; Mattingley, J.B. Brain regions with mirror properties: A meta-analysis of 125 human fMRI studies. *Neurosci. Biobehav. Rev.* **2012**, *36*, 341–349. [CrossRef]
2. Kilner, J.M.; Lemon, R.N. What We Know Currently about Mirror Neurons. *Curr. Biol.* **2013**, *23*, R1057–R1062. [CrossRef]
3. Babiloni, C.; Del Percio, C.; Vecchio, F.; Sebastiano, F.; Di Gennaro, G.; Quarato, P.P.; Morace, R.; Pavone, L.; Soricelli, A.; Noce, G. Alpha, beta and gamma electrocorticographic rhythms in somatosensory, motor, premotor and prefrontal cortical areas differ in movement execution and observation in humans. *Clin. Neurophysiol.* **2016**, *127*, 641–654. [CrossRef]
4. Lapenta, O.M.; Ferrari, E.; Boggio, P.S.; Fadiga, L.; D'Ausilio, A. Motor system recruitment during action observation: No correlation between mu-rhythm desynchronization and corticospinal excitability. *PLoS ONE* **2018**, *13*, e0207476. [CrossRef] [PubMed]
5. Bimbi, M.; Festante, F.; Coudé, G.; Vanderwert, R.E.; Fox, N.A.; Ferrari, P.F. Simultaneous scalp recorded EEG and local field potentials from monkey ventral premotor cortex during action observation and execution reveals the contribution of mirror and motor neurons to the mu-rhythm. *Neuroimage* **2018**, *175*, 22–31. [CrossRef]
6. Fox, N.A.; Bakermans-Kranenburg, M.J.; Yoo, K.H.; Bowman, L.C.; Cannon, E.N.; Vanderwert, R.E.; Ferrari, P.F.; van IJzendoorn, M.H. Assessing human mirror activity with EEG mu rhythm: A meta-analysis. *Psychol Bull.* **2016**, *142*, 291. [CrossRef] [PubMed]
7. Kozelka, J.W.; Pedley, T.A. Beta and Mu Rhythms. *J. Clin. Neurophysiol.* **1990**, *7*, 191–208. [CrossRef]
8. Nitsche, M.A.; Paulus, W. Excitability changes induced in the human motor cortex by weak transcranial direct current stimulation. *J. Physiol. London* **2000**, *527*, 633–639. [CrossRef]
9. Stagg, C.J.; Antal, A.; Nitsche, M.A. Physiology of transcranial direct current stimulation. *J. ECT* **2018**, *34*, 144–152.
10. Nitsche, M.A.; Nitsche, M.S.; Klein, C.C.; Tergau, F.; Rothwell, J.C.; Paulus, W. Level of action of cathodal DC polarisation induced inhibition of the human motor cortex. *Clin. Neurophysiol.* **2003**, *114*, 600–604. [PubMed]

11. Auvichayapat, N.; Rotenberg, A.; Gersner, R.; Ngodklang, S.; Tiamkao, S.; Tassaneeyakul, W.; Auvichayapat, P. Transcranial direct current stimulation for treatment of refractory childhood focal epilepsy. *Brain stimul.* **2013**, *6*, 696–700. [CrossRef] [PubMed]
12. Tecchio, F.; Cottone, C.; Porcaro, C.; Cancelli, A.; Di Lazzaro, V.; Assenza, G. Brain functional connectivity changes after transcranial direct current stimulation in epileptic patients. *Front. Neural Circuits* **2018**, *12*. [CrossRef] [PubMed]
13. Lin, L.-C.; Ouyang, C.-S.; Chiang, C.-T.; Yang, R.-C.; Wu, R.-C.; Wu, H.-C. Cumulative effect of transcranial direct current stimulation in patients with partial refractory epilepsy and its association with phase lag index-A preliminary study. *Epilepsy Behav.* **2018**, *84*, 142–147. [CrossRef]
14. Fregni, F.; Boggio, P.S.; Mansur, C.G.; Wagner, T.; Ferreira, M.J.; Lima, M.C.; Rigonatti, S.P.; Marcolin, M.A.; Freedman, S.D.; Nitsche, M.A. Transcranial direct current stimulation of the unaffected hemisphere in stroke patients. *Neuroreport* **2005**, *16*, 1551–1555. [CrossRef] [PubMed]
15. Zimerman, M.; Heise, K.F.; Hoppe, J.; Cohen, L.G.; Gerloff, C.; Hummel, F.C. Modulation of training by single-session transcranial direct current stimulation to the intact motor cortex enhances motor skill acquisition of the paretic hand. *Stroke* **2012**, *43*, 2185–2191. [CrossRef]
16. Nemanich, S.T.; Rich, T.L.; Gordon, A.M.; Friel, K.M.; Gillick, B.T. Bimanual Skill Learning after Transcranial Direct Current Stimulation in Children with Unilateral Cerebral Palsy: A Brief Report. *Dev. Neurorehabil.* **2019**, 1–5. [CrossRef]
17. Rich, T.L.; Nemanich, S.; Chen, M.; Friel, K.; Feyma, T.; Krach, L.; Nawshin, T.; Meekins, G.; Gillick, B.T. Transcranial Direct Current Stimulation (tDCS) Paired with Occupation-Centered Bimanual Training in Children with Unilateral Cerebral Palsy: A Preliminary Study. *Neural Plast.* **2018**, *2018*, 14. [CrossRef]
18. Furuya, S.; Nitsche, M.A.; Paulus, W.; Altenmüller, E. Surmounting retraining limits in Musicians' dystonia by transcranial stimulation. *Ann. Neurol.* **2014**, *75*, 700–707. [CrossRef]
19. Oldfield, R.C. The assessment and analysis of handedness: The Edinburgh inventory. *Neuropsychologia* **1971**, *9*, 97–113. [CrossRef]
20. Moriuchi, T.; Iso, N.; Sagari, A.; Ogahara, K.; Kitajima, E.; Tanaka, K.; Tabira, T.; Higashi, T. Excitability of the primary motor cortex increases more strongly with slow than with normal-speed presentation of actions. *PLoS ONE* **2014**, *9*, e114355. [CrossRef]
21. Stefan, K.; Wycislo, M.; Classen, J. Modulation of associative human motor cortical plasticity by attention. *J. Neurophysiol.* **2004**, *92*, 66–72. [CrossRef]
22. Qi, F.; Nitsche, M.A.; Zschorlich, V.R. Interaction Between Transcranial Random Noise Stimulation and Observation-Execution Matching Activity Promotes Motor Cortex Excitability. *Front. Neurosci.* **2019**, *13*. [CrossRef] [PubMed]
23. Bisio, A.; Avanzino, L.; Lagravinese, G.; Biggio, M.; Ruggeri, P.; Bove, M. Spontaneous movement tempo can be influenced by combining action observation and somatosensory stimulation. *Front. Behav. Neurosci.* **2015**, *9*, 228. [CrossRef] [PubMed]
24. Bisio, A.; Avanzino, L.; Gueugneau, N.; Pozzo, T.; Ruggeri, P.; Bove, M. Observing and perceiving: A combined approach to induce plasticity in human motor cortex. *Clin. Neurophysiol.* **2015**, *126*, 1212–1220. [CrossRef]
25. Hannah, R.; Rocchi, L.; Rothwell, J.C. Observing Without Acting: A Balance of Excitation and Suppression in the Human Corticospinal Pathway? *Front. Neurosci.* **2018**, *12*. [CrossRef] [PubMed]
26. Vigneswaran, G.; Philipp, R.; Lemon, R.N.; Kraskov, A. M1 corticospinal mirror neurons and their role in movement suppression during action observation. *Curr. Biol.* **2013**, *23*, 236–243. [CrossRef]
27. Kraskov, A.; Philipp, R.; Waldert, S.; Vigneswaran, G.; Quallo, M.M.; Lemon, R.N. Corticospinal mirror neurons. *Philos. Trans. R. Soc. Lond. B Biol. Sci.* **2014**, *369*, 20130174. [CrossRef] [PubMed]
28. Mukamel, R.; Ekstrom, A.D.; Kaplan, J.; Iacoboni, M.; Fried, I. Single-neuron responses in humans during execution and observation of actions. *Curr. Biol.* **2010**, *20*, 750–756. [CrossRef] [PubMed]
29. Okada, Y.; Shibamoto, C.; Osumi, Y.; Asano, C.; Takeuchi, R.; Nabeshima, S.; Morioka, S.; Shomoto, K. Transcranial Direct Current Stimulation Combined with Action Observation and Electromyographic Biofeedback Training in a Patient with Writer's Cramp. *J. Mov. Dis.* **2018**, *11*, 82–86. [CrossRef]
30. Nitsche, M.A.; Schauenburg, A.; Lang, N.; Liebetanz, D.; Exner, C.; Paulus, W.; Tergau, F. Facilitation of implicit motor learning by weak transcranial direct current stimulation of the primary motor cortex in the human. *J. Cogn. Neurosci.* **2003**, *15*, 619–626. [CrossRef]

31. Stagg, C.; Jayaram, G.; Pastor, D.; Kincses, Z.; Matthews, P.; Johansen-Berg, H. Polarity and timing-dependent effects of transcranial direct current stimulation in explicit motor learning. *Neuropsychologia* **2011**, *49*, 800–804. [CrossRef] [PubMed]
32. Stagg, C.J.; Nitsche, M.A. Physiological basis of transcranial direct current stimulation. *The Neuroscientist* **2011**, *17*, 37–53. [CrossRef]
33. Polania, R.; Nitsche, M.A.; Ruff, C.C. Studying and modifying brain function with non-invasive brain stimulation. *Nat. Neurosci.* **2018**, *21*, 174. [CrossRef] [PubMed]
34. Focke, J.; Kemmet, S.; Krause, V.; Keitel, A.; Pollok, B. Cathodal transcranial direct current stimulation (tDCS) applied to the left premotor cortex (PMC) stabilizes a newly learned motor sequence. *Behav. Brain Res.* **2017**, *316*, 87–93. [CrossRef] [PubMed]
35. Zhu, F.F.; Yeung, A.Y.; Poolton, J.M.; Lee, T.M.C.; Leung, G.K.K.; Masters, R.S.W. Cathodal Transcranial Direct Current Stimulation Over Left Dorsolateral Prefrontal Cortex Area Promotes Implicit Motor Learning in a Golf Putting Task. *Brain Stimul.* **2015**, *8*, 784–786. [CrossRef]
36. Antal, A.; Nitsche, M.A.; Kruse, W.; Kincses, T.Z.; Hoffmann, K.-P.; Paulus, W. Direct current stimulation over V5 enhances visuomotor coordination by improving motion perception in humans. *J. Cogn. Neurosci.* **2004**, *16*, 521–527. [CrossRef]
37. Wiethoff, S.; Hamada, M.; Rothwell, J.C. Variability in Response to Transcranial Direct Current Stimulation of the Motor Cortex. *Brain Stimulation* **2014**, *7*, 468–475. [CrossRef]
38. Strube, W.; Bunse, T.; Nitsche, M.A.; Nikolaeva, A.; Palm, U.; Padberg, F.; Falkai, P.; Hasan, A. Bidirectional variability in motor cortex excitability modulation following 1 mA transcranial direct current stimulation in healthy participants. *Physiol. Reports* **2016**, *4*. [CrossRef]
39. Jamil, A.; Batsikadze, G.; Kuo, H.-I.; Labruna, L.; Hasan, A.; Paulus, W.; Nitsche, M.A. Systematic evaluation of the impact of stimulation intensity on neuroplastic after-effects induced by transcranial direct current stimulation. *J. Physiol. London* **2017**, *595*, 1273–1288. [CrossRef]
40. Samani, M.M.; Agboada, D.; Jamil, A.; Kuo, M.-F.; Nitsche, M.A. Titrating the neuroplastic effects of cathodal transcranial direct current stimulation (tDCS) over the primary motor cortex. *Cortex* **2019**, in press.

MDPI
St. Alban-Anlage 66
4052 Basel
Switzerland
Tel. +41 61 683 77 34
Fax +41 61 302 89 18
www.mdpi.com

Brain Sciences Editorial Office
E-mail: brainsci@mdpi.com
www.mdpi.com/journal/brainsci

www.ingramcontent.com/pod-product-compliance
Lightning Source LLC
Chambersburg PA
CBHW062105220926
51850CB00010B/2897
9783039287963

For Bruce D. Sales

Contents

Acknowledgments

Editors at New York University Press obviously have done much to shepherd the development of this text. I am, though, especially grateful for Jennifer Hammer's careful editing and suggestions for revisions. I also am very appreciative of the manner in which she swiftly responded to my initial book proposal and found reviewers who provided insightful comments and wise counsel. I was very pleased to see that all took the manuscript on its own terms and helped develop the arguments I had proposed, rather than asking that I write a book that addressed issues that, although important, were not central to what I was trying to do: examine our understanding of the role of religion in adolescent development and fashion appropriate legal responses that respect adolescents and broader societal (including religious) concerns.

The Psychology, Policy, and Law Program at the University of Arizona also deserves special acknowledgment. The intellectual environment fostered by Bruce Sales has done much to free me, in terms of both time and spirit, to write this text. Engaging students in the psychology program and in the law school, most particularly Tracey Geer, Cordelia Guggenheim, Scotia Hicks, Shannon Locke, and Sean Sullivan, have made teaching so rewarding and appropriately challenging that I found it easy to share and develop many of the ideas presented in this text, even though what we discussed related less to religious issues than to the legal system's role in each of our lives.

My family also deserves special acknowledgment. Helen, my wife and my constant source of support and inspiration, deserves much more than I can ever offer. Although I can never praise her enough, I find great comfort in knowing that she knows and warms my deepest sentiments and that her place in my life makes me a much better person (and scholar) than I ever could have hoped. Our three sons, Thomas, Henry, and Marc, all remind me of the best reasons we have been placed on Earth and why the issues I address in this text are of such significance. My parents, Marc and Isabelle, have been such a powerful force in my life that it renders me speechless when I think of how much I owe them for having helped instill in me respect for religious and spiritual concerns. My brother Yves and my sister, Linda, remain powerful role models as they remind me of the importance of moving beyond one's individual concerns, reaching out to others, and finding one's place in a changing world. Last, my brother Gilles, who died unexpectedly and tragically as I was starting this project, undoubtedly served as a deep source of inspiration and as a reminder of the need to cherish life and to embrace its mysteries.

PART I

Introduction

1

Religion and Adolescents in Changing Times

Adolescents exhibit strikingly high levels of religious beliefs, attitudes, and behaviors. Ninety-seven percent of American teens ages 13 to 17 believe in God (or a universal spirit), 76 percent believe that God observes their actions and rewards or punishes them, 93 percent believe that God loves them, 91 percent believe in heaven, and 76 percent believe in hell (Gallup, 1999). In addition to having spiritual concerns, 80 percent of American teenagers view religion as at least fairly important to them; 93 percent report being affiliated with a religious group or denomination, and more than half of adolescents surveyed regularly engage in religious practices, such as praying alone, attending church or synagogue, frequently reading scriptures, or being involved with religiously affiliated youth groups. Surprisingly, more than one in four teens considers spiritual life to be more important to the teen than it is to his or her parents, indicating that teens are slightly more likely to attend church or synagogue than adults. Equally intriguing, more than three-quarters of adolescents feel confident that they will be more religious than their parents.

Adolescents' religious beliefs, development, and environments clearly are important to them. Yet, the law does not know how to approach adolescents' religious rights and needs. The law often ignores the many intricacies of this transitional developmental period as it typically classifies adolescents either as children or as adults. In some contexts, the law finds the religious concerns of adolescents indistinguishable from those of young children, which allows it to subject them to paternalistic policies based on assumptions of dependence, vulnerability, incompetence, inexperience, and immaturity. In other contexts, it treats adolescents' religious concerns as it treats those of fully mature adults, assuming that teens are competent to make decisions, accountable for their choices, and entitled to no special accommodations. Sometimes, the law essentially remains silent and leaves open to speculation the assumptions and conceptions about adolescents that it uses to determine the nature of the legal rules that affect their religious concerns. In these contexts, great discretion is left to lower courts and to those who implement laws, such as law enforcement officers, school personnel, and social service providers. And, in still other instances, adolescents' rights, abilities, and vulnerabilities do not even factor into the resolution of religious disputes, even when the disputes clearly involve adolescents. These wide disparities regrettably have not spurred the legal system and legal commentators to confront the awkward task of fashioning a more unified, carefully crafted approach to adolescents' religious rights.

The law's attempt to tailor itself to adolescents' needs and to affirmatively foster their development toward positive outcomes has been characterized by a lack of sustained, systematic efforts to address these inconsistencies and polarized tendencies (c.f. Levesque, 2000a, 2002). The challenges that arise in the context of adolescents' religious rights are distinctive, especially in light of the glaring clashes between historical and modern perceptions of religion and the place adolescents occupy in those perceptions. The social sciences, and especially the psychological sciences, historically have taken a highly negative view of religion and its influence on individual development (Donelson, 1999; Breakey, 2001). Likewise, the historical relationship between law and religion also has not been entirely positive. As many have argued, religions typically have resisted the development of important human rights, such as freedom of belief and protection from discrimination (see Gustafson & Juviler, 1999), and historically have fostered massive human rights violations (see Appleby, 2000). Yet, social science research, even within psychology, is now beginning to focus on the positive role that religious beliefs and institutions can play in fostering healthy individual and social outcomes (see Plante & Sherman, 2001; Koenig, McCullough, & Larson, 2001). Likewise, legal systems in the United States and in many other nations have enshrined rights that generously offer protections for liberty of conscience and freedom of religious exercise, for guarantees of religious pluralism, equality, and nondiscrimination, and for other entitlements owed to religious individuals and even religious communities (see Lerner, 2000; Witte, 2001). Yet, even though these laws recognize that religious rights belong to every individual and every group, how and whether adolescents can enjoy the protections, guarantees, and entitlements now seen as fundamental human rights possessed by all remains one of the greatest mysteries and challenges of the new millennium. Moreover, the recent focus on religion's positive influences and on new policies to protect and foster religious convictions pervasively ignore the ways in which religious beliefs, practices, and groups may in fact not always be good for adolescents *and* for broader society. The possible and often real harms linked to religious concerns make urgent an effort to shape, through our laws and policies, adolescents' religious development, attitudes, practices, and environments in ways that foster their fundamental rights, respect the rights of others, and respond to the changing place of religion in civil society.

Law, Religion, and Adolescents

Landmark U.S. Supreme Court cases involving religious concerns illustrate the wide variations in the legal system's approach to adolescents and in its view of the extent to which adolescents can be deemed to possess religious rights. In *Lee v. Weisman* (1992) and *Santa Fe Independent School District v. Doe* (2000), for example, the Court described adolescents as especially vulnerable to religious ideas and therefore in need of protection from peer pressure they would face if prayers were to be offered at public occasions. Yet, in other cases, such as *Board of Education of Westside Community Schools v. Mergens* (1990) and *Good News Club v. Milford Central School* (2001), the Court has portrayed even young adolescents as being as mature and as sophisticated as adults and therefore as able to resist pressures to join or participate in religiously inspired assemblies that would otherwise not occur but for opportunities provided by the public schools. In those contexts, the Court reasoned that parents would be the ones who would feel the coercion but would be capable of resisting it as they decide whether their children should attend proselytizing meetings on public school grounds. Thus, the first group of cases protects adolescents by finding them unable to resist public, state-supported prayers, while the second group of cases allows adolescents to join prayer meetings and attend religiously motivated meetings offered on state-controlled school grounds.

Other cases have placed even greater emphasis on the right of parents to control their adolescents' religiosity. One approach essentially removes adolescents' rights and interests from consideration, while another envisions respect for those rights. The first approach finds expression in the leading cases involving parental rights and religious freedom. In those cases, the law does not even consider adolescents' abilities worth mentioning, framing their rights as clashes between parental rights and governmental obligations. Most notably in *Wisconsin v. Yoder* (1972), for example, the law respected the rights of parents to direct their adolescents' religious and secular education in the absence of direct harm to the adolescents or of any burden placed on society by such harm. The Court essentially deemed it immaterial whether the adolescents in question would (or in fact did) reject their parents' beliefs. Indeed, in this case, it was precisely because the Amish students involved had reached adolescence and thus were susceptible to influences coming from outside their own and their parents' culture that the Court supported the parents' rights. Yet, in other circumstances, parents' assertions of a right to control their children's religious lives have resulted in recognition by the Court that adolescents possess the ability to exercise their own rights. Thus, in *West Virginia State Board of Education v. Barnette* (1943), parents refused to allow their children to salute the United States flag while repeating the Pledge of Allegiance because doing so would violate their religious ban on the worship of graven images. The Supreme Court reasoned that everyone, including the parents and the students themselves, possesses rights that cannot be infringed and that must be protected if individuals are to become citizens who respect fundamental freedoms. Parental claims, then, can and often do result in paradoxical outcomes related to the legal conception of adolescents' rights.

In other contexts where the state's higher interests are at stake, those interests may be so compelling as to override consideration of parents' right to direct their children's upbringing or adolescents' abilities to resist the effects of religious proselytizing or to decipher the state's role in supporting religious beliefs. Again, landmark cases litigated before the U.S. Supreme Court present two very different views of adolescents. One approach recognizes that adolescents can hold profound religious beliefs, an approach that allows the state to hold them accountable for holding and expressing such beliefs. The other approach recognizes that some issues that confront adolescents may involve profoundly religious matters but nevertheless allows the state to restrict adolescents' choices in a way that ignores their firmly held religious convictions.

Cases that deal with the state's interests in the prevention of early pregnancy and poverty exemplify the second approach, essentially ignoring adolescents' religious convictions and the extent to which other individuals' religious attitudes may affect adolescents without their knowledge. In *Bowen v. Kendrick* (1988), for example, the Supreme Court let stand the public funding of abstinence-only sexuality education programs and adoption-only reproductive services provided by religious institutions that do not provide the range of services that secular institutions provide to adolescents. As a result, the Court found that service providers could make decisions on service provision on the basis of religious convictions about, for example, abortion and sexuality, even though adolescent clients might well be unaware of the resulting bias and even though adolescent clients may therefore be less able to exercise their own religious convictions because they are unaware of other options potentially available to them.

The other context allows for taking adolescents' religious beliefs so seriously that the legal system assumes not only that adolescents can hold religious beliefs but that it can punish them for expressing those beliefs. This was the ultimate outcome of cases that have laid the foundation of the Supreme Court's approach to hate crimes, a category that includes crimes that stem from religious hate. The two leading hate crimes cases involved juveniles, even though their minor status had no notable impact on the Court's analyses and rulings. The first case, *R.A.V. v. City of St. Paul* (1992), led the Court to protect hateful speech. This First Amendment rights case invalidated a statute that prohibited the expression of particular content; in this case a juvenile hastily shaped broken chair legs into a cross that he and his friends then burned inside the fenced yard of a neighboring African American couple. In *Wisconsin v. Mitchell* (1993), the Court unanimously supported the state's penalty-enhancement statute for hate crimes; it ruled that the First Amendment did not protect the juvenile's beliefs that led to the crime. This case involved an aggression by an African American adolescent against a white adolescent in response to a highly charged scene in the movie *Mississippi Burning* in which a gang of white thugs viciously assaults an African American child while he is praying. The holding in *Mitchell* allows for punishing more severely those who succumb to religious hatred. Both cases support the view that adolescents can hold deep religious beliefs, either in the form of speech worth protecting or of an action worth punishing.

In yet other contexts, which arguably constitute the vast majority of adolescents' everyday religious concerns and environments, the Supreme Court essentially has remained silent. As a result, adolescents' rights are not clear, and the factors that would weigh in determinations of their rights remain uncharted. One possible issue that may arise in these contexts involves adolescents' levels of maturity and legal competency. If adolescents are deemed potentially mature, legal systems *may* require inquiries into the extent to which their views vary from those of their parents. Given the Supreme Court's silence, legal systems essentially still proceed in a haphazard manner. One of the most notable examples is cases litigated in the lower courts that involve decisions on medical questions made by adolescents. Different results continue to emerge in cases where there is potential conflict between parents' religious beliefs and an adolescent's right to decide whether to accept those religious precepts (for a review of cases, see DiCamillo, 1998). These differences touch only a few of the controversies that go unrecognized and unchallenged; among these are conflicts that emerge as a result of religious beliefs relating to, for example, adolescents' relationships with peers; sexuality; reading materials; educational aspirations; disciplinary rules; and basic views about adolescents' proper roles in their families and in the broader society.

The cases described highlight several key points regarding the legal system's approaches to religious issues that involve adolescents. First, no simple rule guides legal responses to disputes that involve adolescents' religious beliefs and religious environments. Second, assumptions about adolescents' religious beliefs vary considerably, depending on particular social contexts and the claims of others. Third, visions of adolescents' abilities and religious convictions differ greatly; some see teens as having full, autonomous maturity and solidified religious orientations, whereas others ascribe to them essentially total immaturity and an inability to hold firm religious convictions. Last, the law tends to adopt sweeping standards; sometimes it holds visions and assumptions about adolescents' religiosity that do not fluctuate as they otherwise would if religious concerns were not directly involved, while in other situations the nature of adolescents' rights varies depending on their relative age and their exhibited maturity level within the broader category of minors. The cases, then, reveal numerous and often conflicting assumptions in the legal system's efforts that sometimes address and at other times ignore the peculiarities of adolescents' religiosity and religious environments.

Emerging Concerns about Current Legal Responses

The wide disparity in the way in which the legal system approaches adolescents' religiosity in and of itself may not be problematic. It may well be that the system adequately weighs and balances numerous interests and needs. Our legal system's tradition of slowly emerging principles and precedents certainly has proven worthy in other areas of law. Indeed, other observers undoubtedly could examine the same cases we have looked at and highlight far more consistencies than divergences. Most notably, for example, and at a very broad level, these cases do promote the policy that great power should be reserved to parents and to those who act in the place of parents. The cases also hold consistently that the state may serve as the ultimate arbiter of competing religious claims in only the most extreme cases and only with the most compelling reason to interfere. Yet, while consistency may exist, it remains important not to play down or ignore apparent divergences and conflicting outcomes. Quite significant in themselves, the varying outcomes and supporting rationales become increasingly important in view of other developments in adolescent rights, laws, and policies.

Divergences tend to be ignored even as the courts march toward a recognition that adolescents are individuals with rights and that they are capable of possessing and exercising these rights. Although it may not be appropriate to bestow upon adolescents total control over their own rights in every context—no right enjoys absolute protection—it does seem peculiar that, when other rights have been deemed fundamental, the Supreme Court has not recognized and supported adolescents' exercise of those rights when they are, or may be, in conflict with parents' beliefs and demands. This is particularly true of controversies that involve adolescents' religious rights. Religious rights may receive the highest protection; they are deemed so important that they find explicit articulation in the First Amendment. Yet, that recognition and the needed protections have not been applied to adolescents as forcefully as have, for example, other explicit rights (e.g., due process rights; see *In Re Gault*, 1967) and even implicit rights (e.g., the right to privacy, see *Bellotti v. Baird*, 1979) (see Levesque, 1997).

Another concern emerges from the broad jurisprudential transformation frequently conceived as the "new federalism." The Supreme Court's new federalism seeks to elevate the power of state governments and, in part, encourages state governments to pursue their own constitutional rights agendas (Eisgruber & Sager, 1997). What the states have done in the wake of *City of Boerne v. Flores* (1997) illustrates the significance of the trend. In that case, the Court found unconstitutional federal efforts to protect religious freedoms by requiring states to demonstrate a compelling interest before they can burden religious conduct; once the compelling interest is found, the Court decision limits the government's efforts to further that interest to the "least restrictive means" (*City of Boerne v. Flores*, 1997). In legal parlance, the challenge involved the extent to which religious conduct would enjoy "strict scrutiny." This means that if a right is fundamentally important, the state must have a compelling interest in order to interfere with that right, and the interference must be tailored narrowly to achieve that interest, intruding on the right to the least degree possible. Although the case has been interpreted as evidence of the Court's hostility to religion, the case also has been interpreted in the opposite way. The Court explicitly avoided a ruling that protects religious conduct against state and local laws; as a result, several states have enacted statutes that protect religion in ways deemed unconstitutional at the federal level (see Gildin, 2000).

Mandates that require minimal restrictions on religious rights directly impact adolescents. Their effect likely will be much broader than is envisioned by the many commentators who focus on their effect on efforts to intervene in and to prevent maltreatment, such as the refusal of some religious parents to provide life-saving medical treatments (see Levesque, 2000b). The compelling interest of the state in the support of adolescents justifies the use of a wide panoply of remedies to provide the fullest protection for the individual adolescent, as well as for the public fisc. The implications of this support are most obvious in efforts to obtain child support from noncustodial parents. Given the state's interest in providing support for children, states use a range of enforcement tools, ranging from criminal contempt citations for nonsupport to civil child support orders to withholding of wages and sequestration of property to enforcement of garnishment orders through contempt citations. These tools, however, are considered unacceptable by courts that support the view that the states must use the least restrictive means of enforcing child support when the enforcement would burden religious freedom (see Bruno, 1999). These legal developments and interpretations of the need to protect religion, then, may further the worthy goal of protecting religious beliefs at the expense of the state's ability to foster healthy development by providing adolescents with basic financial support from their religious parents.

The manner in which social policies are increasingly becoming faith-friendly provides another cause for concern about the failure to address adolescents' religious rights more systematically. Rather than maintain a high wall of separation between religion and the government, both federal and state legislative mandates increasingly embrace religious institutions. For example, the federal government's most dramatic welfare reform overhaul measure, the Personal Responsibility and Work Opportunities Reconciliation Act of 1996 (1996), contains Section 104, the so-called Charitable Choice provision, which encourages states to utilize "faith-based organizations in serving the poor and needy," requires that religious organizations be permitted to receive contracts, vouchers, and other government funding on the same basis as any other nongovernmental provider, and "protects the religious integrity and character of faith-based organizations that are willing to accept government funds." As enacted, Charitable Choice covers each of the major federal antipoverty and social welfare programs; congressional efforts to expand its scope to other social service programs and other federal policy domains are under way. Although initially meant to increase access to social services for those living in poverty, the new orientation extends to other government-supported services. Most notably, it helps account for an increase in the number of primary and secondary school students opting to attend private, sectarian schools at taxpayers' expense. It also is the dominant rationale for school reform efforts that offer parents government resources, in the form of tax incentives and vouchers, to send their children to religious schools (Todd, 2000). These are extraordinary developments. Rather than simply allowing religious institutions to exist and to supplement government services, the new mandates actively support (albeit indirectly) parts of their religious missions. Despite claims that these efforts violate First Amendment freedoms, the Court has increasingly shown a tendency to support faith-friendly efforts. Indeed, the Supreme Court has overruled earlier decisions that prevented the government from providing funds for instructional equipment to be used in parochial schools (*Mitchell v. Helms*, 2000), and it has let stand programs that use state funds to send public school students primarily to sectarian schools (see *Jackson v. Benson*, 1998). These developments demonstrate that legislative efforts that treat religious institutions as full partners in social service provisions find support from policymakers, and the final arbiter of those efforts' legitimacy—the Supreme Court—seems poised to respect the expression of such efforts.

Given the broad transformations that are occurring in the place of religion in law and policy, it is not unreasonable to suggest that the changes in the very nature of public service provision can lead to dramatic change in adolescents' lives. These changes already impact how schools address adolescents' ethical, moral, and spiritual concerns. They affect the balance that child protective services and juvenile justice services must seek as they strive to respect the religious rights of parents, communities, and adolescents themselves. These changes also affect families' ability to dictate the religious development of their adolescent children and alter religion's impact on adolescents' emotional and social development. Religion's influence on adolescent development has led leading authorities to encourage adolescents, families, and communities to turn to religion and to spiritual resources to buffer adolescents against social pathologies that contribute to youth violence (see, e.g., Garbarino, 1999) and to address adolescents' religious development in order to help adolescents to thrive and to actively contribute to their communities (see, e.g., Youniss, McLellan, & Yates, 1999). These leading scholars of adolescent development join an increasingly strong contingent of policy researchers who also champion an even more deliberate turn to religion to address social ills, especially the many problems associated with adolescent development (DiIulio, Larson, & Johnson, 1999). Among the most aggressive and persistent policy reform proposals are school reform efforts that champion character education in public schools and provision of greater access to sectarian schools so that education can nourish adolescents' spirituality (see Milligan, 2000), a point frequently reinforced by analyses of influential political campaigns by activist religious groups that aim to recast public schools into allies for fundamentalism (see Detwiler, 1999; Lugg, 2000). Leading scholars, researchers, and policy analysts, then, increasingly embrace the need to reconsider the place of religious institutions in the healthy development of adolescents and society.

Effective youth policy-making must involve an effort to find ways to increase the chances that approaches to matters of faith will contribute to positive adolescent development. The success of efforts to harness religious beliefs actually remains debatable, however, despite an emerging consensus that only good results from religious influences (c.f. Hamilton,

M., 2000; Donelson, 1999). In truth, adolescent religiosity and religious institutions do not always serve adolescents and society well.

There is, admittedly, a vast body of research that reveals adolescent religiosity's link to positive outcomes. Compared to their less religious peers, religious adolescents exhibit a concern for others and for behaving in ways that do not place a burden on them. Most notably, religious adolescents appear less likely to engage in behaviors that compromise their health (e.g., carrying weapons, getting into fights, drinking and driving) and are more likely to behave in ways that enhance their health (e.g., proper nutrition, adequate exercise and rest) (Wallace & Forman, 1998). Religious adolescents also engage in lower rates of delinquent activity (Benson, Donahue, & Erickson, 1989; Donahue, 1995; Evans, Cullen, Dunaway, et al., 1996; Free, 1994); are more likely to delay the initiation of sexual activity and intercourse (for a review, see Levesque, 2000a); are more attached to school and earn better grades (Montgomery & Francis, 1996; Zern, 1985); are more prosocial and concerned about others (Donahue, 1995; Perry & McIntire, 1994); score higher on tests of moral development (Kedem & Cohen, 1987); reveal more "wisdom" in the choices they make (Furrow & Wagener, 2000); and even report higher levels of happiness (Francis, Jones, & Wilcox, 2000). Although some may dispute the nature and robustness of religion's impact on healthy development and prosocial orientations, the links seem indisputable.

While the links between religion and prosocial adolescent behavior increasingly appear undeniable, they do not negate the reality that adolescent religiosity may not lead to unequivocally good outcomes—at least, "good" as defined by broadly accepted social standards. For example, our nation's history is replete with hate activity and prejudice linked to religious beliefs, and newly recognized hate crimes have become a subject of great public concern (see McPhail, 2000). The growing public awareness of hate crimes and hate crime policy, however, has been fueled by social movements that pervasively ignore the significant place of adolescents in hate activity (see, e.g., Jenness & Broad, 1997). The most recent statistical evidence reveals that adolescents commit the vast majority of hate crimes and are the most frequent victims of such crimes and that religious hate constitutes the second leading category of hate crime offenses (Bureau of Justice Statistics, 2001). Although the reasons adolescents turn to hate activity remain complex and are often disregarded by researchers and policymakers, we do know that the search for personal meaning, including the need to belong, have hope, and feel secure, impels attachment to hate activity (see Levin & McDevitt, 1993)—dynamics that parallel those that, according to researchers, make adolescents especially susceptible to recruitments by cults or extremist groups (see Curtis & Curtis, 1993). The link between hate activity and religion, however, seems even more evident in the remarkable extent to which religious convictions play important roles in some groups (e.g., World Church of the Creator) identified as purposefully intolerant and even terroristic (see Dobratz, 2001; Sharpe, 2000) in ways deliberately framed to lure adolescents (Blazak, 2001). Importantly, even if cults and other nontraditional religious groups themselves may not be problematic, adolescents' involvement with them still may contribute to negative consequences for their mental health and relationships (see Robinson & Frye, 1997). Likewise, most adolescents who commit hate activity tend not to be affiliated with hate groups, and most typically "age out" of their criminality (Levin & McDevitt, 1993), which serves only to reinforce the emerging conclusion that the roots of hate are much more broad, normative, and even unremarkable than we may realize and that policymakers must aim to address the everyday manifestations of hate (see, e.g., Franklin, 2000). Given these findings, it remains rather peculiar that leading scholars with interests in youth policy have not voiced much vocal opposition to the increasing tendency among policymakers to embrace religiosity and religious institutions.

This point becomes even more significant when we consider the extent to which recent jurisprudential and legislative developments ensure that individuals may not receive the protections we generally and properly see as basic human rights. Although many challenges abound, most illustrative is the persistent failure to address private wrongs. Legal commentaries, for example, focus almost exclusively on the need for national and state governments to place proper limits on religious expression and to curtail the state's role in supporting religions. Although important, that focus leads to a tendency to ignore ways that the law still may allow private citizens to respond to religious mandates in ways deemed inappropriate and problematic. Two examples illustrate both the significance of the challenge and the general failure of the law to respond more appropriately. The first example involves the frequently recurring controversies around the place of religion in public schools. In *Santa Fe Independent School District v. Doe* (2000), for example, the Supreme Court reaffirmed its ban on efforts by school officials to organize prayer in public schools. The Court's finding that student-led prayers at high school football games were unconstitutional ignited efforts by private citizens to organize student prayers without explicit school sponsorship. This result is related to but much less overtly violent than the frequent persecution, harassment, and intimidation reported by those who challenge efforts to advance religions in public schools (see Alley, 1996; Ravitch, 1999). Thus, the law prohibits school-sponsored public prayer in schools, largely because it may harm those who refuse to participate in it; yet, the law permits similar harms when individuals commit them of their own volition. The second example involves the frequently ignored controversy surrounding the ability to hold faith-based service providers liable for violating fundamental rights. Indeed, the failure of faith-based organizations to provide certain services that are contrary to their faith, such as certain medical services, are not necessarily defined as rights violations because the individual does not have enforceable rights unless the entity that acts to deny the services is the state itself or a quasi-governmental entity (White, 1999). Courts and legislatures have yet to untangle the thorny issues of whether faith-based service providers are private when supported by government funds and the extent to which, even though they might remain private, they can be held liable when they discriminate in their service provision. Although these examples may not seem particularly troublesome in that the state cannot be expected to protect individuals from all harms, they do reveal the failure of the legal system to respond to private harms and to provide clear mandates when private harms may occur.

Related to this concern is the question about the manner in which religion impacts broad social justice issues. These issues involve the fundamental nature of society and the extent to which society properly addresses its problems. How society addresses religious issues fundamentally relates to how it addresses issues of prejudice, the extent to which it promotes respect for liberties of conscience, how it responds to adolescents' needs when these needs have the potential to conflict with familial concerns, and how it addresses issues of community and personal health. For example, states increasingly contract with faith-based drug-treatment programs, encourage churches to mentor low-income families, work with private-sector housing programs, and implement voucher systems for training and job-placement services (Rees, 1999). The provision of these services through faith-based institutions raises many unaddressed questions. When religious organizations are the only ones that provide certain social services that otherwise would be provided by a secular entity, the risk of coercive proselytization and indoctrination increases, even though laws may require the availability of alternatives. For example, federal laws require faith-based providers to make available "sensitive" reproductive and life-sustaining services, yet these services are not necessarily provided when certain faith-based organizations are the sole providers of health services (see White, 1999). Innovative policy developments, then, may lead to effective service provision, but they also may allow for overlooking rights that people would retain were the state to provide the service itself.

The ultimate question regarding the emerging policy directions is how the law's response to religious issues that affect adolescents can contribute to positive outcomes for individual adolescents and for civil society. That we may want to guide adolescents' religious development toward prosocial ends may strike many who cherish religious freedom as odd, misguided, wrong, and simply immoral. Yet, even though such goals may be anathema to many, the law clearly supports such goals, and the law always has been willing to intervene to protect adolescents' and society's interests against religious practices—and even against religious beliefs—deemed harmful. Despite that willingness, as we already have seen, precisely how society intends to balance religious interests with those of adolescents and society is unclear, changes appear to be lacking any principled consistency. In this book I urge a turn to social sciences, especially to the sciences that examine adolescent development, to guide the laws and policies that regulate adolescents' religious development and environments.

Why and How the Developmental Sciences Matter

Ascertaining how a policy is to achieve valued ends necessarily requires an understanding of developmental influences. Only by understanding the influences of peers, families, communities, and governmental services on adolescent life can we envision the best ways to approach adolescents' religiosity and religious environment. And only an understanding of religiosity's influence on adolescents and their beliefs and actions can effectively guide our efforts to imagine a policy response that takes adolescents' experiences and position in society seriously enough to generate informed and effective youth policy.

How can the developmental sciences influence legal analyses? Their use would seem troublesome, given that many observers properly view developmental ideals as contentious. That concern is particularly fitting for our inquiry, given our society's immense religious diversity and the necessary commitment of religions to determining the nature of worthy ideals. In the United States alone, more than one thousand different religious sects and denominations pursue their own visions of truth (Witte & Martin, 1999). Individuals amplify that diversity; people who practice the same religion do not necessarily share religious perspectives, since perspectives necessarily grow out of a mix of irrational, nonrational, and rational beliefs that depend partly on distinctive locales, traditions, faiths, and personal experiences. The multivocal character of religious traditions and truths expresses itself even in the diverse ways in which members of fundamentalist and evangelical groups approach their religions (Smith, 2000). Partly because of this reality, the analysis that follows aims to adopt and champion the legal system's fundamentally democratic values, rather than simply adopt social science ideals and designate them as worth developing. The analysis understands the basic ideals enshrined in our legal system as those that seek to develop responsible, democratic citizens who respect the rights of others and who balance those needs with the dictates of their own consciences and values (c.f. Levesque, 2000a, 2002). The analysis simply looks to social sciences to help fashion a legal system that fosters those civic values and the consciences to abide by them.

Importantly, research on adolescent development already has investigated values consistent with civic ideals and the environments that support those ideals. Lessons learned from that research undoubtedly will inform our understanding of religious issues that affect adolescents. For example, important research studies characteristics that can constitute a model character, an ideal marked by a sense of empathy, altruism, and cooperation and a commitment to promoting and respecting others' welfare (see Killen & Hart, 1995; Berkowitz & Grych, 1998). Equally significant research has found that a sense of personal agency (a belief that one's life is intrinsically driven) (Csikszentmihalyi & Rathunde, 1993; Ryan & Deci, 2000), and a sense of competency and resilience (a belief that one concertedly and effectively engages with the constraints, rules, challenges, and complexities found in the environment) are critical for optimal development (Masten & Coatsworth, 1998; Levesque, 2002). The research urges us to focus on adolescents in their social contexts if we are to understand their lives and promotes the view that successful development derives from individuals' active engagement with their surroundings. As we will see, these and other rich findings from the broader study of adolescence offer themes that can help guide our understanding of religious issues as they relate to adolescents.

Our reliance on the legal system as a starting point and our espousal of the use of a broad social science approach to shaping legal responses to adolescents' religious development are significant for two reasons. The first reason is that researchers studying religious development and religious influences have yet to develop a robust empirical analysis like that developed by researchers investigating other areas of adolescent life. Much research on religious concerns remains correlational and fraught with numerous limitations. The most effective way to approach those findings, I argue, is to examine them in light of our broader understanding of adolescence and the place of adolescents in law and society. My view diverges from that of the only other social scientists to examine (briefly but directly) religious issues in the context of the Supreme Court's use of the developmental sciences (Bersoff & Glass, 1995), who argued *against* the use of current social science evidence as it relates to adolescents, mainly because the research addresses legal issues inadequately. Waiting for more evidence seems problematic. The law does not allow for full experimentation, and social scientists already know much about the adolescent period that may be of assistance in efforts to address legal assumptions and pressing social needs (see Buss, 2000). The second reason the approach reaches significance relates to the extent to which the ideals revealed by both social science research and the law coincide with those of religious traditions. As others have noted, each major religious tradition espouses a number of the basic concepts seen as cardinal both in human rights discourse and in religious philosophy (e.g., notions of compassion, dignity, conscience, reason, liberty, equality, tolerance, love, openness, responsibility, justice, mercy, integrity, righteousness, and accountability) (see Witte, 2001; Thierstein & Kamalipour, 2000). This profound theological commitment to fostering human rights mirrors what the social sciences understand as necessary for fostering healthy personal and social development. However, again for practical reasons related to the immense diversity of religious traditions in the United States, that argument is not made here, since we see it as beyond the purview of this analysis. Instead of glossing over important differences in religious beliefs and dictates, and realizing that some religious traditions and those who live by them cannot easily accommodate competing views (see, e.g., Detwiler, 1999; Appleby, 2000), our analysis centers on how the legal system, policymakers, institutions, and those who interact with adolescents can structure society in ways that respect and foster adolescents' religious rights in our pluralistic, civil society.

The social sciences potentially have much to contribute to legal and policy movements that support adolescents' religious rights, but the reality is that laws have only slowly adapted to changing realities and in some cases have not changed at all. Numerous factors account for the reluctance to adapt social and legal policies to our empirical understandings of adolescent life. First, symbolism guides the law. The general belief is that parents determine their children's upbringing, despite strong evidence to the contrary (see, e.g., Harris, 1995). Second, our society exhibits a profound cultural attachment to privacy and to family life. Americans value individualism, self-sufficiency, and independence, and those values are clearly reflected in the venerable tradition of noninterference with the rights of parents within the family; governments are therefore wary of involving themselves in family conflicts (Levesque, 2001). Third, social sciences and legal systems use and offer conflicting paradigms (see Haney, 1980). For example, the Supreme Court must base its legal decisions on legal principles founded on historical precedents, a process that frequently is hostile to the social sciences. Indeed, it is in the context of religion and adolescents' rights that the most robust psychological findings from leading journals on adolescent development have led some Supreme Court justices to see other justices' attempts to use psychological science as a ploy involving amateur psychology that has no place in constitutional jurisprudence (*Lee v. Weisman*, 1992, p. 636). Fourth, results have not been uniformly positive in cases where social science evidence has been used, for example, in efforts to support greater intervention in family life. History is replete with innovations gone wrong, and those who inform policy-making are not free from their own political agendas (see Levesque, 1995). Fifth, the innovations demand a reconception of rights. Adolescents are in the unique position of being both dependent and independent, and accommodating such obvious uncertainty remains problematic given the law's insistence on narrow conceptions of rights. Last, religion's many roles in public and private life figure prominently among the most constant and contentious issues in American society, and that contentiousness works against the consensus needed for legal reform.

Although many reasons militate against the use of findings from the social sciences in determining legal, political, and religious policy, this does not mean that such research should be ignored. The law frequently emphasizes psychological factors in deciding cases that deal with adolescents' rights in religious contexts. The need to understand the extent to which adolescents are mature enough to reject their parents' religiously based medical decisions provides a clear example of the need to consider social science evidence (Levesque, 2001). Moreover, courts and policymakers require assistance in evaluating whether any compelling basis exists on which to develop and follow legislative prescriptions. Most notably, in the controversies surrounding the funding of religious schools through vouchers, only sound research can evaluate legal procedures and policies to ensure that they remove rather than create unintended negative effects. Social scientists can also assist adolescent clients and professionals in their interactions with one another. Those interactions are ubiquitous in adolescent life; for example, psychologists interact with adolescents as clinicians, as counselors in schools, as mental health center experts, in health care settings, and as experts who guide parents. Professionals who interact with adolescents continue to commit themselves to the promotion of human welfare, including the need to respect privacy, protect the civil rights of their clients, and even challenge social injustice (see Levesque, 2000a). Thus, although the law often clashes with social science realities, the social sciences still have an important role to play in the assessment of policy approaches to adolescent life. As we will see, regardless of one's views of the utility of the social sciences for guiding legal analyses and policies, such analyses can make invaluable contributions by facilitating more useful and intelligent discussions of complex matters that reach the heart of society and our humanity.

Plan of the Book

The impact of changes in the regulation of religion will likely be complex, far-reaching, and often paradoxical. In light of the rapidly changing legal terrain and existing social science findings, this investigation seeks to examine key concerns, identify emerging themes, and offer possible ways to address religion's role in adolescence. To do so, the text proceeds as follows. Chapter 2 begins our analysis with a close look at the nature of religiosity and its transmission. The analysis begins by noting the need to differentiate among the numerous constructs used to investigate adolescents' religiosity and their religious environments. As we will see, research generally focuses on indexing religiosity, rather than on religious environments, and analyses often are merely efforts to determine whether adolescents belong to an organized religion, practice a religion, hold particular beliefs, or simply are spiritual. We then examine the current understanding of adolescents' religious orientations and place this in the context of developmental science, an analysis that uncovers many limitations in our understanding of religious orientations but still reveals important conclusions and themes. That discussion serves as the critical basis from which we explore the transmission of religiosity, highlight factors that influence adolescents' religiosity, and detail the implications of existing developmental science.

Chapter 3 examines the social science evidence relating to the impact of religion and religiosity on adolescents. The chapter focuses on the link between religiosity and adolescents' positive mental health and prosocial behavior, as well as religion's role in a wide range of harms. After discussing religion's possible influences on adolescents, the chapter details the many ways in which religion actually does influence adolescent development. This analysis focuses on the roles of families, peers, and communities and on adolescents' own dispositions. The discussion reveals several key misperceptions that ignore religion's multivalent roles and demonstrates how an understanding of those roles would help policymakers better achieve the outcomes they desire, especially the frequently espoused goals of reducing intrusion in religious domains and respecting religious freedom.

Chapters 4 and 5 move away from social science analyses and toward legal analyses. Lessons learned from the previous discussions guide our legal analyses beyond traditional concerns. As a result, the chapters focus on detailing both traditional and emerging trends in policies that affect adolescents' religious development and religious environments. Chapter 4 highlights the nature of major doctrinal shifts in jurisprudential and legislative efforts related to religions' role in public and private life. It serves as background for our understanding of how the legal system approaches religion and serves as the foundation for the legal analyses that follow. Chapter 5 examines legal developments and how they directly apply to adolescents. The focus is on the way in which the law regulates the social forces (parents, communities, and peers) that impact adolescents' experiences and developmental outcomes. The discussion also highlights limitations in current regulations, such as the general failure to address private actions that impact religious consciousness. The chapter ends with an examination of the extent to which the legal system can make use of the social sciences to help it redirect its efforts to achieve more positive developmental outcomes for adolescents and society.

Chapter 6 concludes by highlighting the need to recognize the role religion plays in adolescent life and by charting the necessary components of a more effective youth policy. The discussion first details the ways in which laws and policies that affect adolescents and their religious environments are at a critical juncture. For the first time, those who evaluate adolescents' religious rights may refer to new legislative mandates, to a changing jurisprudential landscape, and to a firmer empirical understanding of adolescence and of religion's place in human development. The chapter explores policy responses that accord with social science observations as they acknowledge adolescents' own realities while effectively balancing respect for adolescents' rights with communal needs. The analysis emphasizes how responses that affirm adolescents' rights further the deep, historically tested rationales for regulating adolescents' religious development and environments. The chapter ends by envisioning the essential components of a youth policy that would benefit from social science research and emerging legal reforms and that would prepare both adolescents and their social environment for an increasingly changing and challenging world.

PART II

Religiosity's Role in Adolescent Development

2

Adolescents' Religious Development

Adolescents search for meaning and purpose in life. Their new cognitive, social, and emotional abilities allow them to begin formulating a system of beliefs that will guide them through the remainder of their lives. The most dominant social institution specifically dedicated to meaning-making—religion—thus potentially occupies an important place in adolescents' experiences. Religions provide structure, cohesion, and support as they foster religious and spiritual development. Even when formal religious institutions prove unsatisfactory, adolescents still engage in spiritual quests and attempt to develop their unique perspectives and to make sense out of disparate experiences. These quests may not be influenced directly by formal religious institutions, but they necessarily are influenced by teens' own dispositions and by other socializing agents, most notably their parents, peers, and communities. Importantly, those socializing agents also may be infused with religious beliefs, so that religious concerns are particularly influential on adolescents' development. Some adolescents may move beyond the frameworks of organized religions and dabble in the mysteries of cults and the occult, but in all of them the striving runs deep and is influenced by social environments that serve their need to arrive at a valued purpose in life.

Little controversy surrounds the proposition that adolescents search for meaning and even engage in spiritual quests. In fact, the proposition seems so self-evident that few researchers empirically investigate the nature of adolescents' quests and the development of their religious and spiritual orientations. The paucity of research is matched only by the availability of research that examines the factors that influence religious orientations. The state of research is rather remarkable, especially given the apparent significance so many attach to religion and the almost universal extent to which even adolescents understand themselves as concerned with finding meaning and purpose in life. The quest to find meaning in life and the potential role of religion in that quest provide the most fundamental impetus and rationale for an effort by researchers to better understand adolescents' religious development.

Although available research may be limited, we can learn much by culling the available research on religious development and placing it in the context of what we know about the nature of adolescence and the forces that shape it. Two concerns make the following analysis necessary and of considerable significance. Understanding how adolescents internalize religious beliefs serves as a first step toward understanding how religion influences their religious and spiritual orientations and how religious concerns impact other aspects of their lives. With that understanding, we can create policy reforms that address religious issues. Understanding the nature of religious development and the forces that operate on it, then, provides the proper background for understanding the potential impact of many policy changes involving religion and for conceptualizing changes in policies in ways that appropriately respond to adolescents' religious orientations.

Conceptual Challenges in the Study of Adolescent Religiosity

Individual research projects vary considerably by how they define religion, what they mean by religious involvement, and what they consider important for understanding religious development. These differences, chiefly resulting from the many methods and orientations used to study religious concerns, make understanding religious issues difficult and perplexing. Given what we know about other areas of research on adolescent development that face similar definitional issues, it remains unlikely that even future investigations will agree on one single definition or conceptualization of religion, religious development, religious orientation, or religious salience. This lack of agreement on definitions, however, does not invalidate efforts to review the data and to draw general conclusions; it simply highlights the need to adopt a much more tentative, nuanced, and inclusive approach. To appreciate the need for such an approach, it is important to understand the nature and the relative merits of the manner by which researchers typically conceptualize religious issues.

Investigators tend to use the term "religiosity" interchangeably with one's institutional religion (see Hill, Pargament, Hood, et al., 2000). Researchers often focus on whether individuals belong to a particular religion or exhibit characteristics of those who belong to a particular religion. By far the most common measure simply focuses on the subject's denominational affiliation (e.g., Catholic, Baptist, Jewish, Muslim, other, or no religious affiliation). Recognizing that such variables do not address adolescents' religiosity, other studies use a single item of church or synagogue attendance to index religiosity; still others use items that measure religious salience (e.g., importance of religion or God in one's life). Other researchers focus on respondents' actions as a way to measure religious salience; they may study the degree to which subjects deem prayer an active and/or meaningful part of their lives, the frequency with which they read or study sacred scriptures, or the extent to which adolescents participate in various formal and informal religious activities. Research, then, typically defines religiosity as the way an individual expresses his or her relationship to a higher power or God through either his or her belief system or communal rituals (see Mahrer, 1996).

Some researchers approach religious issues by focusing on spiritual concerns, although this is much less common. Unlike efforts to measure religiosity or religious affiliation, research on spirituality generally takes one of two different directions. The first concerns itself with individual experiences, such as personal transcendence, supraconscious sensitivity, and meaningfulness. It generally focuses on adolescents' concerns about the existence of some kind of higher power, whether a personal deity or an impersonal force, and the manner by which adolescents turn to this higher power for inspiration, guidance, replenishment, and comfort. For example, research has examined the impact of experiences of bereavement on adolescents' spiritual quests (Batten & Oltjenbruns, 1999) and has examined the spiritual basis for the treatment of chemically dependent adolescents (Hackerman & King, 1998). The second set of studies focuses on individual dimensions by conceptualizing spirituality as a worldview. This approach views spirituality as a framework for answering fundamental questions of existence, including issues about the nature of reality, one's origin, identity, mortality, and relationship to others and the world itself. In addition to organizing one's world, this framework plays a key role in the development of a code of attitudes and behaviors, both personal and collective (see Balk & Hogan, 1995, Fowler, 1981; Sperry & Giblin, 1996). This code relates to spirituality in the sense that it emerges from the process of surrender to a higher power and the answers that this process provides. A person's ability to maintain these convictions relates directly to the strength of his or her relationship to this higher power and/or to a religious institution. These two approaches differ greatly from research that studies religiosity by, for example, measuring religious affiliation. The focus on individual experiences, though, deemphasizes the more communal aspects of religious matters, a rather profound limitation given that religions focus as much on behavior in this world as on what may transpire in the next one or on what already may have transpired in the prior one.

Other research simply integrates the approaches described with religiosity and spirituality and conceptualizes religion as a search for significance in ways related to the sacred (Pargament, 1997). The approach is remarkable for its comprehensiveness in that it incorporates substantive and functional aspects of religion. The definition allows for a focus on beliefs, practices, and feelings directed toward spiritual beings and also comprises questions that relate to the ultimate meaning and concerns of religion. Although it achieves greater comprehensiveness by integrating the first two approaches, even this approach remains limited to the extent that research limits the scope of what constitutes religion; for example, it pushes aside concern for the less sacred, as Satanists may claim. It also fails to include those who simply do not believe in the sacred, as atheists may claim. Likewise, the approach does not address the inclusion of those who do not involve themselves in religious experiences, such as nonpracticing individuals who identify themselves as belonging to a religious denomination. There is thus a need for an even more comprehensive approach.

Another way to investigate religious issues has not been used in previous research on adolescents but nevertheless is the approach adopted here. This approach focuses on religious orientation as it constitutes a person's worldview and on religion's role both for particular adolescents and for their social world. This approach to understanding religiosity includes research that expressly investigates religion, spirituality, and matters of faith, including rejection of religious and spiritual beliefs. Taking a more comprehensive and inclusive approach offers several advantages. First, it allows us to include more research that examines the place of religion in adolescent life. Second, it allows us to examine both behavioral dimensions, which may be public or private (e.g., denominational affiliation and attendance, prayer and devotional reading), and subjective dimensions, which may include attitudes, beliefs, experiences, and self-perception, as well as attributions that involve religious or spiritual content, such as religious identity and feelings of closeness to God. Third, an inclusive approach seemingly relates to individuals' perceptions of religion: individuals tend to take a comprehensive approach to religion as they approach the sacred through the personal, subjective, and experiential path of spirituality that often includes organizational or institutional beliefs and practices (Zinnbauer, Pargament, Cole, et al., 1997). Fourth, and as we will see more clearly in the chapters that follow, a broad approach mirrors the way the legal system conceives of religion and of individuals' religious concerns. Last, such an approach allows us to be less judgmental, a necessity given that achieving a consensus on how to define, describe, or conceive of ideal developmental attainment remains one of the chief difficulties that besets studies of religious development. For the purposes of investigation, then, this analysis adopts a broad approach to religiosity.

Adolescents' Development of Religious Orientations

Studies of how adolescents understand religious issues and the place of religion in their lives generally adopt a developmental approach. This approach has its roots in early empirical efforts to understand adolescents' religiousness, efforts that were launched when investigators began to analyze individuals' religious functioning across the stages of human development. Not surprisingly, that research generally adopted theoretical models of the psychosocial, ego, cognitive, and moral stages of development. As a result, the study of adolescents' religious understanding tends to focus on the process of religious or faith development and the driving force behind it. Although many researchers now offer their own conceptualizations of religious or faith development, and none necessarily focus solely on adolescents (see Oser & Reich, 1996; Streib, 2001), three of the best-known approaches to understanding religious development illustrate well the focus, thrust, significance, and limitations of developmental research and its implications for our understanding of the development of adolescents' religious orientations.

The first approach centers the understanding of religious orientation and development on beliefs relating to an ultimate being. These researchers generally suggest that a system of rules regulates how we entertain relationships with the ultimate. Religious consciousness emerges and is actualized each time individuals interpret life experiences, pray, study religious texts, and take part in religious community life (Day & Naedts, 1995; Oser, 1991; Tamminen, 1994). Research in this area generally understands religious judgment in terms of the way we cognitively organize religious reality (Oser, 1991). Some posit that we develop a deep "mother-structure" that constitutes the basic religious-cognitive structure. This fundamental, universal structure underlies whatever specific religion or even atheistic position we adopt for addressing religious concerns. Contextual factors, such as culture-specific and time-specific content and different forms of religious socialization, shape the mother-structure, at least in terms of the influence these features have on the rate of development of certain individuals or of groups of individuals in a given society (Day & Naedts, 1995; Oser & Reich, 1990; Reich, 1993; Tamminen, 1994). The diversity and the contextual complexities of religious belief, however, lead some to conclude that this notion of a universal deep structure of human religious reasoning is unsatisfactory (Day, 2001). Differences in emphasis lead some researchers to dismiss religious diversity as insignificant in comparison to the universal characteristics that are found across human religious experience, while they lead others to emphasize that the way individuals conceive of religion, religious experience, and religious decision making varies considerably by gender, social class, ethnicity, and religious affiliation. Although context driven, the structure (or what some view as the narrative individuals use to understand their experiences) addresses fundamental religious concerns, such as the search for the meaning of life, hope, freedom, transcendence, and eternity. From this view, religious judgment proceeds through a series of qualitative transformations (but not necessarily universal stages) influenced less by the accumulation of new religious contents than by a progressively more complex and integrated cognitive framework for understanding and approaching one's world (see Oser, 1991). It is this cognitive pattern that allows for more complicated thinking about fundamental concerns and that allows for a more mature relationship with the ultimate being.

A second approach understands religious development as involving matters of faith (see Fowler, 1981, 1991). This view posits that faith development essentially entails a process of making meaning out of life, a process in which people create loyalties and commitments to centers of value. Faith, and the development of faith, evolves from a three-part structure: the self, the self's significant others, and the ultimate other or the center(s) of value and power in one's life (Fowler, 2001). Rather than focusing only on how individuals understand their relationships to ultimate beings, this approach understands matters of faith as involving what people center their lives around, which may be God or religious beliefs but which also may be devotion to family or financial success or some charismatic personality. In this sense, faith may not always be religious; the quest for meaning occurs whether the person embraces or repudiates religion. Faith nevertheless still serves as a motivating power in ordering priorities in one's life. In addition to the trust one may have in the object of faith, the process creates for the individual a "mastery story" that guides the person in making choices and underlies how he or she interprets events in life. This approach to faith development also emphasizes cognitive development and capacities and generally views cognitive development as a necessary, although not always sufficient, component of a more complex system of faith. Research in this area, buttressed by research on the stages of cognitive development, suggests that cognitive development necessarily relates to the sophistication or complexity of faith structures. Faith structures result from the individual's process of "meaning-making" (Fowler, 1981). The notion that meaning-making depends, to some degree, on cognitive development highlights well how this model assumes and builds on the interrelationship between faith development and normal human development, especially the stages of cognitive and psychosocial development. These links also gain significance to the extent that they become problematic as a tide of challenges aims to debunk the reliance of faith and religious development theories on stage models of human development (see Streib, 2001; McDargh, 2001). Yet, recent conceptualizations still maintain, despite broad and deep criticisms, that faith develops in a sequential, invariant, and universal manner and that different types of faith (or worldviews) develop differently, depending on the cultural environment (Fowler, 2001).

Another leading model adopts a similar focus on faith and cognitive development but also highlights the importance of contextual considerations. This approach views individual adolescent's religious reasoning and beliefs as depending upon both exposure to religious experiences and the interpretations attached to them (see Elkind, 1971). It thus moves away from a purely cognitive model and highlights how others influence adolescents' interpretations of religious experiences. Research and theory in this area place particular emphasis on parents' responses to religious experiences and focus more on the important role of internalization, the means by which children acquire their values and morals (see Elkind, 1971). The approach conceives of the process as entailing more than simply learning, memorizing, or imitating models. Internalization involves a deep incorporation of religious beliefs into one's cognitive substructures. Adolescents abstract their beliefs, attitudes, and behaviors from those of individuals to whom they are attached. Moral and religious beliefs, then, become the understanding of one's responsibilities to others and to God, through years of active construction of events with significant others. These beliefs are both persistent and pervasive in adolescents' interpretations of experiences. Given that teens constructively and continuously interpret (and reinterpret) their experiences, their interpretations vary over time as their experiences confirm their abstractions and realities (both imagined and real). These experiences lay the groundwork for religious development, which occurs through two separate but interacting paths: institutional religion, which encompasses formal religious instruction, rituals, and observances, and spontaneous religion, the feelings, concepts, and attitudes that arise from one's interpretation of institutional religion.

These models place adolescents' religious and faith development within the context of the entire lifespan, an approach that offers unusual benefits but that also has limitations. Most notably, existing approaches tend to view development as a general unfolding that begins with the highly egocentric religiosity of childhood and that, if conditions are favorable, culminates in the well-integrated and autonomous religiousness of middle adulthood. As with other theories based on structuralist, cognitive-developmental perspectives, the models tend to assume a universally sequential and hierarchical development. This view of cognitive development proposes a series of structural transformations in cognitive capacities that determine the extent to which we move from one stage of development to the next. This concept of development remains problematic to the extent that it implies moving from a less valued to a more valued state, such as is suggested in the terms "underdeveloped," "developing," "developed," and "highly developed." Importantly, recent reviews probe notions of religious styles and do attempt to move beyond a single view of what would be the proper outcome of religious or faith development (Fowler, 2001; Streib, 2001), but how they can do so while still holding onto (or at least not losing the benefits of) cognitive-structuralist models remains unclear. Difficulties also emerge to the extent that the theories tend to presume autonomous thinking and autonomous social interactions and also tend to posit sequential stages devoid of theological content. As we will see, adolescents are far from autonomous in their thinking and social interactions, and there is great diversity in researchers' views about an end state of religiosity.

Again, researchers have attempted to respond to these concerns, and they increasingly pay attention to contextual and theological issues (see, e.g., Fowler, 2001; McDargh, 2001). Studying religious issues developmentally and in theological context may help counter the difficulties with more limited research approaches, since the focus typically relates to individuals' views and relationships with ultimate beings, parents, peers and communities. But, approaching issues related to religiosity and faith in that manner seems problematic, given that some approaches aim for a broader view of religious issues; for example, spirituality may be present and even more pure at birth, as evidenced by the recognition by many that children are born with a spirit that marks their very humanity and closeness to the eternal or divine (see Nigosian, 1994). The current understanding of religious development, then, has important limitations, but we must not forget that these limitations tend to reflect those of developmental approaches in general and do not detract from the validity of the propositions properly evaluated on their own terms and through their own theoretical postures.

Despite setting themselves up for considerable challenges, models of religious orientation and development that seek to describe a general, normative process of development offer much to our discussion. The contributions of existing research and models become quite apparent when coupled with current understandings of human development. First, significant variation exists in people's abilities to conceive of religious constructs. Although much clearly relies on individuals' cognitive capacities, much also relies on their social circumstances and surroundings; most notably, many nonbelievers view religious beliefs as magical, devoid of reality, or simply immature, while believers argue that those beliefs represent higher thinking. Second, the study of faith development helps us to understand religious diversity. Although much heated debate centers on the existence of universal stages, those debates do not detract from the claim that the formal structuring of stages reveals only part of one's worldview; environments provide structuring and deconstructing powers and may either stifle or enrich development. Third, individuals actively construct their own religious worldviews. Religious development unfolds through a process determined by what adolescents bring to their environment and what the environment provides them. Last, although individuals may reject formal religion, the social nature of development necessarily includes matters of faith, trust, hope, and other orientations that people adopt as their general approach to their world. Individuals continuously reconstruct their religious orientations as experiences impact how they define their place and purpose in life. Matters of faith, spirituality, and religiosity all affect people's worldviews as people deal with both existential and everyday matters. These four considerations, and the variety of religious beliefs and contexts, suggest that it makes practical sense to approach religious knowing and commitment as an aspect of human development in general and to understand that religious development requires attention to other individualistic, developmental and social factors.

Religious Orientations' Place in Adolescents' Identity Development

Adolescents appear particularly sensitive to lacunae—the gaps, incompleteness, and insufficiencies in their lives and in their social worlds. Their peculiar sensitivity to these issues derives from a mixture of inexperience, biological readiness, marginal status, and managed experience and from the cultural confusion that accompanies the adolescent condition (see Violato & Travis, 1995). The awareness also arises from teens' need to commit themselves to others. That need involves adolescents' desire to go beyond themselves and to commit themselves to people and causes for the latter's sake and not as mere means to personal benefit (Eisenberg, 1991). This self-transcendence provides a source of meaning that gives life its worth and individuals the feeling that they are actualizing or fulfilling themselves (see Magen, 1998; Levesque, 2002). The awareness of social lacunae, the need to fill the void, and the need for commitment outside the self all make religious issues particularly salient for adolescents.

The significance of identity issues to adolescents makes obvious the salience of the search for meaning, regardless of whether individuals construe the search as religious. Few doubt that adolescents' central developmental task involves developing a firmer sense of identity. Although a sense of self emerges early in infancy and adults also renegotiate their identities, developmentalists most commonly designate and understand the period of adolescence as a period in which individuals most thoroughly and self-consciously develop the integrated awareness of their own tastes, values, skills, and behaviors that constitutes a sense of identity. During the adolescent transition, one's sense of identity becomes noticeably less childlike and assumes its adult outlines, particularly in the form of competence supported by a recognizable set of values and beliefs.

Given the focus on personal meaning and values, it is difficult to discount the importance of the role of religion in adolescents' identity construction. Teens' acquisition of the ability to think abstractly, their desire to confront important questions, and their overall engagement in the process of identity formation incline adolescents to address issues of religion and spirituality (Markstrom, 1999). Adolescents' search for comprehension, inspired by their new skills, and their concerns lead them to turn to religious issues for answers (Worthington, 1989). Research on religious commitment that reveals that major crises and challenges trigger religious concerns (Balk, 1999) further supports the significance of religious issues during adolescence. Many researchers suggest that spiritual development often takes place when an individual must examine, assess, and reconstruct values and beliefs and then act autonomously on them (Butman, 1990). These descriptions precisely parallel how leading theorists have described the task of adolescence, a transitional period marked by challenges that many view as a series of crises that require at least temporary resolutions (Erikson, 1959/1980). The need to address these concerns is more than well established and does more than simply exist; the need to address those issues attains urgency to the extent that the failure to respond to them arguably contributes to destructive outcomes, such as a sense of despair and apathy (see Csikszentmihalyi, 1984). The intertwining of identity and religion is of particular interest to researchers who examine the nature of adolescents' identity development and find that adolescents at different levels of identity development approach religious issues differently and that teens with "achieved" and healthy identity formation tend to explore religious issues more deeply (see Hunsberger, Pratt, & Pancer, 2001; Markstrom-Adams & Smith, 1996; Fulton, 1997). By offering explanations for existential issues, connections to society, and a sense of belonging through rites and rituals of faith, religion can play an important role in helping adolescents develop a secure identity and self-concept (Blaine, Trivedi, & Eshleman, 1998).

Adolescents' cognitive abilities and affiliative concerns drive not only their new self-evaluations and their personal values but also their new evaluations of their world. Even though adolescents' identity formation process has been associated with deep turmoil, selfishness, and character flaws, the process also simultaneously and typically involves positive commitment beyond the self. The process fundamentally involves expectations that adolescents will develop a balance between their unique individual goals and values and an acceptance of societal norms and expectations. Moral development research characterizes the period of early adolescence as a time when adolescents evince concern with maintaining, supporting, and justifying the social order, embodied in the rules of their families, groups, or nation, and with doing their duty and acting in socially approved manners (see Gilligan, 1982; Kohlberg, 1981). Although the concern seems most obvious during early adolescence, it continues well into adolescence, as revealed by teens' increasing sensitivity to peer and other sources of social evaluation.

Adolescents' affiliative concerns highlight the typical conceptualization of adolescence as a time when one's relational focus shifts from parents to peers. As a result of this relational shift, much identity development occurs within the social, cognitive, and emotional dimensions of peer relations. Most notably, close peer relationships allow adolescents to reflect on and experience their own identities through ties with people other than parents and allow exploration of options viewed as departures from parental wishes. Although we may often characterize the adolescent period as a movement away from parents, the complexity of that move often leads to mythic exaggerations and popular misunderstandings. The identity formation process does not lead adolescents to reject parents or their values. Emotionally close parent-child relationships actually are critical to healthy identity formation. Adolescents typically reveal much attachment to their parents, and parents continue to exert the greatest influence on teens' long-term decisions and basic values. Despite popular fears of negative peer influences, parents do remain most influential, and peers' influence does seem limited in scope and duration (see Fasick, 1984); furthermore, peer influences appear at least as likely to be prosocial as antisocial (Levesque, 2002). Peers, however, wittingly or unwittingly, may influence the process of exploration that produces deeper questioning of values inherited from parents and consideration of how those values fit with the adolescent's sense of self. In the identity formation process, then, parental and peer social expectations become salient to adolescents, and those expectations inspire the identity formation process as much as the more obvious biological and cognitive changes.

The significance of social expectations finds clear expression in the affiliative dimensions of religious identity formation. Although religious identity development clearly involves a complex balance among peer, parental, and individual social and cognitive changes, broader communal concerns also play significant roles in the identity formation process. The affiliation-based aspect of identity, which involves religion, appears much earlier than adolescence (see Van Ausdale & Feagin, 2001). Firmly rooted by early childhood, this aspect of identity seems fairly well established by midadolescence. Although group identity has received the most attention in the context of ethnic and racial identity, research does reveal that it also includes religious affiliation and is particularly strong in religious groups that keep themselves apart from the rest of society (Ozorak, 1989) or that try to adjust to dominant cultural dictates (Bankston & Zhou, 1996; Brega & Coleman, 1999).

The highly context-dependent nature of the identity formation process also helps account for the manner in which religious identity varies according to familial, gender, ethnic, racial, and religious concerns (Markstrom-Adams, 1992; Kooistra & Pargament, 1999). For example, ethnicity shapes adolescents' exploration and commitment processes with respect to the identity domains of occupation, religion, politics, and gender roles (Rotheram-Borus & Wyche, 1994). Some ethnic groups with particularly strong communal orientations may discourage identity exploration because it is a process characteristic of individuation. However, it does seem that even adolescents who grow up in cultures that expect them to assume roles and values identical to those of their parents and their community must internalize those values, a process that makes possible fulfillment of their contextually dependent role.

The influence of contexts should not lead us to conclude that identity development is a passive process. Successful identity development necessarily involves the process of active exploration. This process involves gathering information about oneself and one's environment so that one can make commitments about one's life choices (Grotevant, 1992). As mentioned earlier, ground-breaking research in this area pictured the necessity of confronting decisions and making choices as crises involving development along different axes: crises, in the sense of doubt and indecision accompanied by awareness of competing possibilities, and commitments, viewed as resolving doubts and selecting among multiple possibilities (see Marcia, 1966). The value placed on crisis experiences suggests that exploration allows individuals to acquire a self-understanding that equips them to make decisions driven by genuine personal preference, rather than by externally imposed obligations and expectations (see Kroger, 1996). The value of crisis may be rejected by some who argue that membership in cohesive social groups with rituals vitiate the need for them, but the very need for rituals, especially violent ones, demonstrates that societies recognize the need for, and even create, crises to guide adolescent development (see Herdt, 1981). Identity formation, with or without intense, socially controlled rituals, involves revisiting outcomes; the absence of rituals simply makes more obvious the manner in which identity development consists of a series of multiple choices and changes made during an ongoing exploration, rather than a terminal decision made in late adolescence following a finite period of information collection. As adolescents explore, initial outcomes lead to an initial identity consolidation; the adolescent's continued evaluation of these outcomes determines whether and how he or she will proceed to explore further and achieve consolidation and development.

Influences on Adolescents' Religious Orientations

No one doubts the existence of a religious identity, or an identity that responds to religious issues even in irreligious ways. Few also dispute religion's significance as a socializing agent. Parents, peers, communities, and adolescents' own experiences and dispositions all influence adolescents' religiousness. A realistic understanding of adolescents' religious orientations must look beyond whether the teens appear to be in certain stages, are exploring religious concerns, or possess certain dispositions that drive their identity development. It must focus on variations within and influences on the activities of exploration and commitment to religious orientations. As an important area of identity, however, religion has yet to excite much research interest. Despite the paucity of studies that directly examine these issues, existing research does reveal important insights and trends.

We already have seen that parents have a powerful influence on adolescents' identity development. Research suggests that parents similarly impact adolescents' efforts to shape their religious orientations. Yet, despite the widespread belief that parents directly control their children's religious development, research has yet to support this claim. Several investigations report that parents' religious activity does not appear to be a particularly strong direct influence during the adolescent years (see Erickson, 1992; Cornwall, 1988; Benson, Donahue, & Erickson, 1989;

Ozorak, 1989). This does not mean that parents do not impact adolescents' religious orientations. But, they seem to do so in subtle, indirect ways.

The most obvious way parents influence their children's religious orientations relates to the manner in which they channel their children's experiences (see Cornwall, 1988). This involves parents' subtle influence on the contexts in which their children socialize, where they spend their free time, and their exposure to social models. For example, home religious behavior appears quite influential. Adolescents who pray, study the Bible, help the poor, think about religious issues, and so on profess strong personal religious beliefs, even though much of this homebound behavior almost certainly proceeds under the direction of parents (Erickson, 1992). In addition, parents' impact on whether children attend religious education classes or participate in religiously supported activities reveals how parents can remain powerful socializing agents. As expected, the more parents find matters of religion and faith important, the more likely they are to engage in efforts to socialize their children in ways that foster religious faith (Danso, Hunsberger, & Pratt, 1997; Dudley, 1999).

Parents also indirectly influence their children's religious orientations as they communicate and transmit their own worldviews. Parents help provide adolescents with ways to give meaning to life experiences. Research has long established that infants' primary experiences form the basis for their internal working models—the child's mental representations of the self, of significant others, and of the child's relations with them (Ainsworth, 1989). This internal representation forms the basis for later representations of the self and the world and guides the child's interactions with others (Bretherton, 1985). Although some changes do occur later in life, attachment theory suggests that these internal working models tend to resist change. This resistance derives from the way internal working models operate outside conscious awareness and from the fact that they guide the interpretation and assimilation of new experiences (Bowlby, 1988). Despite the tendency toward continuity in attachment patterns, the development of working models does not seem subject to critical periods or critical experiences (Bowlby, 1988). Changes may occur over the course of a lifetime, including during adolescence, when internal working models seem especially vulnerable to damage from negative experiences or to benefits from positive experiences (see Hamilton, C., 2000). Although research has yet to link our understanding of internal working models to adolescents' religious identity development, the two concepts seem clearly intertwined. By influencing the cognitive structures their children use to understand and cope with the world, parents help to ensure that their children will find social environments, including religious beliefs, compatible with their own cognitive schemes. Arguably, even when adolescents rebel, it is within the framework of their parent-given social, emotional, and cognitive structures. It is in this way, rather than through overtly coercive measures, that parents direct their children into religious beliefs and practices compatible with, if not identical to, their own. This research helps make obvious reasons for the pervasive and familiar belief that parents reap what they sow.

Parents also influence their children's religious orientations by their adoption of certain parenting styles. Religious parents may have parenting styles that directly contribute to adolescents' own religiosity. For example, parents with higher levels of religiosity tend to engage in more desirable parenting practices (Strayhorn, Weidman, & Larson, 1990), and church attendance may be more predictive of parenting styles than church denomination (Alwin, 1986). Parents in Christian denominations value parental control of children, as well as children's intellectual autonomy (Ellison & Sherkat, 1993a). High levels of parental control and monitoring, along with support and open communication, are correlated with a reduced frequency of adolescent problem behaviors (Barnes & Farrell, 1992; Barnes, Farrell, & Banerjee, 1994). This reduction may in turn reduce the frequency of conflicts related to parental rule setting, including requirements for church attendance and for participation in other religious activities. Adolescents who perceive their parents as supportive and caring report higher levels of religious participation, religious orthodoxy, and divine support (Bjarnason, 1998). These perceptions may be the result of the greater supportiveness of religious parents (Ellison, 1992), or the experience of a supportive family environment may predispose children and adolescents toward a favorable view of religion and religious activities (Hertel & Donahue, 1995). Caring and supportive parents may generate the perception of God as caring and supportive (Hertel & Donahue, 1995) and at the same time foster an emotional attachment to society. This view receives considerable support from well-established research indicating that different parenting styles create different emotional climates and that these climates give meaning to parent-adolescent relationships (see Steinberg, 2001).

Although the family arena remains very important for adolescents' religious development, teens do tend to move toward spending more time with peers, who take on an increasing saliency in their lives. As we have seen, one of the most important relational changes that occurs during the adolescent period involves a shift in relational focus from parents to peers. While parents maintain a strong influence over the content of children's religious experiences and over their commitment to religious beliefs, peer interactions also matter greatly. What we know about peers' powerful roles in adolescent development (see, e.g., Harris, 1995; Brendgen, Bowen, Rondeau, & Vitaro, 1999) suggests that peer interactions facilitate the development of religious orientations along social, cognitive, and social dimensions. Peer interactions may facilitate the ability to engage in self-reflection about identity, including religious identity. As they do for any other issue that they see as important to finding a place for themselves in society, adolescents evaluate religion as they use their relationships with peers to scrutinize themselves and their experiences. In addition, peer relationships may provide emotional support that helps adolescents move from parental dependence to more autonomous religious thinking and behavior. As adolescents practice separating from their parents by attaching themselves to peers, they can distance themselves from their parents' ideas as they gain independence. By offering transitional support, in part by providing opportunities for close relationships with individuals other than parents, peer relationships allow adolescents to explore options, both those that are contrary to parental wishes and those that conform to parental preferences.

Discussions of peer influences on adolescents' religiousness frequently elicit comparisons with the role of parental influences. Both sets of relationships highly influence adolescent development, but parents' impact seems more determinative (see Brendgen, Vitaro, & Bukowski, 1998); when teens must make important, far-reaching choices, their parents' views seem most salient. Parents' roles seem more significant than those of peers in transmitting religious values, and these values help adolescents through important life transitions (de Vaus, 1983). To the extent that adolescents take direction from their peers, that influence tends to be limited in both scope and duration. That the influence is time limited, however, does not diminish its significance. Everyday decisions have the potential to exert a long-term influence. Research on religion and adolescent sexual behavior highlights this point: having friends who are deeply religious highly influences adolescents' sexual behavior and general level of delinquency, an influence seemingly more predictive than parents' own religiosity (Levesque, 2000a). Likewise, relationships with peers seem to have greater impact on many indices of adolescent adjustment (e.g., depression, sympathy, and aggression) than do relationships with parents (Laible, Carlo, & Raffaelli, 2000). Even that research, however, focuses on short-term effects and recognizes the bidirectionality of influences, with both parents and peers fostering adolescent adjustment and beliefs that, in turn, facilitate the formation and preservation of relationships that help sustain behaviors, attitudes, and values.

Communities also impact adolescents' religious development and orientations. Precisely how communities do so, however, remains to be investigated directly. But, research relating to parenting and adolescent development does shed some insight. Community practices and beliefs impact parenting attitudes and beliefs, and a religious community's beliefs and attitudes about child-rearing impact the way parents approach their child-rearing decisions. For example, conservative Protestant fathers are considerably more likely than their nonevangelical counterparts to engage in paternal supervision and affective parenting (Bartowski & Wilcox, 2000). Religions also impact beliefs about the roles that families and communities play in nurturing adolescents, exemplified by the important role churches play in helping minority communities address the negative effects oppression has on adolescents (Billingsley, 1999; Taylor, Ellison, Chatters, et al., 2000). Sectarian education also may provide an important source of influence. Education offered in religiously inspired settings has a strong impact on adolescents' religiosity, with massive amounts of research highlighting the effectiveness of some religious schools in contributing to academic and prosocial development (see Bryk, Lee, & Holland, 1993; Youniss, McLellan, & Yates, 1999) and of religious involvement in producing parallel outcomes in public schools, even when race, gender, and socioeconomic factors are taken into account (Regnerus, 2000). But, interestingly, the extent to which even sectarian schools can achieve religious goals depends on the values of the surrounding community and its broader cultural values; even fundamentalist Christians schools accommodate to American popular culture and draw from the culture of professional education (Wagner, 1990). Simple demographic factors reveal important variations. For example, compared to adolescents in other regions of the country, adolescents in the South attend church more often, are more likely to say that religion is very important to them, and are least likely to say that they do not belong to a religious group (Wallace & Williams, 1997; Donahue, 1995; Benson, Donahue, & Erickson, 1989; Bahr & Hawks, 1995). Likewise, compared to white and Hispanic adolescents, black adolescents are more religious across all of the three indicators of religious salience just mentioned (see Donahue, 1995; Wallace & Williams, 1997).

In addition to broader community influences, experiences with local religious organizations play a significant role on adolescent religiosity. By far the most critical factor relating religious communities to adolescents' levels of religiousness is the relevance of its activities, beliefs, and practices to the adolescents' experiences. For example, adolescents are less likely to drop out if the religious community offers relevant and interesting religious programming, allows adolescents to be themselves, rather than having to conform to an expected persona, provides opportunities for them to take active leadership in various aspects of community life, and engages in dialogue with the aim of making the religious experience relevant to their lives (see Dudley & Muthersbaugh, 1996). Although important, these findings are not surprising; individuals respond more favorably to organizations that recognize them and meet their needs.

In addition to social forces outside the adolescent, it is important to examine how adolescents themselves influence their own religious development. It appears clear that adolescents idiosyncratically interact with their environment and shape their own religious development. They may interpret religious issues differently. As we already have seen, adolescents' ethnic and racial minority status impacts their religiosity; similarly, adolescents' sex also impacts their religiousness. On average, adolescent girls seem to be more religious than boys (Amoateng & Bahr, 1986; Forliti & Benson, 1986), as measured by reported religious importance, attendance, and affiliation variables (Wallace & Williams, 1997; Benson, Donahue, & Erickson, 1989; Bahr & Hawks, 1995). The most compelling research reveals that gender differences arise from differential socialization; girls are more likely to be submissive, passive, obedient, and nurturing than boys—all characteristics that are also predictors of religiosity *within* each sex (Thompson, 1991). Adolescents also seek out religious beliefs and practices differently. Adolescents who have suffered personal crises, such as the loss of a parent or sibling, tend to struggle more with religious issues (Batten & Oljenbruns, 1999). These findings are supported by a long line of research indicating that fears of death, desires to control nature, and need for comfort and hope during times of crisis lead to increased religiosity (see Miller & Hoffmann, 1995). Likewise, although adolescents' age does not appear to relate strongly to the importance that they ascribe to religion or to the likelihood that they will be affiliated with a religious denomination, older adolescents attend religious events less frequently than younger adolescents (Wallace & Williams, 1997). Adolescents may also elicit different responses from their environments in certain special circumstances. Adolescents who become ill and whose parents accept faith-healing traditions, for example, may find their lives more profoundly impacted by religious precepts than do healthy adolescents (Hartsell, 1999). A less visible but more common example is a court decision to require adolescents who have been adjudicated delinquent to participate in some form of faith-based services (e.g., the equivalent of Alcoholics Anonymous for adolescents) and in other group programs that may be, but do not appear to be, religiously affiliated (e.g., the leading drug and alcohol treatment model for adolescents, known as the Minnesota Model, which adopts a disease approach, requires abstinence, and mandates participation in AA-type programs; see Hohman & LeCroy, 1996). Like any other individuals, then, adolescents may idiosyncratically approach and evoke religious responses from their social environment.

Religious orientations do not emerge from a vacuum. Religion necessarily plays a significant and even organizing role in adolescents' formation of a sense of self, and numerous social and individual forces influence religious orientations and identity development. The influencing factors vary widely, and research clearly suggests definitive influences on adolescents. The impact of these influences appears quite remarkable, according to the research, and exists outside particular religious traditions. Whether content and quality of faith messages (from parents, peers, communities) and adolescents' interpretations of those messages do influence religious orientations and identity development remains an empirical issue that has yet to be investigated. Even without that research, however, it does seem fair to conclude that, even though the influence of specific religious traditions may vary, the influence of religion in general is felt and integrated through families, peers, communities, and individual dispositions, all of which determine the extent to which adolescents develop and maintain certain religious orientations.

Critical Implications of the Developmental Sciences

The development of a religious orientation is an important part of adolescents' overall identity formation process. Religious orientations involve the core of adolescents' beliefs, values, practices, and affiliations.

They do much more than become part of the identity formation process and one's core sense of self; they have profound social implications. Our current understanding of religious orientations and the ways they influence identity raises important concerns. Efforts to accord appropriate respect to adolescents' religious and spiritual needs could benefit by taking social science findings seriously.

As part of the larger identity development process, adoption of a religious orientation appears to require active exploration. As we have seen, engaging in critical exploration seems to be a significant component of achieving a healthy sense of identity. Whether and how to help adolescents engage in exploration is thus a critical concern. Intrinsic religious orientations, which value religion for itself, tend to be healthier than extrinsic orientations, which see religion as a means to an end (Paloutzian, 1996). Research reveals that faith in transcendence, altruism, compassion, love for others, and other characteristics associated with mental health dominate the worldview of the intrinsically oriented; on the other hand, more extrinsically oriented individuals appear more likely to use faith and religious practice as a means to achieve personal and social goals that lie outside the teachings of religion and, as a result, are more likely to display narrow, negativistic attitudes toward themselves and others. Moreover, active exploration constitutes a critical component of adolescents' religious identity development. Those who embrace religion out of an unreflective sense of obligation to achieve approval or to avoid feelings of guilt manifest more negative mental health outcomes (Ryan, Rigby, & King, 1993; Markstrom-Adams & Smith, 1996) and also reveal more problematic social orientations (e.g., they are more likely to be prejudiced; see Fulton, 1997) than those who experience their embrace of religious beliefs as volitional, self-determined acts assimilated into an integrated sense of self. Fostering active exploration and examination of religious concerns through, for example, respectful and tolerant investigations of religious values (see, e.g., Leicester, Modgil, & Modgil, 2000) seems central to fostering healthy identity and social outcomes.

If we conceptualize adolescence only as a period of moving away from parents, we may fail to recognize teens' enduring bonds to their parents; yet, adolescence undoubtedly involves separation to the extent that it involves preparation for functioning more actively in the broader society. Separating one's own religious identity from one's parents' is as important as any other separation that occurs during the identity formation process. This separation may be especially important. Since parents generally care most about reproducing their religious identities in their children, religion may be the area that parents try hardest to instill in their children, especially since religion itself may fuel a sense of obligation to instill particular beliefs and values (see Danso, Hunsberger, & Pratt, 1997). The potential centrality of religious beliefs in individuals' lives may result in deep tension between adolescent and parents if the former embraces a religious identity that differs from that of her parents. Thus, the issue of how to balance and respect the needs of adolescents, parents, and communities is of central concern.

The significance of peer relationships in adolescence raises yet another set of critical concerns. The extent to which parents control the nature, extent, and composition of peer interactions is likely to impact adolescents' religious development. A growing body of research demonstrates that parents actively manage their children's peer relations (see Parke & Bhavnageri, 1989). By controlling with whom their children associate, parents control the development of their children's group identification, spiritual beliefs, and value systems. When parents limit the pool of peers to those raised in the parents' tradition, those peers most likely will help point the child in the direction of the parents' religious community. This form of parental management suggests that the effects of peer influence may actually result from indirect parental influence; parents help children form friendships within the religious community, and these friendships in turn help keep them there.

Consideration of parental and peer influences on religious development highlights the critical need to adopt a developmental perspective if we are to grasp the intricacies of the various influences on religiosity. The extent to which adolescents become apostates provides a clear example of the nature of parents' roles. Even though parents may not control their responses to their peers, adolescents tend to self-select on the basis of sameness: left to their own devices, they associate with peers who are like them (Brown, 1993). This self-selection derives from a combination of community and educational segregation, self-selection in activities, and a preference for familiar social interactions. As a result, friends are more similar to each other than to other peers on a variety of attitudes, goals, and behavioral characteristics (see Cairns, Cairns, Neckerman, et al., 1988). Similarly, monitored and appropriately structured interactions with different groups, as opposed to unstructured interactions, significantly affects not only perceptions of the other but also of the self. Evidence for this process derives from efforts to increase tolerance toward adolescents from different cultural groups; the evidence suggests that simply heightening cultural awareness does not increase tolerance. Theory and research suggest that, as members of different ethnic minority groups in a society come into contact, they become more tolerant of one another only if the contact occurs in an environment perceived to be fair (Pellebon, 2000). Interaction with peers is valuable not only for promoting adolescents' own development; it is also valuable for developing effective citizens. As a result, decisions on how to manage peer relations—whether to limit or foster interactions with certain peer groups—and determinations of who should manage these relations is critical; parents, individual adolescents, their peers, and society may be seeking different outcomes.

Because of their influence on adolescent development overall, communities must play an active role in adolescents' religious development. As we have seen, adolescents invariably focus on community and social concerns as they negotiate their way through the interpersonal demands of the adolescent period. Adolescents can increasingly base their actions on feelings of sympathy and caring for others as they become increasingly able to give weight to the good feelings that they may gain from making others feel good (see Eisenberg, 1991). They engage in the critical task of integrating their views of themselves with their personal inclinations and with the opportunities afforded by society as they construct a sense of who they are and what they will become. Adolescents' capacities, however, do not always develop. Whether or not they do develop in full depends on the existence within the community of appropriate supportive opportunities. Without appropriate opportunities to work out identity issues, adolescents may turn to gurus, demagogues, or totalitarian leaders with easy answers to complex problems (see, most notably, Erikson, 1959/1980). Given adolescents' concern for belonging and the many different ways this can play out, society may need to take more active steps to provide adolescents with appropriate responses, opportunities, and supports that address their religious needs. Such steps, however, raise complex issues, since our society has taken the official position that it will not intervene in individuals' private religious concerns.

A last group of concerns involves the many challenges faced by adolescents. Adolescence is a period marked by critical changes and life events that are often experienced as stressful: the onset of puberty, increased autonomy from parents, new relations of intimacy, transition to different schools, and career decisions. These developmentally mediated stresses—stresses that arise from normative changes that accompany the transition from childhood to adulthood—are exacerbated for many adolescents by other circumstances. In addition to normative changes, many adolescents also experience conditions that may jeopardize a healthy transition, for example, marital disruption, economic disadvantages, and discrimination based on race, sex, or sexual orientation (see Levesque, 2000a). Group differences, however, do not necessarily predict the manner in which adolescents will respond to challenges (see, e.g., Coll, Lamberty, Jenkins, et al., 1996). In addition, the small variance explained by group variables such as race and ethnicity, income, and family structure offers little predictive power at the individual level, and there is tremendous variation in outcomes within racial and ethnic groups, income strata, and family structures (see Blum, Beuhring, Shew, et al., 2000). These differences highlight the need to examine neighborhood, family, school, peer, and individual characteristics and to study how those characteristics interact within various demographic groups if we are to truly understand the dynamics that contribute to any aspect of adolescent development. Adolescents necessarily confront typical and atypical issues as they adjust to biological, cognitive, and social changes that accompany puberty. Recognizing that adolescents are not all the same necessarily complicates our efforts to derive simple resolutions to complex issues.

Conclusion

Important themes emerge from our effort to understand the nature of the various influences on adolescents' religious orientations. At their core, religious orientations simply are part of adolescents' worldviews. As such, it is not surprising that families, peers, communities, and adolescents' own dispositions influence adolescents' religious orientations. Adolescents are active participants in their religious development, a development that arises from their cumulative interaction with the whole spectrum of environmental influences, both direct instruction and more subtle, symbolic messages and attitudes. Although not surprising, this conclusion is significant.

The impact of broader social forces, in addition to parental and individual adolescent dispositions, on religious orientations means that policies that address religious issues must take these forces into account. Although society treats parents with great deference in matters relating to the development of their children's religious orientations, parents clearly do not act alone in guiding their children's religious development. The family's religion and formal religious instruction are only two elements of this development. Religious communities and the social environments in which adolescents find themselves also play important roles. This finding is of considerable significance. Policies may not want to leave religious matters exclusively to family life and communities, since it is clear that other factors influence religious development. In fact, it is clear that policies already impact religious dispositions to the extent that they necessarily affect the social forces that in turn impact religious orientations. There is room for more deliberate policymaking.

Understanding that many influences shape adolescents' religious development, however, does not mean that we necessarily want to influence religious orientations in a more deliberate manner. The social precepts that guide societal responses to religion are based on the claim that society protects the right of individuals to believe what they wish and that there exists a deep need to protect freedom of conscience. That recognition, however, does not live up to empirical scrutiny. Our analysis reveals that social factors necessarily do influence religious dispositions. Given the significance of social influences and the common interest in seeking a responsible balance among those interests, society may have a stake in rethinking its approach to religion, a stake that derives from the impact religious orientations have on adolescents' development and social dispositions and the impact of those dispositions on society. The analysis, then, necessarily must turn to understanding how religious orientations and religious environments influence adolescents' dispositions.

3

Religiosity's Potentially Paradoxical Influences

Massive amounts of data now detail how religion influences and is influenced by other aspects of one's personal and social life. Most notably, research continues to examine religion's effects on mental health and its numerous pernicious social dispositions, especially religion's role in fostering intolerance, prejudice, and bigotry (see Batson, Schoenrade, & Ventis, 1993). Recent research also explores religion's impact on more prosocial dispositions, namely the extent to which religion increases concern for others in need and the desire to relieve those needs (Wilson & Janoski, 1995). Many reviews report significant trends and even suggest conclusive links between religion and specific psychological and social dispositions (see Koenig, McCullough, & Larson, 2001).

Despite important findings that have emerged from the study of religion's place in peoples' lives, the nature of religion's influences on some periods of human development is ill explored and frequently overlooked (Trusty & Watts, 1999). Research has yet to focus systematically on the relationship between religion and adolescents' developmental outcomes (c.f. Wallace & Forman, 1998). The general failure to examine the potential peculiarities of religion's roles in adolescents' lives does not mean that research has ignored religion's role in adolescence altogether. When researchers do examine religion's roles, however, their analyses tend to assume positive effects—that is, they highlight how religious participation, experiences, beliefs, and values produce beneficial outcomes, such as reduced rates of delinquency (See Benda, 1995; Johnson, Jang, Larson, & Li, 2001; Baier & Wright, 2001). The focus on positive outcomes is not accidental; the empirical study of adolescence emerged as the study of religion turned its attention to religion's positive effects. Much of the study of adolescence devotes itself to identifying and preventing dispositions deemed harmful in adulthood. The broader study of the impact of religion on psychological and social dispositions suggests, however, that the focus on religion's salubrious effects on adolescent development presents an incomplete picture. Religion may also negatively shape communal, familial, and individual experiences.

Although it may seem important to consider religion as both a positive and a negative influence, the trend toward focusing on its positive impact makes it difficult to examine religion's negative effects. Much more controversy surrounds research that identifies negative outcomes than attaches to research that finds positive links. More important, however, there is much less research on religion's problematic influences on adolescent outcomes than on its benefits. As a result, a comprehensive analysis of religion's role in adolescents' experiences must rely heavily on literature on adolescent development that has not traditionally been classified as belonging to either the psychology or the sociology of religion, the two dominant fields that empirically study religion. Broader research on adolescence and the influences on it, then, must serve as the organizing context for understanding the nature of religion's influences on adolescent development. Although the approach may be necessary given the state of the field, even if there were more research on the specific influence of religion on adolescents, the research would be understood best in the context of our broader understanding of adolescence.

Religiosity's Salutary Influences

Studies that examine religious influences on adolescents most consistently focus on adolescents' own religious orientations, rather than on religious influences on the social environments in which adolescents find themselves. That research reveals that, compared to their less religious peers, religious adolescents seemingly not only exhibit a heightened concern for others but also behave in ways that do not place undue burdens on themselves. Given that analyses focus on adolescents' own religiosity, the findings appear quite powerful and lead many to conclude that addressing social ills requires increasing adolescents' religiosity (DiIulio, Larson, & Johnson, 1999). Although such conclusions may be premature, especially given that what precisely constitutes positive adaptations and healthy social orientations remains controversial (see Levesque, 2002), this research orientation has produced at least four major areas of adolescent research that highlight the tendency among religious adolescents to adopt more prosocial and socially desirable orientations.

Delinquency research most often provides evidence of religious adolescents' concern for themselves and for abiding by societal prescriptions. Adolescents who hold religious beliefs and participate in religious activities tend to engage in less criminal and delinquent activity (Stark, 1996; Welch, Tittle, & Petee, 1991; Rodell & Benda, 1999). For example, research typically finds an inverse relationship between religiosity and use of numerous types of drugs, ranging from cigarettes, alcohol, marijuana, and amphetamines to depressants (Bahr & Hawks, 1995; Cochran, 1993; Free, 1994; Newcomb, Maddahian, Skager, & Bentler, 1987; Bahr, Maughan, Marcos, & Li, 1998). The most robust findings emerge from research that indexes the multiple aspects of religiosity and religious affiliation. On average, adolescents are more likely to turn away from drug use if they frequently attend religious services, report that religion is important to them, and affiliate with religious denominations that explicitly prohibit drug use (Miller, Davies, & Greenwald, 2000; Wallace & Williams, 1997; Amey, Albrecht, & Miller, 1996; Cochran, 1991)—all consistent relationships found across many religious denominations (Amoateng & Bahr, 1986; Barnes, Farrell, & Banerjee, 1994). Importantly, studies also report negative relationships between religiosity and many other specific forms of delinquent activity, such as decreased likelihood of carrying weapons, getting into fights, and driving under the influence of alcohol (Wallace & Forman, 1998; Donahue, 1995; Evans, Cullen, Dunaway, et al., 1996; Free, 1994). Generally, then, adolescents identified as more religious are less likely to place their health at risk by engaging in delinquent behavior.

Research on adolescents' sexual risk taking and romantic relationships also reveals that religiosity increases adolescents' tendencies to abide by conventional rules and to place themselves at less risk for negative developmental outcomes. Most notably, highly religious adolescents initiate sexual activity later, have fewer sexual partners, and engage in sexual intercourse less often than their less religious peers (Thornton, 1985; Cooksey, Rindfuss, & Guilkey, 1996; Forste & Heaton, 1988; Kahn, Rindfuss, & Guilkey, 1990; Studer & Thornton, 1987; Thornton & Camburn, 1989). Likewise, religious adolescents tend to hold different sexual values; although simply viewing oneself as religious may not affect some sexual attitudes (see Donnelly, Duncan, Goldfarb, & Eadie, 1999), intrinsically religious adolescents—those whose religion shapes their everyday actions—do hold more conservative sexual attitudes (Reed & Myers, 1991). As a result, for example, these adolescents are less likely to approve of premarital sex, abortion, and contraceptive use—all of which decreases the chances that they will engage in these behaviors (Marsiglio & Shehan, 1993; Boggess & Bradner, 2000; Brewster, Cooksey, Guilkey, & Rindfuss, 1998). Religious participation and orientations, then, tend to have significant net effects on adolescents' sexual attitudes and outcomes in ways deemed more healthy for individual adolescents and society.

The extent to which religious adolescents adopt less risky social orientations also finds clear expression in their approach to schooling and academic achievement. Religious adolescents generally are more attached to schools and earn better grades (Montgomery & Francis, 1996; Zern, 1985). Religious commitment also improves students' chances of living up to their ability levels, as measured by the Scholastic Aptitude Test (SAT) and the Graduate Record Examination (GRE) (Zern, 1987, 1989). The results hold true within specific religious and social groups. For example, many studies report a positive correlation between the degree of religious commitment and academic achievement among Christian high school students (Koubek, 1984; Brody, Stoneman, & Flor, 1996); others find that religious private school students generally outperform the general population not only because of the religious orientations of those schools but also because of the religious commitment of the students (Sander, 1996). The general rule, then, is that more religious students outperform less religious students in academic achievement; these results seem unaffected by gender, socioeconomic status, ethnic or racial minority status, or attendance at religious schools (Jeynes, 1999).

The finding that religious students adopt more socially exemplary orientations gains support from research that examines adolescents' moral development and prosocial activities. Religious adolescents tend to score higher on available tests of moral development (Kedem & Cohen, 1987). Likewise, religious adolescents tend to act in ways that reveal more prosocial commitment and concern for others (Donahue, 1995; Perry & McIntire, 1994). Most notably, parochial school students who report positive attitudes about religion consistently have been found to engage in more noncompulsory volunteer work (see Forliti & Benson, 1986). In a comprehensive, nationally representative survey of adolescents, 74 percent of high school seniors who said that religion was important to them also engaged in community service on a monthly or more frequent basis, while only 25 percent of their nonreligious peers engaged in similar service (Youniss, McLellan, & Yates, 1999). Thus, students who say religion is important in their lives are almost three times more likely to do service than students who report that religion is insignificant to them. For those interested in fostering civic and political involvement, these findings are quite significant. The development of a service-orientation in adolescence increases the chances that the orientation will continue in adulthood, and religious participation by adolescents has been shown to predict the extent to which parents and communities can foster that orientation (Smith, 1999).

Although this line of research has yet to examine other areas in which religiosity and religion can impact adolescents' experiences and outcomes, it does seem fair to conclude that research that focuses explicitly on adolescents' religiosity and developmental outcomes supports the popular assumption that religion can be a protective factor against numerous negative dispositions. Many adolescents who do not engage in risk-related behaviors also report living by their religious beliefs and accepting the proscriptions of their faith. On average, adolescents who exhibit religious behavior and find religion important are less likely than other adolescents to engage in problem-related, high-risk behaviors and are more likely to engage in behaviors that promote long-term well-being. That religion may present an important protective factor against risk is not surprising, given the current understanding that adolescents' high or low risk-taking tendencies characterize their general approach to social interactions. Socially concerned adolescents who respect authority and tradition tend to engage in less problem behavior. The converse also is true: problem adolescents tend to exhibit multiple problems and engage in many risky behaviors (e.g., delinquent adolescents tend to engage in sexual risk taking, do poorly in school, and reject some forms of authority, perhaps including religious traditions) (Levesque, 2002). Adolescents' religious beliefs and attitudes, then, seemingly infuse adolescents' social orientations in many ways deemed beneficial to themselves, their close relationships, and their communities.

Religiosity's Problematic Influences

Some commentators and nonbelievers express negative attitudes toward particular religions or religious beliefs. Few, however, empirically explore how adolescents' religious views and practices may effect negative developmental outcomes. Despite researchers' failure to provide rigorous empirical data, the available evidence does support claims that religious beliefs and practices can produce destructive consequences. These consequences typically arise when adolescents become involved in nonmainstream religions and when mainstream religions' practices produce unexpectedly harmful effects. Understanding these negative consequences generally requires moving beyond adolescents' reported religiosity and focusing on religious environments themselves, an approach that some believe to be more objective (see Batson, Schoenrade, & Ventis, 1993).

Two examples help illustrate the manner in which adolescents involved with nontraditional religions may be at risk for unhealthy outcomes. The first derives from the increasing interest in adolescents' involvement in witchcraft, the occult, and Satanism. Researchers associate these experiences with adolescents' religious and spiritual needs and link these nonmainstream religious experiences to problem behavior. For example, research has examined the extent to which adolescents admitted to mental health or juvenile justice institutions engage in some form of Satanism and has found that from 5 percent to 10 percent of adolescents accept and practice these beliefs (Burket, Myers, Bradford, & Carrera, 1994; Damphousse & Crouch, 1992). Although studies of adolescents in need of mental health and other services have found negative correlations between adolescent health and religious beliefs, some authors view teens' involvement in these religious beliefs and practices more as signs of rebellion than as true religious experiences (Emerson & Syron, 1995). They understand these ritualistic and spiritual quests less as pathology or criminal activity than as responses to the dominant society's religious beliefs (see Lowney, 1995). Yet, the general belief is that involvement in religious cults that indoctrinate, isolate, and alienate individuals from family and community members fosters negative mental and physical health (see Curtis & Curtis, 1993; Koenig, McCullough, & Larson, 2001). Regardless of one's interpretation of the role of occult experiences in adolescence, it does appear that those experiences at least reveal the possibly negative consequences of adolescent religiousness and, equally important, the challenges adolescents face in their responses to more mainstream religions.

In addition to Satanistic and other occult experiences, nonmainstream religions may champion beliefs and actions that outsiders can see as abusive. Most notably, new religious groups frequently have been accused of committing acts of physical child abuse (Ellison & Bartowski, 1995; Singer, 1996), including ritualized child sexual abuse (Richardson, 1999). Abusive practices, however, are not limited to fringe groups; mainstream religions also have been charged with perpetuating beliefs that result in the sexual abuse of children, as seen in controversies regarding female circumcision (see Levesque, 2001) but more obvious in the literature that finds complex correlations between certain religious beliefs and the rate of domestic violence (Ellison, Bartowski, & Anderson, 1999; Ammons, 1999; Burris & Jackson, 1999) and claims that church patriarchies increase the risk that leaders will abuse children (Kennedy, 2000). Equally controversial are differences in religious groups' use of spiritual healing rather than more generally accepted medical practices. Most notably, a leading and highly publicized study reported a high incidence of child deaths in families that eschewed medical care in favor of faith healing (Asser & Swan, 1998). Although many outsiders may view the latter practices as abusive, many others argue that they are not, a point supported by the existence of many child welfare law exemptions for spiritual healing (see Levesque, 2000b). As expected, these findings remain quite controversial, but their existence at least suggests that religious beliefs and practices may evoke debatable results.

Despite the focus on new and nonmainstream religions, it is important to consider the extent to which dominant religions also may foster potentially harmful parenting practices. As suggested earlier, mainstream religious teachings and practices may harm adolescents. By far the most controversial example, one yet to be resolved, is the conflict between the approval by contemporary fundamentalist Protestants of corporal punishment and the known destructive consequences of physical punishment (see Greven, 1991; Capps, 1992). Researchers have associated key tenets of evangelical theology with parental support and use of corporal punishment (Weihe, 1990; Ellison, Bartoski, & Segal, 1996; Ellison & Sherkat, 1993b; Gershoff, Miller, & Holden, 1999) and have also linked corporal punishment to family violence, in the form of violence against children and, when the children themselves reach adulthood, violence against adults (Straus, 1994; Straus, Sugarman, & Giles-Sims, 1997). The extent to which rationales for supporting corporal punishment remain linked to conservative religious values, even when research indicates that many religious groups support corporal punishment and many conservative Protestant groups spurn it (Ahn, 1994), suggests that certain religious beliefs can gain and retain social dominance despite their links to potential harm.

In addition to instilling particular views of parenting, certain religious groups have been seen as problematic because they foster certain views of morality and seek to impact adolescents' personal relationships, those of nonadherents. Efforts by religious groups to influence education in public schools provide a clear example. Few question that religious groups that espouse moral traditionalism have had an impact on the education of adolescents, as revealed by their highly contentious support for the banning of books they view as immoral in public schools (see Tamney & Johnson, 1997). More recently, however, these views have become even more controversial as certain religious groups have vigorously opposed comprehensive sex education programs offered by public schools in response to trends in adolescent sexual behavior. Many commentators who have reported on the public schools' inability to address adolescents' sexual needs (especially those of sexual minorities) blame the failure of sex education programs on the opposition of conservative religious groups (Simson & Sussman, 2000; for a review, see Levesque, 2000a). Less obvious negative outcomes are evident from the investigations of religiosity's impact on sexual relationships. As we have seen, less religious adolescents report more sexual activity. Although these findings seem to point to a benefit of religiosity, the research actually identifies a less healthy outcome, since adolescents who are more religious are *more* likely to engage in risky sexual activity when they do engage in sexual activity. For example, many researchers have noted that strong religious beliefs relate to a decreased likelihood of practicing "safe sex," particularly the failure to use condoms effectively to prevent diseases and unwanted pregnancies (see Zaleski & Schiaffino, 2000). Whether religious and nonreligious adolescents engage in the same risky practices (given that religious and nonreligious adolescents present differences in age of initiation of sexual activity and in the nature of their relationships), religiosity apparently presents a risk factor for unsafe sexual practices among sexually active adolescents. Despite the controversies that surround the imposition of particular religious beliefs on educational efforts, then, it seems fair to conclude that religious beliefs and traditions impact sexual activities and may place adolescents at risk for negative outcomes.

Religious organizations' views of what constitutes appropriate moral behavior and relationships also may impact adolescents negatively to the extent that they shape models of achievement. There are clear limits to the many positive links between religiosity and education. Two examples are again illustrative. The first example is the rise in the home schooling movement as a result of parents' religiosity and in the increasing popularity of sectarian schools (Levesque, 2002). Important commentaries suggest that these forms of schooling may improperly constitute a process of indoctrination that instills prejudice and intolerance in children and produces adults ill equipped to compete in a secular marketplace or to enter higher education (see Peshkin, 1986; Buss, 2000; Dwyer, 1998). Whether there is empirical support for these commentaries remains to be determined, especially given some sectarian schools' abilities to foster academic success (Bryk, Lee, & Holland, 1993) and prosocial commitments (Youniss, McLellan, & Yates, 1999). Regardless of the empirical findings, however, it does seem that some religious groups may foster prejudicial beliefs, and some adolescents—and, eventually, society—may suffer from their not having had particular social experiences (Minow, 1999; Buss, 2000).

The second example of religious organizations' impact on adolescents' achievement relates to the manner in which some religious views stifle educational opportunities. For example, conservative Christians hold two views that could limit adolescents' educational attainment: they may be opposed to certain secular forms of education, especially those based on scientific and humanistic values (see Ellison & Musick, 1995), and they may hold traditional gender role values, including the belief that women should restrict their activities to the home (c.f. Peek, Lowe, & Williams, 1991). These two views impact adolescents' schooling and their eventual educational attainment (Darnell & Sherkat, 1997). The relative impact of these values, however, varies according to the depth of the adolescents' own fundamentalism (see Sherkat & Darnell, 1999). When both parents and children adopt fundamentalist values, parents' fundamentalism positively impacts adolescents' academic achievement, regardless of whether the adolescent is male or female. Adolescent males who are fundamentalists, however, receive greater boosts than females. An even greater difference arises when the adolescents do not hold fundamentalist values: male adolescents are not significantly hampered by their parents' views, while girls are clearly negatively impacted by having fundamentalist parents (Sherkat & Darnell, 1999). Religious views, then, impact adolescents' educational pursuits, in terms of both the nature and the eventual level of education attained.

Although the bulk of studies that find negative influences of religion on adolescents' development tend to focus on religious institutions themselves, research that focuses on individuals' own religiosity also reports problematic links. Research on coping provides clear examples of ways in which religiosity may increase the risk of problematic outcomes. Many religious coping styles have been identified as maladaptive—responses that seem to match poorly the features and requirements of specific stressful situations. Most notably, coping in collaboration with a divine other seems healthy, but passively leaving the responsibility for resolving crises entirely up to divine intervention seems problematic (Pargament, 1997). Likewise, those who develop "righteous anger," pray for divine vengeance, and feel abandoned by divine forces evince problematic responses to crises (see Pargament, 1997). Problematic results also may arise when social pressures within religious congregations foster social norms that may increase the negative consequences of certain stressors. For example, many religious groups frame problem behavior as sin and ascribe responsibility for events or conditions to an individual's flawed character, rather than to the interplay of contextual or other factors. Such a view can adversely impact the individual's health to the extent that it fosters feelings of guilt and shame, erodes feelings of competence, self-worth, and hopefulness, and distracts the person from more productive coping responses (Ellison, 1994). Likewise, some groups may encourage or tacitly condone the withdrawal of community support, a factor clearly related to negative outcomes. These two possibilities were seen in one study as the reason that unmarried adolescent mothers who were more religiously active suffered higher rates of depression than those who were less religiously active (Sorenson, Grindstaff, & Turner, 1995). Religion, then, may foster feelings of guilt or shame, erode feelings of hopefulness, competence, and self-worth, and encourage the withdrawal of community support from those who do not conform to social norms. Although correlations between religious issues and negative personality outcomes are only beginning to receive empirical scrutiny, these links have led the American Psychiatric Association to recognize "religious or spiritual problems" as a new diagnostic category in the Diagnostic and Statistical Manual for Mental Disorders (DSM-IV) (American Psychiatric Association, 1994).

One of the most widely reported negative impacts of religiosity is the role religion can play in fostering prejudice. Much research has examined the extent to which religious individuals become prejudiced toward certain groups, behaviors, and attitudes. Although many religions place a focus on tolerance, forgiveness, and assisting those in need, they may also seemingly foster the opposite, especially with regard to those seen as transgressing their moral dictates or otherwise outside the select group (see Reimer & Park, 2001). Importantly, many of the more problematic aspects of religiousness may derive less from problematic religious beliefs than from individuals' attitudes toward religion, whether religiosity indicates an internalized commitment to religion or a more self-serving commitment adopted for utilitarian reasons (see Allport, 1966). Most notably, adolescents who internalize their religion as a means to some other end, such as a way to achieve social gain or personal protection, display more prejudice toward stigmatized groups than do adolescents who adopt a religious orientation that emphasizes the positive value of doubt and open-ended search (Fulton, 1997). Other research confirms that the nature of individuals' commitment to religion determines the extent to which they adopt positive social orientations; intrinsically motivated individuals reveal high levels of tolerance, especially toward groups viewed as acceptable by their religious community, and individuals who value a more tentative, complex, and skeptical view of their religion reveal even higher levels of tolerance and sensitivity to others' needs. Extrinsically motivated individuals, those who adopt religious beliefs for self-serving ends, tend to evince more prejudice and intolerance and less caring attitudes toward others (Batson, Schoenrade, & Ventis, 1993).

Efforts to identify the problematic influences of religion and religiosity on adolescents reveal important points. First, adolescents' own orientations toward religion determine the extent of their religion's influence. Second, adolescents' religiosity and religious environments may contribute to negative outcomes, although current empirical research fails to identify negative outcomes from religiosity, at least from mainstream religions and traditional religious involvement, as frequently as some observers would expect. Third, understanding the role of religion in adolescents' experiences requires a close look at the community, peers, and families in which adolescents find themselves. We now turn to efforts that attempt to understand the nature of the relationship between religion and adolescent outcomes to clarify our current understanding of *how* religiosity and religion impacts adolescents' experiences.

Understanding How Religion Influences Adolescents

The research we have reviewed presents two contrasting images of adolescents and the role of their religious environments in fostering developmental outcomes. One image portrays religiously oriented adolescents as individuals who attend services, believe in the supernatural, trust authority, engage in civic affairs and community service, get involved in school activities, and avoid the usual risks that many adolescents experience as a matter of course with alcohol, drugs, tobacco, and sex. This view typically focuses on adolescents' own religiosity more than on their religious environments. The other image focuses more on adolescents' environments than on adolescents' own religiosity. This approach reveals a much more negative view of religion's role in adolescents' experiences and social orientations. Indeed, this second view in many ways essentially presents adolescents as victims of religion. Research, then, typically sees positive outcomes as deriving more from adolescents' own religious views and sees negative outcomes as deriving more from religious environments affected by religious views.

Understanding the existence of these two seemingly contradictory images sheds considerable light on the manner in which religion actually impacts adolescents' experiences. Religion undoubtedly plays a complicated role in adolescents' lives. That role suggests that religion influences adolescents in numerous ways. That is, numerous interrelated factors, notably parents, communities, and peers, simultaneously influence adolescents' religiousness and developmental outcomes. The extent to which developmental outcomes take positive or negative turns necessarily depends on the contexts in which adolescence experience religion and the influence of other social factors on those contexts. Understanding religion's impact on the adolescent experience, then, requires understanding how different contexts, and religions' roles in them, impact adolescents.

The way that families influence adolescent development and religion plays a role in family life demonstrates the need to consider religious contexts. The relationship between parent-child relationships and adolescent substance use has numerous possible ties to religious beliefs and practices. Research suggests that the influence of parent-adolescent bonds on religiosity and its outcomes may be indirect, affecting variables such as adolescents' religiosity and educational commitment (Huizinga, Loeber, & Thornberry, 1995; Litchfield, Thomas, & Li, 1997). From this view, religious involvement may be encouraged by parents, and the religion may reinforce the family's teachings and thereby impact adolescent outcomes. Longitudinal studies reveal that positive parent-child bonding and monitoring during early adolescence relate to adolescent religiosity, which has a negative later association with adolescents' use of alcohol and drugs (Litchfield, Thomas, & Li, 1997). Importantly, studies that control for the amount of parent-child bonding find that religiosity negatively correlates with alcohol and marijuana use. These findings support earlier research that indicated that adolescent drug use appears to be most frequent in families in which parents regularly attend services and adolescents do not (Burkett, 1977). These findings imply that adolescents who are withdrawn or alienated from their parents are most likely to exhibit problem behaviors and that parental control of and involvement with adolescents play important roles in promoting healthy adolescent development. Both parent-adolescent bonds and parental monitoring associate positively with levels of religiosity and decrease the likelihood of adolescent drug use and general delinquency (see Levesque, 2002).

These findings are not surprising, given what we know about factors that protect against delinquency and other problem behavior. Researchers have long noted that parenting style impacts adolescents' outcomes. Most notably, researchers distinguish between two relationship styles that yield divergent outcomes. Authoritative styles demand age-appropriate, mature behavior from their children and simultaneously foster children's autonomy in a warm and supportive environment. These relationships encourage children to participate actively in discussions of decisions that affect them, but parents retain the final say. Authoritarian parenting, in contrast, places an emphasis on children's conformity to parental rules in the context of low parental support. Authoritarian parents expect their children to obey without questioning or reflecting on the specifics of given situations. The former style continues to be viewed as most healthy: adolescents from homes marked by authoritative child-rearing styles achieve more in school, report less depression and anxiety, score higher on measures of self-reliance and self-esteem, and are less likely to engage in antisocial behavior, including delinquency and drug use (Steinberg, 2000). Many studies that assess the values and parenting behaviors of specific religious groups conclude that religiosity engenders authoritative parenting (Brody, Stoneman, Flor, & McCrary, 1994). These findings are even more pronounced for religious or ethnic minority groups, for whom religion or spirituality serves as a central cultural value (Hill, Soriano, Chen, & LaFromboise, 1994), but they generalize to the majority culture and to individuals who view themselves as committed to religious beliefs (Gunnoe, Hetherington, & Reiss, 1999).

The impact of families on adolescents' religiosity and outcomes seems even more obvious in families marked by violence, dysfunction, and problematic adolescent involvement with religion. A seminal study of 143 adolescents, for example, found that these adolescents' involvement in Satanism reflected the dynamics of the total family system (Emerson & Syron,1995). In a circular way, the family's dysfunctional patterns influenced the young person's involvement with satanic groups; this involvement, in turn, affected the entire family. As adolescents' involvement in Satanism increased, parental dysfunction also seemed to increase, so much so that, in more than half the families observed, the adolescents directly controlled their families through either overt or covert terrorism. Other research indicates that familial responses to adolescents who break their involvement with alternative religions largely determines the extent to which those adolescents will experience negative outcomes (Robinson & Frye, 1997). Families, then, can play a particularly influential role in adolescents' religious experiences and in affecting the impact of those experiences.

Although families may provide important social foundations for both positive and negative outcomes, adolescents occupy a much broader world and are highly influenced by extrafamilial forces. Most notably, as has been discussed, peer associations figure prominently in adolescents' experiences and impact adolescents' outcomes. Two examples document the link between peer networks and problem behavior. The first area involves adolescent drug use (Ary, Tildesley, Hops, & Andrews, 1993; Bahr, Marcos, & Maughan, 1995; Coombs, Paulson, & Richardson, 1991; Hawkins, Catalano, & Miller, 1992; Needle, McCubbin, Wilson, et al., 1986; Orcutt, 1987). Adolescents who associate with peers who use drugs, not surprisingly, initiate drug use at higher rates than do adolescents whose primary peer associates shun drugs (Elliott, Huizinga, & Ageton, 1985; Huizinga, Loeber, & Thornberry, 1995; Kaplan, Martin, & Robbins, 1984). Peers usually introduce a drug and encourage its use; adolescents rarely use drugs if none of their friends do (Khavari, 1993). These findings, however, do not negate the significance of parents, at least to the extent that adolescents who use drugs tend to choose friends who use drugs. That is, longitudinal data indicate that socialization and selection effects are about equal in strength (Kandel, 1980). It seems, for example, that parental monitoring and drug use by family members influence the choice of peers, which impacts the level of adolescent drug use (Bahr, Marcos, & Maughan, 1995; Chassin, Pillow, Curran, et al., 1993). Involvement in a religion provides a network of support and friendship that seems to insulate adolescents from opportunities to use drugs (Bahr, Maughan, Marcos, & Li, 1998).

The second example relates to sexual activity. Spiritually interconnected adolescents may participate in voluntary faith assemblies, which can act as powerful referent groups and influence behavior and promote abstention from risky behaviors. The importance that adolescents give to religion and to spiritual interconnectedness is inversely associated with a history of voluntary sexual activity (Holder, DuRant, Harris, et al., 1999). In multivariate analyses, however, only spiritual interconnectedness, particularly spiritual interconnectedness with friends, independently and inversely associates with a history of voluntary sexual activity (Holder, DuRant, Harris, et al., 1999). It is therefore important to consider how peers, whose selection may be influenced by parents, affect the extent to which religious beliefs influence adolescents' experiences.

Communities also clearly impact adolescents and shape their religious and secular experiences. Religious communities serve as resources through, for example, their community development, mentoring, and outreach efforts. Religious groups also play a central role in the delivery of social services. While governments may design and fund social programs, nonprofit organizations tend to administer them and increasingly are becoming the local organization that offers health, medical care, food, and educational services (Salamon, 1995). In fact, faith-based organizations provide the greatest source of service provision for adolescents at risk (DiIulio, Larson, & Johnson, 1999; Queen, 2000).

In addition to providing direct services, religious communities support socializing institutions that provide opportunities for adolescents to reach positive developmental outcomes. Three examples are illustrative. First, the structure of religious communities may enhance the welfare of individual community members (Ellison, 1994; Ellison & George, 1994; Welch, Tittle, & Petee, 1991). Emerging research on how adolescents cope with racial and economic discrimination reveals that church attendance, independent of other factors, helps make young black males from high-poverty neighborhoods substantially more likely to escape poverty, crime, and other social ills; religion's impact is even more powerfully predictive than such variables as peer influences (Johnson, Jang, Li, & Larson, 2000). The links between strong integration into a church community and positive outcomes for African American adolescents seem indisputable (Brega & Coleman, 1999). Other research reveals similar findings for other groups, including immigrant Vietnamese adolescents in low-income neighborhoods, whose ethnic religious communities provide social supports and constraints that encourage behavior likely to lead to upward mobility (see, e.g., Bankston & Zhou, 1995). Research generally concludes that religious participation promotes social adjustment to the host country by providing a solid base of ethnic identification (Bankston & Zhou, 1996). The potential of church going and other religious influences to improve the life prospects and the psychological adjustment of minority youth varies, in part, as a function of the influence of these factors on how young people spend their time, the extent of their engagement in positive structured activities, and the degree to which they receive support from responsible adults (Levesque, 2002). This view renders almost obvious, but still quite significant, the recent finding that the effects of church involvement on adolescents' educational outcomes actually may not vary much by ecological context; church involvement seemingly predicts equally well (and positively) for students in poor neighborhoods and for those from middle-class or wealthy neighborhoods (Regnerus, 2000). These findings confirm that religious institutions and activities help to integrate individuals into social groups and direct individuals' behavior by making them subject to group demands and expectations. Deeper involvement in a religious group means greater group integration and thus greater susceptibility to the group's influence.

The second example involves the influence of religious beliefs and faith-based services on adolescents who are not religious and on the secular services they receive. Throughout U.S. history, a wide variety of religious movements have sought public influence, political movements have appropriated religious resources and symbols, and denominations and sects have alternately sought a public voice and condemned the same (Casanova, 1994). Not surprisingly, research has revealed how religious communities, such as Christian fundamentalist and evangelical groups, have exerted often contradictory influences on U.S. social reform and social policies (see Schaefer, 1999), focusing on issues ranging from capital punishment, abortion, and sexual orientation legislation to welfare reform and divorce law to tax breaks (Regnerus & Smith, 1998; Berg, 2001; Peach, 2002). For our purposes, the increasing involvement of such groups in national and local politics and their growing success at framing social issues, such as sexuality education, in moralistic terms provides a clear example of how religious communities may take influential sociopolitical stances and influence even adolescents who are not members of their communities (Levesque, 2000a). In addition, services offered by religious communities differentially impact adolescents, even adolescents who do not belong to the religious groups providing the services. Most notably, for example, adolescents who view themselves as nonreligious benefit the most academically from Catholic schools (Sander, 1996). Religious communities, then, do more than provide individuals with certain worldviews; religious communities provide institutions that reflect and foster those worldviews.

The role of religion in helping adolescents cope with their ethnicity provides our third example of the powerful influence of religious communities. Research on religion's role in minority adolescents' ability to deal with their stigmatization provides a clear example. As noted, African American adolescents who show a strong commitment to their religious beliefs and to their church internalize to a lesser extent the negative attitudes and stereotypes others hold about them and show a more positive adjustment (Brega & Coleman, 1999). These findings complement a long line of research that documents a host of psychosocial benefits among adolescents and individuals who hold a strong sense of ethnic identity (see Roberts, Phinney, Masse, et al., 1999). Given those findings, it should not be surprising to find that research that examined how adolescents deal with belonging to minority religions also provides an important example of the role religious communities play in adolescents' outcomes. It seems self-evident that belonging to a minority religious group would cause additional stressors and challenges to adolescents already struggling with other normative developmental tasks. Most notably, it would seem that adolescents from minority religious groups would face tensions between cultural assimilation and maintenance of a religious tradition on the one hand and stresses that arise from belonging to a minority religion in a larger Christian culture on the other. Although adherence to minority religious groups undoubtedly engenders tensions, it also provides important coping resources. For example, religiously identified Jewish adolescents are more likely to cope with stress by seeking God's direction and support, seeking Jewish cultural and social support, and engaging in a spiritual struggle (Dubow, Pargament, Boxer, & Tarakeshwar, 2000). That is, adolescents who are more embedded within Judaism have a more accessible and more compelling set of resources for coping with their ethnic concerns than do Jewish adolescents who are less identified with their ethnic or religious identity. Importantly, these findings mirror those of research into the influence of religious participation on minority adolescents' outcomes. Religious activities provide more than arrangements of symbols and norms; they also provide concrete social networks that can occupy a central place in the life of a community (see Bankston & Zhou, 1996). Belonging to a religious community seemingly has its benefits; not belonging leaves one to one's own devices.

Although religious communities themselves may influence adolescents' experiences, broader community factors also may impact both adolescents' religiosity and the influence of that religiosity on their behavior. In addition, religiosity may impact adolescents differently depending on their place in society. Sociologists have long noted the existence of racial differences in whether religious affiliation affects fertility-related behaviors. African American teens with no religious ties are at greater risk of nonmarital first intercourse than are those with a professed denomination (Brewster, 1994). This finding has been repeated with rural (Lock & Vincent, 1995) and middle-income African American adolescents (Murry-McBride, 1996). There is, however, no evidence to support claims that religion influences the risk of nonmarital intercourse among white teens (Brewster, Billy, & Grady, 1993). Importantly, race appears to have the opposite effect on contraceptive use; denominational affiliation differentially influences contraceptive use at first intercourse for white teens, but not for black teens (Kahn, Rindfuss, & Guilkey, 1990). Minority status, then, influences adolescents' experiences differently, depending on adolescents' religiosity, religious affiliation, and outcomes under study.

Although religion may impact adolescents through familial, peer, and broader community influences, those influences do not detract from the possibility that adolescents benefit or suffer from religiosity simply because religion itself has positive (or negative) effects. Individuals participate in "divine relations" as well as social relations, and divine relations may approximate the intensity of concrete social relationships. Studies now show that divine interaction may enhance individual well-being, including a feeling of empowerment caused by consultation with the divine in everyday life and an expansion of one's role set beyond the secular social world (Pollner, 1989). Likewise, religious participation may not merely extend one's secular support network but may lead one to see God as a supportive partner in coping (Pargament, Ensing, Falgout, et al., 1990). In this view, it is a personal relationship with one's God, rather than one with fellow believers, that increases individual well-being. Although the perception of divine support may be rooted in the concrete experience of having other people caring and supportive, perceived divine support does seem to play an important role in helping believers cope with the trials of everyday life (Hertel & Donahue, 1995). Again, however, not all the links are positive. A belief in original sin may be problematic to the extent that it is inversely linked with self-esteem, but this negative effect may be offset by a belief in or a perception of divine grace and forgiveness (Watson, Morris, & Hood, 1988). The possibility clearly exists that we may be influenced by matters some of us simply do not comprehend.

Several authors suggest that the key to the effect of religion on well-being lies in the sense of coherence, comprehensibility, and meaningfulness of reality that religious beliefs, religious communities, and perceived divine support can engender (Ellison, 1994; Pollner, 1989; Williams, 1994; Bjarnason, 1998). This interpretation neatly meshes with recent studies that present religious communities as tightly knit support groups that buffer individuals from negative social and psychological outcomes. Frequent churchgoers report that they have larger social networks, have more contacts with network members, receive more types of social support, and have more favorable perceptions of their social relationships—all of which tend to be associated with positive health (Ellison & George, 1994; Wright, Frost, & Wisecarver, 1993). These benefits are compounded by the manner in which religious communities place moral constraints on behavior in ways that facilitate good physical health, positive family and interpersonal relations, and ethical work conduct and financial dealings and inhibits stress-inducing lifestyle choices (Ellison, 1994; see also Welch, Tittle, & Petee, 1991). Commitment to a religious organization and its goals, such as proper conduct, also may provide a meaning to life and may constitute the type of commitment that makes problem behaviors, such as drug use, less attractive (Merrill, Salazar, & Gardner, 2001). The power of religious matters, then, may derive from the very reason religions exist: to help us understand, and live by, our proper place in the world.

Perhaps the most likely mediator of religiosity and religious environments on adolescents' experiences is adolescents' acceptance of the family ideology or identification with the religious community. Adolescent religiosity tends to be similar to parental religiosity (Francis & Gibson, 1993; Gecas & Serf, 1990), and membership in a religious community appears to encourage the development of ties to the conventional order and the adoption of norms and behaviors congruent with social responsibility (D'Antonio, Newman, & Wright, 1982; Hirschi, 1969; Payne, Bergin, Beilema, & Jenkins, 1991). Such norms, explicitly encouraged within most mainstream religions, include self-control, avoidance of substance abuse, empathy, and service to others (Gunnoe, Hetherington, & Reiss, 1999). The greater the salience of social forces that foster adherence to religious prohibitions relating to healthy outcomes, the greater the likelihood of a correlation between personal religiosity and healthy outcomes for adolescents. Several important lines of research support such hypotheses. A number of studies indicate that religiously committed teens are less likely to become involved in drug and alcohol abuse (Bahr, Hawks, & Wang, 1993; Brownfield & Sorenson, 1991). Other studies indicate that religiously committed teens are less likely to engage in sexual behavior or to become pregnant when they are still teenagers (Beck, Cole, & Hammond, 1991; Holman & Harding, 1996; Miller & Olson, 1988). Young women with ties to established denominations are less likely than their nonreligious counterparts to become sexually involved during adolescence and prior to marriage (Forste & Heaton, 1988; Thornton & Camburn, 1989). Moreover, adolescent women involved in fundamentalist religions are substantially less likely to use contraceptives at first intercourse, even though their initial sexual experiences occur at a later age (Cooksey, Rindfuss, & Guilkey, 1996; Forste & Heaton, 1988; Kahn, Rindfuss, & Guilkey, 1990). Adolescents who engage in noncompulsory volunteer work also report positive outcomes, highlighting the significance of other relationships to religiosity. Adolescents who volunteer more report more positive attitudes about school and religion and experience more positive parental involvement in their lives (Forliti & Benson, 1986; Youniss, McLellan, & Yates, 1999). Last, religious students tend to avoid behaviors often regarded as undisciplined and harmful to academic achievement, suggesting a relationship between religious commitment and academic achievement (Brody, Stoneman, & Flor, 1996). These research findings all highlight the different roles families and religious communities play in fostering well-being and the way these institutions interact to prevent disruptive outcomes, as well as to heal and empower individuals in ways beneficial to them and to the broader society (see Maton & Wells, 1995).

The fact that a high level of religious commitment is correlated with other positive effects may lead one to suspect that a high level of religious commitment by adolescents would benefit them. For example, research has shown a correlation between religious commitment and physical health (McIntosh & Spilka, 1990). Religious commitment can enable people to handle stressful events more productively (Pargament, 1990; Seligman, 1991), develop social competencies (Thomas & Carver, 1990), and deal with traumatic loss, such as the loss of a loved one (Balk, 1983; Palmer & Noble, 1986). Likewise, a strong relationship exists between religious commitment and family stability (Filsinger & Wilson, 1984; Shrum, 1980). These findings all help support the conclusion that religion is linked to types of environments that predispose individuals toward particular ends.

Although this analysis again suggests that religious institutions and personal religious beliefs lead to positive outcomes, there still is cause for concern regarding how personal and social influences affect adolescents' responses to religion. Clearly, social and intrapsychic processes are not mutually exclusive, although religion and religiosity are very personal matters, they are in large part a product of social influence.

Implications of the Developmental Sciences

Although our review of the literature may suggest that we know much about the influences of religiosity on adolescents' development and dispositions, we must be careful in light of the important methodological limitations of the research. Although the studies vary in quality, at least four concerns arise repeatedly. The data are almost always based on self-reports, which may bias responses toward more socially approved responses. This may be particularly problematic for research that focuses on adolescents' self-reports of religiosity and its self-reported outcomes—the methodology that contributes the most data on religiosity's link to positive outcomes. Self-reports may be especially problematic in this area of research because the studies often deal with matters beyond conscious awareness. In addition, virtually all of the studies are correlational. With correlational studies, causal inferences are unsound; this problem is exacerbated by the fact that adolescence is a short time period, making it more likely that the research measures simultaneously the two variables under study (e.g., religiosity and mental health factors), eliminating the potential for finding causality and suggesting that actually measuring variables at different times may not ever yield results that meet high standards of rigor. Even if causal links were identified, it is important to recognize that statistical findings make broad generalizations that may not apply to particular individuals. Thus, finding statistically significant population-level protective effects of religion does not mean that religion benefits everyone's development, that religion benefits most people's development, that religion exerts harmful or null effects on development, or that we understand how and/or for whom religion produces certain outcomes. Likewise, even if we had more longitudinal research that explored predictive links, that research would not be without its own limits, given the typical use of "unique variance" approaches. Such approaches try to identify factors that are sufficiently unrelated that researchers can test how each will predict an outcome independently of other predictors. In reality, however, factors are related to each other; for example, most predictors of delinquency—such as lack of appropriate connections to families, schools, and community groups—are related. Last, the vast majority of studies also tend to use samples from unusual populations (e.g., one religious denomination or sect), which makes it unclear whether findings can be generalized. Despite the massive amount of research that exists, then, we find that not all of it provides useful information on adolescents' religiosity or on religious institutions and other factors that impact adolescents' religious experiences.

These limitations suggest that we need better research, but they do not mean that the existing research fails to contribute important and useful conclusions. First, negative findings and inconclusive findings are just as important as positive results: not finding a correlation leads to the appropriate conclusion that no causal relationship exists under examined conditions. Second, establishing a correlational link between religion and another variable means that the evidence is consistent with the possibility of causation, including the possibility that some other variable is causing both religiosity and the second outcomes or that these outcomes cause religiosity. Last, as with other areas of research, findings on religion must be carefully evaluated in a manner that dissects their complexities, rather than make sweeping conclusions. A more comprehensive and particularized review provides the basis for a broader, more differentiated set of conclusions.

Although no analysis has managed to address the different dimensions needed to understand religion and religiosity's impact on adolescent development, we have seen that several conclusions can be drawn from the existing research. Religion impacts adolescents through a multitude of relationships, including communal, cultural, peer, familial, and even intrapsychic. Precisely how these relationships impact adolescents still remains to be determined, but these relationships reveal three fundamental points.

First, efforts to draw lessons from existing social science evidence must distinguish different ways of being healthy and of interacting with others and must recognize the possibility that individuals may be healthy in some ways but not others. This translates into the need to examine religion's potential for effecting both positive and negative outcomes. For example, religiosity closely relates, in part, to risk taking, which itself seems to be part of a general disposition resulting in general behavioral practices and cognitive decision making. The finding is not surprising, given the current understanding that adolescents have either high or low risk-taking tendencies, which impact the manner they approach social interactions. As a result, being socialized to be religious may mean being socialized to avoid risks. We already have seen that religious adolescents tend to engage in less risk-taking behavior, for example, delinquent activity. It is important to note, however, that religious doctrine may not necessarily promote risk aversion: many religions encourage risk taking in the form of missionary work, proselytizing, or holding beliefs often at odds with secular or dominant social norms (see Miller & Hoffmann, 1995). Religions' multivalent roles offer much for consideration. The different outcomes of risk taking suggest much more than the conclusion that religions may help channel behaviors. The different outcomes also mean a need to consider the influence of religious-based social service delivery and advocacy and to consider how religious issues infiltrate adolescents' everyday interactions.

Analyses also must take into account the different ways of being religious. Religiosity is likely to impact adolescents' outcomes differently simply because different dimensions of religiousness relate to different outcomes; some people use religion for self-serving ends, while others have a more intrinsic motivation, viewing religion as part of one's quest for meaning. Differentiating the broad concept of religious involvement into different dimensions highlights links to different outcomes. The extrinsic dimension exhibits a rather pervasively negative relationship to many aspects of mental health and social dispositions, whereas intrinsic religious dispositions tend to be positively associated with many positive outcomes. Viewing religion as a quest characterized by a sense that one is seeking and questioning beliefs as part of an effort to commit to a belief system also is correlated with positive outcomes. Although the research may be seen as tentative, the general emerging conclusion suggests that the last two religious orientations provide the healthiest results, both personally and socially (see, e.g., Hunsberger, Pratt, & Pancer, 2001).

Analyses must also attempt to take into account the complexity of adolescents' lives. Adolescents have multiple identities, occupy many status positions, participate in a range of role relationships, and move from one context to another on a daily basis. Efforts to understand the role of religion in adolescent life necessarily must consider adolescents' vast array of relationships. Research has long indicated that it appears to be healthier to have access to, and to move between, a number of independent contexts and role relationships and to embrace separate self-identities (see Simmons & Blyth, 1987) so that failure or loss in one role can be buffered by participation in the other roles. The more positive the environments in which adolescents interact, the more positive the adolescents' life experiences and outcomes. The social environment in which adolescents find themselves plays a significant part in determining adolescents' outcomes.

We need to move beyond the belief that parents, peers, communities, or religions alone dictate adolescents' developmental experiences and outcomes. Like all complex social processes, social forces may eventually yield either positive or negative outcomes. Thus, given the multiple influences on adolescents, it makes sense to worry less about religious beliefs and practices per se and to worry more about the social context in which those practices occur. Although we can draw important conclusions from the literature on religious adolescents' dispositions, we should not glibly interpret the research to mean that religious adolescents are inevitably compassionate, committed moral beings who do not demonstrate the self-centered thinking characteristic of adolescents. Although reports of positive links between religiosity and emotional health are important in their own right, it remains important not to overstate the findings. A focus on beneficial outcomes without considering other social forces most likely will produce such links, run the risk of overinterpretation, and lead to erroneous assumptions about the place of God, faith, religion, and spirituality in adolescent development. In reality, different social forces foster different dispositions and exert different effects.

Conclusion

More important than documenting associations between religiosity and developmental outcomes is understanding how religion influences adolescents' experiences. Although understanding these mechanisms requires navigating considerable complexities and uncharted areas of inquiry, the existing evidence suggests important considerations and conclusions. The amount of empirical support for the impact of religion on adolescents is somewhat surprising, and the consistency of the findings is quite remarkable given the diversity of religions and the diversity of individuals' experiences within those religions, a diversity compounded by adolescents' cultural and individual differences. Despite the massive differences in individual experiences, it does seem that religion influences adolescents' general social orientations. Adolescents who strongly identify with religious teachings and traditions are less likely to engage in risk-related behaviors that result in disruptive outcomes; adolescents who fail to respond appropriately to their religious impulses appear to exhibit more negative outcomes. Adolescents who abide by conventional social rules and who live in environments that support adolescents and beneficial conventional social rules exhibit salutary influences.

Just as we now know more about the consequences attached to a failure to abide by religious prescriptions, we can draw important conclusions about the general manner in which religion impacts adolescents' experience. Religious beliefs and orientations may be personal concerns, but they are in large measure a product of social influences. Accounting for the links between religiosity and adolescents' experiences means essentially accounting for the nature of healthy socialization and the nature of environments that produce effective outcomes. Although some variation may exist, depending on social conditions, effective environments address adolescents' needs, provide them with a voice in matters that affect them, and offer hope. The extent to which religions foster positive environments partly determines religion's impact on adolescents; the extent to which religions in themselves constitute positive environments determines the extent to which particular religions, beliefs, and practices have the most positive impact on adolescents.

Although we have identified important social science findings, it requires a leap of faith to take them seriously. Although research can identify trends, it is certainly not the main consideration for policymakers. Policies, especially those that relate to adolescents and religion, must consider public opinion and existing legal mandates. Although policies must take into account numerous and often competing forces and values, our analysis of the ecology of religious influences on diverse outcomes provides the necessary background for an informed policy analysis that can lead to better policy alternatives in light of public and legal concerns. The next two chapters address these legal concerns, and our last chapter offers a way to integrate them with social science findings.

PART III

Regulating Adolescents' Religious Orientations and Environments

4

Shifts in the Regulation of Religion

We live in a land of religious freedom, but our legal system in reality highly regulates religion. The legal foundation of the United States, the Constitution, contains the First Amendment's "religious clauses," which affirm that "Congress shall make no law respecting an establishment of religion, or prohibiting the free exercise thereof. . . ." These two clauses delineate a spectrum of possible relations among the government, religious institutions, and individual religious believers (and, by implication, nonbelievers). In general, the Free Exercise Clause may require the government to exempt religions and religious believers from general laws, while the Establishment Clause may prevent the government from providing funding or special preferences to religious entities. These two poles delimit a gray area where government may acknowledge religion without being required to do so. Determining what governments properly do within these two poles remains quite challenging, controversial and even somewhat ephemeral.

The Constitution's statements on religion may appear self-explanatory, but how we are to remain faithful to them has seemed far from obvious, required judicial interpretation, and evolved to face new challenges. Indeed, the wide reach of the First Amendment's religion clauses emerged only quite recently, and it was only after the U.S. Supreme Court established its own expansive reach that the clauses were essentially revived from nearly one hundred fifty years of dormancy to contribute to much heated judicial, scholarly, and public debate about religion's role in public life. As suggested by the phrase "Congress shall make no law . . ." the clauses actually regulate only the federal government, and they were applied only to acts of Congress until the 1940s. Through case law and judicial interpretation of historical accounts, the Supreme Court made the Free Exercise Clause applicable to the states by "incorporation," an interpretive process by which the Court extended to the liberties protected from federal encroachment by the First Amendment similar protections from state encroachment through the Fourteenth Amendment's Due Process Clause. The case in question, *Cantwell v. Connecticut* (1940), involved a Connecticut law that prohibited door-to-door solicitation, a law declared unconstitutional because it violated the Free Exercise rights of Jehovah's Witnesses. It was seven years later that the Court similarly extended the reach of the Establishment Clause. In that case, *Everson v. Board of Education* (1947), the Court, while supporting a very high standard of church-state separation, held constitutional a New Jersey statute that reimbursed parents for the cost of busing their children to parochial school. From a constitutional and historical perspective, then, the mandate for states to follow the Constitution's provisions regulating religion appears quite momentous, even amazing, when we recognize that this interpretation of the religion clauses, catalyzed by the two cases, emerged largely as a phenomenon of the second half of the twentieth century, a period that itself witnessed, largely due to legal changes, dramatic transformations in the role of religion in public life.

The cases that extended the Constitution's reach to state and local governments have significance beyond the actual law they settled. The cases created an important breakthrough for understanding the legal regulation of religion. Given that most religion cases arise as a result of state rather than federal judicial action or legislation, the cases instrumentally fostered litigation to which the Supreme Court could respond in efforts to guide and help address religious issues confronted by all levels of government. Since incorporation, the federal courts have largely determined national, state, and local policy on the proper role of religion in public life. Such enormous federal power seems especially dramatic in view of the cogent arguments put forth asserting that the incorporation cases actually repealed the Constitution's provision that Congress may make no laws regulating religion—that the clauses regulating religion amounted to a decision by the federal government *not* to address questions concerning the proper relationship between religion and the government (see Smith, 1995). This approach would mean that there would be no national law, and thus no constitutional law, theory, or principle prescribing the proper relationship between government and religion and that the regulation of religion would rest solely with the states. Given the latter possibility, it is not surprising to find that considerable judicial and political work still continues the task of interpreting the Constitution's clauses as applicable to states and local governments (see Glendon & Yanes, 1991).

Litigation constantly challenges where and how the law can draw lines to separate the government, religion, and individuals, and the resulting jurisprudence draws and implements those lines with important opportunities to respond to changing conceptions of religion's role in public and private life. That legal institutions can respond to religion's role in society is evidenced by the Court's significant doctrinal shifts in its regulation of religion and by the federal and state legislatures' adaptation to those shifts. Given these shifts, analyzing how the law determines the role of religion in adolescents' lives necessarily must start with an understanding of how and why the legal system regulates religious institutions, practices, and even beliefs.

Free Exercise Jurisprudence

The Supreme Court's Free Exercise jurisprudence defines the limits on how much government regulation may intrude into citizens' religious practices. The Court must define that limit, given that the Court itself has interpreted the Constitution's Free Exercise Clause as providing religious believers with the right to exemptions from laws that infringe on their religious practices. The right to exemptions, by implication, raises the question of the extent to which the Constitution requires that citizens be protected from laws that regulate religious beliefs (important attempts have been made to distinguish beliefs from practices and the legitimate right of the state to regulate either or neither). To understand how the Court balances these concerns, one must understand how it has developed relevant jurisprudential standards and how it envisions further developments in the application of those standards.

The Court's first major free exercise case set the standard for nearly a century. That case, *Reynolds v. United States* (1878), involved the constitutionality of a federal law that prohibited bigamy by male Mormon settlers, who claimed that they had a religious duty to marry more than one woman. While the Court acknowledged that the Free Exercise Clause prevented the government from prosecuting citizens for their religious beliefs, it distinguished that from the enforcement of laws governing conduct and held that Mormons could not claim a constitutional exemption. The Court interpreted the Free *Exercise* Clause as, in actuality, a freedom of *belief* clause that merely protects religious opinion: "Congress was deprived of all legislative power over mere opinion, but was left free to reach actions which were in violation of social duties or subversive of good order" (*Reynolds v. United States*, 1878, p. 164). As a result, the Court found that the state had a legitimate right to apply general laws that could infringe on religious adherents' sincere practices. The Court's rigid deference to congressional will resulted, in this case, in particularly onerous intrusions, given that Mormon doctrine required male members of the Church to practice polygamy when circumstances permitted and that the penalty for failure to do so would be "damnation in the life to come" (*Reynolds v. United States*, 1878, p. 161). From the outset, then, the legal system arrogated to itself considerable power—it would judge how people practice their religions and could even require individuals to act against their faith.

Although *Reynolds* continues to be the subject of considerable commentary and controversy, for our purposes the case provides two fundamental lessons. From its first analysis of laws that impact religion, the Court assumed that the legal system possessed the authority to regulate religious conduct in the name of advancing social welfare. That authority meant that the Court had to defer to the state's regulatory power. Equally important, in the Court's first analysis of these laws, it took the opportunity to limit the freedom to follow the dictates of a religious practice. In fact, even though the Court had asserted that the Free Exercise Clause may require a religious exemption from a general law that regulates conduct when legislation conflicts with religious precepts, the Court never granted such an exemption until it shifted its standard of analysis in the 1960s. Until that time, the Court's reasoning in *Reynolds* helped decide a long line of cases that allowed the government to prohibit religious practices even when individuals had acted under the loftiest spiritual motives as required by their religious beliefs.

It was during the socially and politically tumultuous years of the 1960s that the Court, rather than simply relying on the government's interests, developed important standards to protect citizens making free exercise claims. Two cases involving observance of the Saturday Sabbath illustrate the major shift. In *Braunfeld v. Brown* (1961), the Court upheld Sunday closing laws that burdened a Jewish businessman's ability to compete, since he observed a Saturday Sabbath. The Court found no unconstitutional burden on the plaintiff's religious practice, since nothing in the plaintiff's faith required him to run a business and the legislature could not operate effectively if every law that increased costs to some religious adherents were unconstitutional.

The Court shifted its analysis in a case that followed two years later. That case, *Sherbert v. Verner* (1963), involved a Seventh-Day Adventist who lost her employment because she refused to work on her Sabbath day, Saturday. Her refusal to violate her religious beliefs meant that she did not qualify for state unemployment benefits after her discharge because she had "refused suitable work" (*Sherbert v. Verner*, 1963, p. 401). The Supreme Court departed from rulings in prior cases by applying strict scrutiny to the state law, a level of scrutiny that seeks to determine whether the challenged regulation burdened religious beliefs and whether the state had a compelling interest in enforcing a narrowly tailored regulation. As a result, the Court held that the state must justify the rule by showing not only a neutral governmental intent but also a compelling government interest. Under this rule, the Court discerned that the unemployment compensation law unconstitutionally burdened the plaintiff's religious practice and found no sufficiently compelling government objective that would justify that burden. In addition, the Court found that permitting the plaintiff to receive an exemption from unemployment compensation laws did not interfere with the operation of the state unemployment compensation system, because the state always used a case-by-case analysis to evaluate individuals' qualifications for benefits. Thus, the Court reasoned that the right was so compelling that it justified burdening the state, especially since the state could adapt appropriately and address the religious objector's claims without excessively burdening itself.

A decade later, the Court used this "strict scrutiny" test to establish the high-water mark for religious protection under the Free Exercise Clause. In *Wisconsin v. Yoder* (1972), the Court sided with Amish parents who refused to comply with the state's compulsory school attendance statute after their children had completed the eighth grade. In an effectively unanimous opinion, the Court recognized that enforcing the statute would "gravely endanger if not destroy the free exercise of respondents' [the Amish parents'] religious beliefs" (*Wisconsin v. Yoder*, 1972, p. 219). Moreover, the Court fully rejected the outmoded distinction between belief and action, stating that, at least in the case of the Amish, "belief and action cannot be neatly confined in logic-tight compartments" (*Wisconsin v. Yoder*, 1972, p. 220). Most important, the opinion demonstrated the *Sherbert* approach's potentially strong protective stance toward religious liberty. When subjected to "strict scrutiny," Wisconsin's interest in preparing students for participation in modern life could not stand against the success and agrarian self-sufficiency of the Amish people. Religious freedom would be secured against undeniably neutral legislation that served highly laudable goals. Quoting *Sherbert*, the Court in *Yoder* indicated that, to be "compelling," the interest a state regulation must address would have to involve "some substantial threat to public safety, peace, or order" (*Wisconsin v. Yoder*, 1972, p. 230). Given that the Amish traditions posed no such compelling threat to the general public, the state had to accept and provide an exemption.

While the strict-scrutiny standard introduced in *Sherbert* and affirmed in *Yoder* promised extensive protections for religious believers burdened by state and federal laws, the cases' significance was not as profound as it might appear. Although the robust "strict-scrutiny" test developed in *Sherbert* and *Yoder* greatly influenced lower-court decisions, the Supreme Court itself appeared markedly less hospitable to subsequent Free Exercise claims. Virtually all of its later decisions, while often invoking strict scrutiny, ruled in favor of government regulations and rejected exemptions to accommodate religious claims. In fact, religious adherents' claims for exemptions from generally applicable laws remain for the most part unsuccessful. Thus, in *Lyng v. Northwest Indian Cemetery Protective Association* (1988), the Court permitted the U.S. Forest Service to build a road through government land that was located on areas sacred to Native Americans; in *Goldman v. Weinberger* (1986), the Court upheld a military dress code that barred the plaintiff, an ordained rabbi, from wearing a yarmulke while in uniform; in *Bob Jones University v. United States* (1983), the Court denied an exemption from antidiscrimination laws to a religious university that claimed that its religious beliefs required racial discrimination; and in *United States v. Lee* (1982), the Court required an Amish employer to pay social security taxes, despite his religious objections. The newly applied strict-scrutiny standard, then, hardly resulted in a wholesale protection of religious rights and claims.

The Court's move away from protecting free exercise claims, despite its rigorous scrutiny of alleged infringements, became even more obvious in the context of unemployment, the very context in which *Sherbert* introduced the high-scrutiny standards. In this context, the Court's approaches again reveal dramatic shifts in thinking. At the end of the 1980s, the Court closely followed *Sherbert* in resolving disputes involving eligibility for unemployment compensation. Most notably, in *Frazee v. Illinois Department of Employment Security* (1989), the Court upheld the right to compensation of a worker who objected to working on Sunday, even though he did not belong to a particular Christian sect. However, a year later, even that context no longer benefited from the strict-scrutiny approach. The Court formally abandoned the strict-scrutiny standard in *Employment Division, Oregon Department of Human Resources v. Smith* (1990) and upheld the state's decision to withhold unemployment benefits from Native American plaintiffs terminated from their positions as drug rehabilitation counselors because they had made sacramental use of peyote in one of their religious ceremonies. While the Court acknowledged that, like the law in *Sherbert*, the Oregon law forced the plaintiffs to choose between upholding their religious beliefs and accepting state benefits, it concluded that "to make an individual's obligation to obey such a law contingent upon the law's coincidence with his religious beliefs . . . contradicts both constitutional tradition and common sense" (*Employment Division, Oregon Department of Human Resources v. Smith*, 1990, pp. 885–890). *Smith* explicitly limited *Sherbert* to the employment compensation context where "the State has in place a system of individualized exemptions" (*Employment Division, Oregon Department of Human Resources v. Smith*, 1990, p. 884). Equally important, the Court did not even require a compelling government interest before denying the free exercise claim. The Court explained that the only cases that justified free exercise exemptions from generally applicable laws were those where there was more than one constitutional right at issue (*Employment Division, Oregon Department of Human Resources v. Smith*, 1990, p. 881), such as a combination of free exercise claims and claims involving the right to free speech or freedom of the press. The Court noted, for example, that it would uphold exemptions for religious exercise combined with the right of parents to direct the upbringing of their children, as in *Wisconsin v. Yoder* (1972). Essentially, the Court held that, as long as religious exercise is free from deliberate political persecution, the Free Exercise Clause does not require the Court to provide heightened "strict scrutiny" to laws that burden the exercise of religion.

It was in the wake of the *Smith* decision that Congress swiftly acted to pass the Religious Freedom Restoration Act (RFRA) (1994). RFRA explicitly restored the strict-scrutiny test of *Sherbert*, which required narrowly tailored laws to achieve the government's compelling interest when the law "substantially" burdens a religious belief or practice. Congress argued that it had the power to pass RFRA because, under § 5 of the

Fourteenth Amendment, it was required to protect citizens' liberty interests in practicing religion unimpaired by government interference. However, in *City of Boerne v. Flores* (1997), the Supreme Court found RFRA unconstitutional on the grounds that it violated the separation of powers between the legislative and the judicial branches of government. In that case, zoning authorities denied Flores, the archbishop of San Antonio, a building permit to expand a church in Boerne, Texas. The archbishop challenged the permit denial, claiming that the authorities could not restrict the church building plans, even though the church was located in a historic preservation district. The Court, applying the *Smith* standards, decided that the archbishop could not obtain an exemption from this neutral law of general applicability by relying on the invalid test prescribed by RFRA. According to the Court, RFRA impermissibly violated the Establishment Clause, since it gave special benefits to religious believers that nonreligious citizens and organizations could not share. The Court explained that applying the *Sherbert* and the RFRA compelling-interest test would create "an anomaly in the law, a constitutional right to ignore neutral laws of general applicability" (*City of Boerne v. Flores*, 1997, p. 513).

Although the Court took a seemingly strong stance against efforts to elevate religious rights, religious rights still enjoy considerable protection. The likelihood of protection is particularly great when the laws that impact religion are neither neutral nor general in nature, as when a law specifically targets certain religious practices. The Court's approach in these contexts is highlighted well by *Church of the Lukumi Babalu Aye, Inc. v. City of Hialeah* (1993). In that case, the Court examined several city ordinances that effectively prohibited the ritualistic sacrificial killing of animals, a practice of the Santeria religion. The Court's majority first noted that, "[a]lthough the practice of animal sacrifice may seem abhorrent to some, 'religious beliefs need not be acceptable, logical, consistent, or comprehensible to others in order to merit First Amendment protection'" (*Church of the Lukumi Babalu Aye, Inc. v. City of Hialeah*, 1993, p. 531). It continued by agreeing with the Church's assertion that animal sacrifice is an integral part of its religion and that it cannot be deemed bizarre or incredible. Given the centrality of the prohibited sacrifices to the Church's religious beliefs, the Court proceeded to address the constitutional claim. In its analysis, the Court first returned to *Smith* for the proposition that "a law that is neutral and of general applicability need not be justified by a compelling governmental interest even if the law has the incidental effect of burdening a particular religious practice" (*Church *of the Lukumi Babalu Aye, Inc. v. City of Hialeah*, 1993, p. 531). The Court noted, however, the existence of an important interrelationship between "neutrality" and "general applicability" and added that failure to satisfy one suggests the failure to satisfy the other. As a result, it concluded that "[a] law failing to satisfy these requirements must be justified by a compelling government interest and must be narrowly tailored to advance that interest" (*Church of the Lukumi Babalu Aye, Inc. v. City of Hialeah*, 1993, pp. 531–532). Because the ordinances aimed to suppress the central element of the Santeria worship service, the Court found that the ordinances were neither neutral nor of general applicability. In its defense, the city advanced two interests it had sought to target with the ordinances: protection of the public health and prevention of cruelty to animals. The Court found the ordinances to be underinclusive on both counts because they failed to prohibit nonreligious conduct that was equally threatening to the public health or equally likely to cause cruelty to animals. Having determined that the laws were neither neutral nor of general application, the laws had to undergo the most rigorous of scrutiny under the compelling interest standard. The Court easily found the ordinances to be both overbroad and underinclusive; it concluded that "[t]he absence of narrow tailoring suffices to establish the invalidity of the ordinances" (*Church of the Lukumi Babalu Aye, Inc. v. City of Hialeah*, 1993, p. 546). The Court then noted that the city had failed to demonstrate that its governmental interests were compelling, finding that "[w]here government restricts only conduct protected by the First Amendment and fails to enact feasible measures to restrict other conduct producing substantial harm or alleged harm of the same sort, the interest given in justification of the restriction is not compelling" (*Church of the Lukumi Babalu Aye, Inc. v. City of Hialeah*, 1993, pp. 546–547). By addressing only Santeria sacrifices and allowing similar and similarly motivated killings and risks to public health, the city had inappropriately infringed on Santeria religious believers' free exercise rights. *Hialeah*, then, affirms the high protection religions receive when laws affecting them are neither neutral nor of general applicability; failing to satisfy these requirements, the law must be justified by a compelling governmental interest and must be narrowly tailored to advance that interest. Religions, then, can receive the Court's highest level of protection.

The Court's enormous power has meant that it routinely struggles to envision appropriate standards upon which to judge and guide governmental efforts that could limit individuals' practice of their religion. In fact, the modern Court has modified the separationist and protective stance it took in the 1960s and 1970s, and its most recent doctrinal move, coupled with highly split decisions (a high number of dissenting and concurring opinions urging different approaches to resolving issues but nevertheless achieving similar outcomes), indicates that the future path of the law may be unclear. Nevertheless, recent cases reveal that the Court does not take a strong stand on free exercise rights when they are pitted against governmental interference. The Court now articulates a fairly low governmental standard for infringement of free exercise rights. Laws that incidentally burden religious practice are permissible if they are neutral laws of general applicability. Despite this apparently permissive approach to regulating religion, the Court still imposes more demanding obligations on the state to justify its laws and to narrowly tailor its intrusion on religious practices if the laws also impact another important right or if they are not meant to be neutral in their application. In these circumstances, the Court will look favorably upon free exercise claims when governments can accommodate those claims without excessively burdening themselves. Despite these considerations, the right to practice one's faith remains not as strong as many would believe.

Establishment Jurisprudence

While the Free Exercise Clause determines the minimum accommodation that the government must provide for religious believers, the Establishment Clause sets a maximum amount of assistance that the government may offer. The express language of the Establishment Clause presumably prohibits a governmental body from advocating, proselytizing, or taxing citizens to promote a religion or an article of religious faith. This extreme view of the establishment prohibition becomes highly problematic in that practically any proposition, even the most uncontroversial, can be alleged to implicate religion if a church or sect has adopted it (or its opposite) as an article of faith. To prevent unintended and paralyzing results, the Court has subjected the provision to numerous refinements and explications that mirror the dramatic shifts in free exercise jurisprudence.

The Supreme Court took nearly 150 years before it availed itself of the opportunity to consider directly the nature and reach of the Establishment Clause. As we saw earlier, that case, *Everson v. Board of Education of Ewing* (1947), involved a law that reimbursed parents for their children's bus transportation to private, religious, and public schools. In its analysis, the Court invoked Thomas Jefferson's "wall of separation" between church and state as it expressed concern about government assistance to children who attended religious schools. Even as it emphasized that the wall "must be kept high and impregnable" (*Everson v. Board of Education of Ewing*, 1947, p. 18), the Court decided to allow the extension of these state subsidies to religious schools. The Court characterized the subsidies as standard state benefits that only indirectly touched religious schools. The cautious Court warned that citizens should not be denied these regular benefits because of an "overzealous" enforcement of the Establishment Clause (*Everson v. Board of Education of Ewing*, 1947, p. 18). Rather, the Court's ruling focused on the fact that the policy assisted in the transportation of children who attended any accredited school and did not directly fund or support religious schools. On the basis of these facts, the Court held that the policy was neutral toward religion and did not violate the Establishment Clause. As a result, the most far-reaching impact of the *Everson* decision was the Court's endorsement of government neutrality toward religion.

The Court's decision to endorse government neutrality was significant, but it did not signal the crumbling of Jefferson's wall. Most notably, the Court re-emphasized the strength and great height of the wall one year after *Everson*, in *Illinois ex rel. McCollum v. Board of Education* (1948). In this case, the Court had to evaluate a public school program that made use of privately paid religious instructors who conducted religious instruction in the Protestant, Catholic, and Jewish religions. The Court's chief concerns were that the school board was supporting religion by letting facilities that were built with tax funds be used for religious instruction and that, although only those students whose parents gave permission attended the religious classes, the state used its "compulsory public school machinery" to furnish the pool from which these students came (*Illinois ex rel. McCollum v. Board of Education*, 1948, p. 212). In considering whether the Establishment Clause permitted such instruction, the Court rested its analysis on the notion that the clause was founded on "the great American principle of eternal separation" (*Illinois ex rel. McCollum v. Board of Education*, 1948, p. 231). On the basis of its concerns and its view of the Establishment Clause, the Court ruled the policy unconstitutional and concluded that allowing clergy to use public school facilities for religious instruction during school hours moved too close to government support of religion.

Although the Court envisioned the separation between church and state as one of the vital safeguards of our constitutional system, the Court soon softened its separationist stance and moved toward a jurisprudence of accommodation. In yet another case involving the place of religion in public schools, *Zorach v. Clauson* (1952), the Court approved a New York program that allowed public school students whose parents had requested their early release to attend religious classes. Although the Court referenced *Everson* in announcing its departure from a strict separationist view, it went much further and accepted a state policy that revealed an accommodationist stance toward religion. In lowering the wall, the Court first noted that only the presence of a genuine tension between religious expression and government indoctrination requires that the separation "be complete and unequivocal" (*Zorach v. Clauson*, 1952, p. 312). In apparent response to the public criticism ignited by *McCollum*, the Court opined that complete separation of church and state would lead to alienation between the two systems, in turn breeding hostility (*Zorach v. Clauson*, 1952, p. 312). The Court found that the policy of accommodating student dismissals for religious education did not engender genuine tension between the free expression of religion and the bar against the government's establishment of religion because the classes were held outside the public school building and without administrative involvement. The Court was well aware that the release time necessarily involved state action that included the force of the state's laws on compulsory attendance and truancy. But the Court deemed the policy a reasonable accommodation that sought to lessen the level of government involvement. By playing down the potential use of public funds to enforce the truancy laws for those who might have violated released time attendance requirements, the Court's analysis allowed for a significant relaxation of a hard-line separationist orientation. Recognizing the need to allow for some state involvement opened the door to the less extreme accommodationist approach to Establishment Clause issues.

Early Establishment Clause jurisprudence, then, presented two approaches, one compelling strict separation between religion and the state and the other supporting accommodation of religious expression and practice. The Court left unclear, however, where permissible accommodation might end and separation might begin. The balance of accommodation and separation fluctuated for nearly three decades, until 1971, when the Court articulated what was then thought to be a clearer standard of analysis in *Lemon v. Kurtzman* (1971). It was in *Lemon* that the

Court synthesized multiple lines of Establishment Clause cases and established a complex standard that would become the source of the most significant jurisprudence involving the Establishment Clause.

In *Lemon*, the Court considered a program that allocated state funds to reimburse the expenses and supplement the salaries of private school teachers in Pennsylvania and in Rhode Island. The Court approached the issue by devising a three-part test in which it forbade any governmental action that (1) has no secular purpose, (2) has a primary effect of advancing religion, or (3) cultivates excessive entanglement between church and state. Though holding that the purpose of the statutes—to improve secular education—was sufficient to survive the first part of the test, the Court struck down both programs by applying the third part. The Court concluded that the programs created excessive entanglement because they presented the potential for administrative entanglement between state and religious institutions, which could in turn lead to political divisions along religious lines. This test, conventionally known as the "*Lemon* test," subsequently was considered the standard for Establishment Clause analyses that seek to erect a high wall between church and state—but it too was scrutinized, refined, and eventually rejected by some members of the Supreme Court.

It was only two years after the pronouncement of the *Lemon* standard that the Court reached its high-water mark in defining how it would limit the government's role in actively supporting religion. The Court applied its most stringent application of the Establishment Clause in *Committee for Public Education and Religious Liberty v. Nyquist* (1973). *Nyquist* involved a tuition reimbursement program in New York that authorized various direct payments to private schools and parents of private-school students, including payments to schools for the purpose of maintaining and repairing equipment and facilities to protect students' health, welfare and safety; reimbursements to low-income parents to help them defray the cost of their children's tuition at private schools; and tax relief to parents who did not qualify for the reimbursements. The Court first found that, because the payments were directly made to private schools (most of which were affiliated with the Roman Catholic Church) without restrictions on usage, the payments had the primary effect of advancing religion. The Court used similar reasoning to invalidate the reimbursement scheme, noting that New York could not guarantee that the funds would be used for secular and neutral purposes. A key element of the Court's analysis was the justices' assertion that, while the payment scheme was an important factor in determining the constitutionality of the scheme, it was not the only factor. Instead, the Court focused on the overall effect of the program, noting that the tuition grants were being offered as incentives to encourage parents to send their children to religious schools by facilitating their ability to afford such schools. By adopting that broader approach, the Court held that the Constitution had been violated, regardless of whether state aid was used for religious purposes or whether parents were free to spend the money as they wished. The Court also invalidated the third part of the program, noting that the program in effect created the same incentive as the reimbursements for sending a child to a sectarian school. Thus, *Nyquist* supported the view that the Court decisively had foreclosed any state-funded voucher program from surviving constitutional scrutiny if the program provided incentives for parents to enroll their children in schools that furthered a religious mission.

In cases that followed *Nyquist*, the Court showed that it had left itself room to accommodate state involvement in religious matters. Indeed, the Court's strict enforcement of the Establishment Clause as pronounced in *Nyquist* diminished noticeably only a decade later. The Court embarked on this landmark jurisprudential shift in *Mueller v. Allen* (1983). The Minnesota law at issue in *Mueller* granted tax breaks to parents for the costs of tuition, textbooks, and transportation, regardless of whether their children were enrolled in public, private, or sectarian schools. The Court distinguished the Minnesota tax relief scheme from the one rejected in *Nyquist* by noting that the Minnesota scheme provided the tax benefit to all parents, regardless of whether their children were in private schools, thus making public funds available to schools on the basis of the private choices of parents. The Court readily admitted that the program provided the same financial benefit to sectarian schools as would direct aid, but the Court asserted that, because the parents were making private choices as to where to apply the aid, the state was not establishing or approving of religion. Thus, and although essentially allowing for the result it had previously forbade, the Court implicitly declined to overrule *Nyquist* as it concluded that the Minnesota scheme did not have the effect of advancing religion.

Mueller stands for much more than a shift in the Court's position. It is the leading source for what the Court now considers when deciding whether a state action constitutes impermissible establishment. The majority in *Mueller* actually made two critical points. First, it called the

Lemon test into doubt for the first time, stating that, while Establishment Clause questions have been guided by the three-part test first announced in *Lemon*, a close look at *Lemon* and its resulting jurisprudence reveals that it merely was offered as a "helpful signpost in dealing with Establishment Clause challenges" (*Mueller v. Allen*, 1983, p. 394). Second, the Court refused to engage in an empirical analysis of the challenged program, which would have shown that parents were using the vast majority of tax benefits to send their children to parochial, not secular, schools. Thus, the Court appeared to abandon the proscription against grants or tax benefits that created incentives for parents to send their children to sectarian schools by making it financially easier for parents to enroll their children in such schools. The Court achieved this result by ignoring evidence that demonstrated a strong incentive on the part of parents in Minnesota to do just that. In the wake of *Mueller*, the Court would begin to apply the *Lemon* test as a mere guiding tool.

The significant shift *Mueller* portended became reality in the Supreme Court's next major Establishment Clause case, *Witters v. Washington Department of Services for the Blind* (1986). In *Witters*, the Court upheld a Washington state program that permitted the payment of an educational stipend for a blind student who was training for the clergy at a religious school. The Court found the program constitutional since the program paid vocational educational grants to enable blind students to attend public, private secular, and private sectarian schools. The Court embraced the incentive standard it essentially had abandoned ealier, determining that the program did not provide a financial incentive for students to choose sectarian educational institutions and that it did not bestow additional benefits on students who chose to attend sectarian institutions. Further, the Court stressed that the same ingredient present in *Mueller* was also present in *Witters*, that the aid was disbursed to sectarian institutions only as a result of "genuine and private choices of aid recipients" (*Witters v. Washington Department of Services for the Blind*, 1986, p. 487). Regarding the *Lemon* test's concern with benefiting particular religions, the Court emphasized that the aid would be used not to promote religion but instead to aid disabled students. This iteration of the *Lemon* test was far less rigorous than that applied in *Nyquist*, or even in *Lemon* itself.

The Court further eroded the *Lemon* test in *Zobrest v. Catalina Foothills School District* (1993). In *Zobrest*, the justices upheld the use of state funds to supply sign-language interpreters to students in sectarian schools. The Court essentially ignored the *Lemon* test and relied more heavily on *Witters* and *Mueller*. The Court revived the neutrality standard, noting that because the program neutrally provided a benefit to "a broad class of citizens without reference to religion," it was not readily subject to an Establishment Clause challenge (*Zobrest v. Catalina Foothills School District*, 1993, p. 9). The Court also reaffirmed the constitutionality of aid that states disburse to sectarian schools through "genuinely independent and private choices" (*Zobrest v. Catalina Foothills School District*, 1993, p. 9). The justices further held that, since the program included both public and private schools, including sectarian schools, it provided no financial incentive for parents to send their children to sectarian schools. Despite these findings, the Court still formally clung to the *Lemon* test, refusing to overrule it outright.

The erosion of the *Lemon* test continued with the most recent case involving state aid to private sectarian schools, *Agostini v. Felton* (1997). In *Agostini*, the Court upheld the provision at government expense, under the Individuals with Disabilities Education Act (IDEA) (1994, 2000), of a sign language interpreter for a deaf student at a parochial school. The Court approved of the program's disbursement of funding to disadvantaged children via the compensation of public teachers who taught disadvantaged children in sectarian schools. In *Agostini*, the Court declared that an injunction against the program, originally upheld by the Supreme Court, had in fact been upheld by an analysis that relied on no longer valid law. The case that was no longer valid law, *Aguilar v. Felton* (1985), had in turn been based on a strict interpretation and use of the *Lemon* test. In *Aguilar*, the Court had applied *Lemon* and had found that the Title I program created excessive government entanglement because public employees would have to be closely monitored by the state in order to guarantee that they did not teach or approve religion. Overturning *Aguilar* is the closest the Court has come to directly overruling the *Lemon* test.

In *Agostini*, the Court focused its analysis, relying on *Witters* and *Nyquist*, on the permissibility of different types of direct governmental aid. The Court compared the challenged program to the use of a government employee's paycheck, a situation in which the employee could sign over part or all of the paycheck to a religious cause or group but "only as a result of the genuinely independent and private choices of individuals" (*Agostini v. Felton*, 1997, p. 226). In overruling *Aguilar*, the Court asserted that Establishment Clause jurisprudence had undergone a change, citing intervening cases that had found no constitutional bar to government-provided aid to handicapped students who attended parochial schools (*Witters v. Washington State Department of Services for the Blind*, 1986; *Zobrest v. Catalina Foothills School District*, 1993). The Court found that the aid in question was not in the form of direct grants; that an empirical analysis of the numbers or percentages of religious school students receiving neutral aid was not relevant; and that the program did not create any financial incentive for parents to send their children to sectarian schools, because funds were disbursed to sectarian and secular school pupils on a neutral basis. The Court further noted that these cases had changed the law by creating a presumption of regularity in the actions of public employees operating in religious settings. This precedent permitted a presumption in *Agostini* that the teachers who administered Title I's services in the parochial setting would not improperly advance religion even if the situation could not be monitored carefully. The entanglement-prong violation of *Lemon* in *Aguilar* thereby disappeared in a presumption.

Mitchell v. Helms (2000) involved a federal program that was a companion to the remedial instruction program reinstated in *Agostini*; the program challenged in *Mitchell*, long known as Chapter 2 of the Education Consolidation and Improvement Act of 1981, distributed federal money to state and local educational agencies, which used the money to purchase educational equipment for loan to private elementary and secondary schools. The statute limited the use of this "equipment," which includes computers, software, library books, VCRs, films, tapes, and other audiovisual material, to programs that are "secular, neutral, and non-ideological" (*Mitchell v. Helms*, 2000, p. 2537). The program included restrictions on the diversion of the loaned equipment for religious use, although these restrictions were not carefully enforced by public authorities. The litigation in *Mitchell* involved a challenge to the program as applied in Jefferson Parish, Louisiana, in which Catholic schools were among the beneficiaries. In a highly contentious opinion, a plurality approached the issue of aid to sectarian schools by establishing two criteria: (1) the aid program must be neutral in aiding sectarian schools and other schools, both private and public, and (2) the government itself must not be engaged in religious indoctrination. By these criteria, the Chapter 2 program easily passed muster. The aid went to a broad array of schools. Moreover, because the aid formula turned on per capita allocation, the plurality found the program neutral in yet another way—the school received government financial aid in proportion to the number of students it attracted, and the program therefore created no incentives to choose sectarian education. Finally, the plurality of the Court rejected the argument that the possibility that the aid would be diverted and used for religious instruction created an Establishment Clause obstacle; any such diversion, though it might violate the governing statute or regulations, would not be attributable to the government and therefore would not violate the Establishment Clause. In its reasoning, the Court squarely rejected the notion that aid to "pervasively sectarian" schools should be treated under rules different from those that govern aid to other schools (*Mitchell v. Helms*, 2000, pp. 2550–2552). Labeling such a distinction offensive, troubling, and anti-Catholic in its "pedigree," the plurality concluded that a doctrine "requir[ing] the exclusion of pervasively sectarian schools from otherwise permissible aid programs . . . [is] born of bigotry [and] should be buried now" (*Mitchell v. Helms*, 2000, p. 2552). The concurring opinion, expressing far more caution than the plurality, still pushed significantly beyond the prior law. *Agostini*, the 1997 decision that reinstated the remedial instruction program, had rested in large part on the finding that the instructors were public employees. Under Chapter 2, that feature was absent; sectarian school administrators and teachers were entrusted with the computers, software, books, and media materials under the general instruction that they be used only for secular and nonideological purposes. That entrustment would have been a fatal policy move in the constitutional jurisprudence of the mid-1970s. But the concurring justices read *Agostini* for the "proposition that . . . presumptions [that sectarian school teachers always and everywhere engage in] religious indoctrination are normally inappropriate when evaluating neutral school-aid programs under the Establishment Clause" (*Mitchell v. Helms*, 2000, p. 2567). Henceforth, according to their opinion, "[t]o establish a First Amendment violation, plaintiffs must prove that the aid in question actually is, or has been, used for religious purposes" (*Mitchell v. Helms*, 2000, p. 2567). Moreover, without joining in the plurality's explicit condemnation of the concept of pervasive sectarianism as the product of anti-Catholic bigotry, the concurring justices joined in essentially undoing the presumption attached to that concept. On the record in *Mitchell*, they concluded that the safeguards against religious uses of publicly financed materials were sufficient and that evidence that the safeguards were being violated could be safely ignored given that it was so sparse and not fully exhibited (*Mitchell v. Helms*, 2000, pp. 2569–2571). *Mitchell* unquestionably broke new ground on crucial issues of money separationism. The case provided the Court with the opportunity to overrule prior law, repudiate crucial principles for blocking aid, and reverse the presumption that sectarian schools cannot be trusted to use public aid for nonsectarian purposes.

Mitchell may reveal the expected and necessary outcomes of evolving jurisprudence on entanglement issues, but Establishment Clause jurisprudence necessarily involves other important concerns as well. Indeed, to the extent that entanglement considerations become less important in determining whether a governmental action impermissibly establishes religion, questions about any challenged statute's secular purpose increase in significance. The Supreme Court has revealed the potential significance of this shift in its response to relentless attempts to display religious texts and to prohibit the teaching of evolution in public schools. In *Stone v. Graham* (1980), for example, the Court rejected the avowedly secular purpose of Kentucky legislation that required the posting of the Ten Commandments in the back of every public school classroom. In an effort to ensure the legislation's constitutionality by furthering a valid secular legislative purpose, the statute had mandated that the bottom of the display state, in small print, that the "secular application of the Ten Commandments is clearly seen in its adoption as the fundamental legal code of Western Civilization and the Common Law of the United States" (*Stone v. Graham*, 1980, p. 41). The Court, however, read the statute's purpose as "plainly religious in nature" because the Ten Commandments are "undeniably a sacred text in the Jewish and Christian faiths" (*Stone v. Graham*, 1980, p. 41). *Stone* does not stand alone in its rejection of stealthy efforts to impose religious views and materials on others. The Court also imputed illegitimate purposes to a legislative act on the basis of its predictable results with regard to efforts to prohibit the teaching of evolution (*Wallace v. Jaffree*, 1985) and to governmental requirement that schools teach evolution only in tandem with creationism (*Edwards v. Aguillard*, 1987). In both cases, the Court found no likely legislative purpose other than to promote religion. As with some previous efforts, the legislative statements challenged in both *Wallace* and *Edwards* claimed that the statutes at issue promoted greater comprehensiveness or promoted academic freedom; the Court nevertheless rejected these secular assertions as mere pretense and easily found language in the legislative history and in the statutes themselves to support the conclusion that the laws sought to encourage religion in the schools. In *Edwards*, for example, Louisiana had enunciated a secular purpose, protection of academic freedom, but the Court reasoned that the statute would effect the opposite. By not encouraging the teaching of all scientific theories about the origins of humankind, the statute, which required a parity between evolution and creationism (i.e., teaching both or teaching neither) had the "distinctly different purpose of discrediting evolution by 'counterbalancing its teaching at every turn with the teaching of creationism'" (*Edwards v. Aguillard*, 1987, p. 589). Clearly, the Court has not given up its role as arbiter of legislative intent.

A look at cases that challenge legislative purposes, however, makes it clear that the teaching of evolution in the public schools is the major exception to the rule that challenges to state regulations that affect religious establishment can easily win their claims. Even in *Stone*, for example, the Court's analysis indicated that some religious displays might pass constitutional muster if the state could mitigate the religious nature of the display; for example, the Constitution permits the integration of the Bible into the school curriculum for the study of history, civilization, ethics, and comparative religion. In fact, the Court has mitigated its approach to government display of religious symbols, and later decisions suggest that the Court is not as likely to question the legitimacy of a stated government reason and to overturn a statute on these grounds alone (see *Santa Fe Independent School District v. Doe*, 2000). As a result, this aspect of constitutional protection typically has not been the most difficult hurdle in Establishment Clause cases. Courts typically allow a generous, deferential reading of the intent of legislative acts.

As issues involving the legislative purposes of statutes lose (or at least change in) significance, so do yet another set of concerns identified in the *Lemon* test: the extent to which the governmental action has a "principal or primary effect . . . that neither advances nor inhibits religion" (*Lemon v. Kurtzman*, 1971, pp. 612–613). Most notably, where the Court has inferred a legislative motive on the basis of predictable effects, the practical analysis and the eventual resolution of this concern closely mimic those involved in the second prong of the *Lemon* test. That these two considerations similarly can duplicate the same decisional factors has prompted some members of the Court to bundle them into an "endorsement" test. In fact, and as we soon will see, in recent cases the Court has by implication abandoned the *Lemon* test and has in its stead adopted the endorsement test to determine the validity of enactments under the Establishment Clause.

The endorsement test invalidates both an intended governmental endorsement of religion and an unintended but reasonably perceived endorsement. The Court uses a subjective standard to assess unconstitutional intentions to endorse. These determinations require the justices to discern and to judge whether the actual intent of the governmental body was to advance one religion over another or religion over nonreligion. An objective test determines the perceived endorsement; it seeks to adopt the point of view of a putative "reasonable observer," an approach similar to the legal fiction of the "reasonable man" particularly common in the area of tort adjudication. The determination turns on the reasonable impression given to a hypothetical reasonable member of the community of possible observers that the Court deems to be the relevant perceivers. This test finds an Establishment Clause violation when government endorses religion, which occurs whenever a reasonable observer would conclude that official activity sends a message to nonadherents that they are outsiders, not full members of the political community, and an accompanying message to adherents that they are insiders, favored members of the political community (*Wallace v. Jaffree*, 1985, pp. 67–79).

Lynch v. Donnelly (1984) best illustrates the way members of the Court apply the endorsement test, which, in effect, combines the first two prongs of the ostensibly abandoned *Lemon* test. *Lynch* involved a challenge to a city's effort to erect a Christmas creche on public property. In *Lynch*, the Court concluded that the relevant observers of the creche were the adult holiday shoppers (significantly, not their children). The reasonable shoppers were determined by the Court to be so inured to commercial settings that the creche (the governmental expression at issue) ought not to strike them as an endorsement of Christianity, regardless of whether testimony in the case revealed that it reasonably did strike an actual observer, certainly the plaintiff, in that fashion (if the impression of the plaintiff had not been reasonable, one expects that the case would have been summarily dismissed). As explained in *Lynch*,

> Government celebration of the holiday, which is extremely common, generally is not understood to endorse the religious content of the holiday, just as government celebration of Thanksgiving is not so understood. The creche is a traditional symbol of the holiday that is very commonly displayed along with purely secular symbols. (*Lynch v. Donnelly*, 1984, p. 692)

The analysis discounts (a) the objective effect that a governmental action in fact advanced or inhibited religion and (b) the subjective perception that it did. These limitations are not explicit in the language of *Lemon* or *Abington School District v. Schemmp* (1963), the cases the Court cited as precedent for the endorsement test. As formulated and applied, the endorsement test appears to embrace a narrower prohibition than prior law, although the Court's discounting of incidental objective effects accords with the spirit and outcomes of earlier decisional law and with legal commentary that tended to reject mere effect-in-fact alone to justify invalidation of a statute.

Efforts to evaluate the impact of recent jurisprudence must proceed quite carefully, but the emerging trend seems well established and its themes reverberate from prior cases. The need for caution arises from the fact that the most recent developments in Establishment Clause jurisprudence generally concerned financial aid to religious schools, and such schools may present a unique context under the Establishment Clause. The recent jurisprudence therefore need not apply in other cases that challenge governmental actions as inconsistent with the Establishment Clause. The developments, though, do appear to reflect tendencies in other jurisprudential contexts. Most notably, the trend toward allowing state funds to be used, albeit indirectly, by religious organizations parallels the use of religious organizations in the provision of social services. That jurisprudence essentially reaches back to 1899, when the Supreme Court first considered the constitutionality of government construction funding for a hospital affiliated with the Catholic Church. In *Bradfield v. Roberts* (1899), the Court found that the hospital itself was not a religious institution, which rendered the Establishment Clause immaterial to the case. Given that almost a century has passed since the Court ruled in this case, government funding of religious entities in and of itself evidently does not violate the Establishment Clause as long as the funds can flow concurrently to both secular and religious providers. In fact, most religiously sponsored nonprofit social service organizations receive funding from the government. In its most recent related case, which involves the provision of social services in an educational setting, the Supreme Court upheld federal grants to religious organizations to provide abstinence education and counseling services to school-age teens and to support research related to adolescent behavior. In that case, *Bowen v. Kendrick* (1988), the Court upheld the Adolescent Family Life Act (1981, 1988). This federal law provides grants to public and nonprofit organizations for services and research in the area of premarital adolescent sexual relations and pregnancy. Among the contested activities funded by the act were counseling and educational services that could be delivered only if they did not promote abortions. The Court reasoned that, although some grantees had institutional ties to religious organizations, on its face, the law did not advance or excessively entangle secular and religious organizations and that the act served a secular purpose. The Court remanded the case to ascertain whether any specific grantees were "pervasively sectarian," which might have had the impermissible effect of advancing religion (*Bowen v. Kendrick*, 1988, p. 621). The funding statute mentioned both religious and nonreligious organizations as potential grant recipients, and the Court found that the law did not direct funding preferentially to religious organizations and thus satisfied the jurisprudential standards used to interpret the Establishment Clause.

Establishment Clause jurisprudence, then, contains several important points. First, the Court permits government actions that entangle religious interests. Second, cases that have been decided tend to be limited to specific contexts, especially to the public schools, so that jurisprudential developments do not necessarily transfer from one context to another. Third, jurisprudence generally permits the neutral provision of public funds, and it especially recognizes the need not to discriminate against religious organizations that offer secular services. The Court allows government benefits to flow to religious institutions that provide secular services; even when the services are not entirely secular, the benefits may still reach sectarian institutions. Fourth, jurisprudence focuses on individual beneficiaries, the Court is more likely to permit disbursement of funds that reach religious institutions when the provisions can be framed as allowing individuals to use public funds for private ends, including supporting services offered by sectarian institutions. This follows from the strong presumption that the provision of governmental benefits to both secular and sectarian institutions is permissible as long as citizens may make their own private choices. This also follows from the strong presumption that religious convictions must be private in their provenance and that religious activity must be the product of the private judgments, convictions, and devotions of individuals and groups. Last, the Court generally lacks interest in empirical investigations to determine the extent of governmental establishment. The Court seems to presume regularity in the action of publicly funded employees in religious settings and emphasizes that law must be decided first and foremost on constitutional, rather than political or statistical, grounds.

Evolving Legislative Mandates

Given that the historical approach to resolving church-state issues has rested on a separationist approach, it is not surprising that legislative mandates, especially federal mandates, pervasively have excluded religious concerns. When religious concerns have been addressed, the legal system has proceeded in a stealthy—but not necessarily pernicious—manner. Religious issues always have been present in the legal process, for example, through the provision of social services supported by government funds (Hamilton, 2001) and the advocacy of religious groups in legislative and judicial branches (Sullivan, 2001). Although religious institutions continue to embrace these efforts, the new approach to religion and religious issues is increasingly variable and visible. Rather than maintaining a wall between religion and the government, the legal system, in recent federal and state legislative mandates, increasingly has become faith-friendly.

The federal government's latest and most dramatic welfare reform overhaul measure, the Personal Responsibility and Work Opportunities Reconciliation Act of 1996 (2001), heralds much more than a shift in the government's distribution of welfare funds. The Act contains Section 104, the so-called Charitable Choice provision, which exemplifies recent efforts to reshape the role of religiosity and religious institutions in public life. The Act encourages states to utilize "faith-based organizations in serving the poor and needy," requires that religious organizations be permitted to receive contracts, vouchers, and other government funding on the same basis as any other nongovernmental provider, and "protects the religious integrity and character of faith-based organizations that are willing to accept government funds." As enacted, Charitable Choice covers each of the major federal antipoverty and social welfare programs.

The apparent success and popularity of faith-friendly measures has spurred congressional efforts to expand their scope to other social service programs and other federal policy domains. In 1999, for example, both the Senate and the House considered numerous Charitable Choice provisions that nearly succeeded in becoming law. Although none of the measures passed that year, they subsequently were reintroduced, passed, and sent to conference to work out differences between the Senate and the House versions. The Charitable Choice Expansion Act of 1999 (1999) sought to provide an umbrella provision that requires consideration of faith-based providers and prohibits discrimination against religious organizations in all programs in which the local, state, or federal branches of government fund private organizations to deliver social services. In addition, The Fathers Count Act of 1999 (1999), offered by the House, sought to permit Charitable Choice in programs to promote responsible fatherhood. In the Violent and Repeat Juvenile Offender Accountability and Rehabilitation Act of 1999 (1999), the Senate sought to permit Charitable Choice in social services related to juvenile justice. In the Youth Drug and Mental Health Services Act (1999), the Senate also sought to expand Charitable Choice in substance abuse and mental health treatment, prohibiting discrimination against religiously based substance abuse programs and allowing such organizations to accept funding without "impairing the religious character of the organizations or the religious freedom of the individuals." Similarly, in the Consequences for Juvenile Offenders Act of 1999 (1999), the House sought to permit Charitable Choice in juvenile justice services. Most of these provisions have now successfully passed the Senate, although some are being held in conference to resolve disputes between the House and the Senate over matters unrelated to Charitable Choice (see McClain, 2001; Kole, 2001; Freedman, 2001). Faith-friendly efforts, then, clearly have become de rigueur.

The new provisions signal a fundamental shift in the relationship between government and religious groups. Although nonprofit, sectarian institutions always have played a dominant role in social service provision, especially to the poor, the new statutes propose a fundamental change in how they are required to function. Traditionally, sectarian institutions that received governmental support essentially were required to act like secular institutions (see White, 1999; Bassett, 2001). The new statutes, in contrast, allow religious groups to receive government monies for social programs without requiring them to censor their religious identity. Faith-based groups that provide state-funded services must refrain only from compelling participants to get involved in religious practices. These mandates leave many issues unresolved, most notably the extent to which individual providers may practice their religions when providing services, such as praying before serving a meal to the homeless, refusing to provide reproductive health counseling to adolescents, and refusing to terminate the life of patients despite advance directives indicating otherwise (see White, 1999; Bassett, 2001). In addition, it remains to be seen whether faith-based services could circumvent significant protections against employment discrimination, the strongest of which come through Title VII of the Civil Rights Act of 1964 (1999). Specifically, the Act prohibits certain employers from discriminating in terms of hiring or firing on the basis of "race, color, religion, sex, or national origin" (Civil Rights Act of 1964, 1999, § 2000e-2). Title VII guarantees, however, are not universal across employment sectors; employment discrimination protections do not apply to any "religious corporation, association, educational institution, or society with respect to the employment of individuals of a particular religion to perform work connected with carrying on by such corporation, association, educational institution, or society of its activities" (Civil Rights Act of 1964, 1999, § 2000e-(1)a). The religious exemption clearly allows only discrimination on the basis of religion, but the resulting impact may be quite broad, as exemplified by a slew of controversial cases permitting religious organizations to discriminate against employees based on their sexual orientation (see *Hall v. Baptist Memorial Health Care Corporation*, 2000) and sexual behavior deemed immoral (such as allowing the dismissal of a single, pregnant parochial school teacher on the grounds that her behavior violated the religious and moral precepts against premarital sex, see *Cline v. Catholic Diocese of Toledo*, 2000). We have yet to understand the full impact of this potentially broad exemption that allows faith-based services to be administered by groups that seek to uphold and practice views that discriminate against others. Although many issues may remain to be settled, legislation already has settled and clearly embraces use of religious groups to supplement (and supplant) governmental efforts to distribute and transform existing social service delivery.

Although initially meant to increase access to traditional social services for those in poverty, the shifting orientation strikingly extends to other government-supported services. Most notably, the change helps to account for an increase in the number of students opting to attend private, sectarian schools at taxpayers' expense. It also provides a rationale for school reform that offers parents government resources, in the form of tax incentives and vouchers, to send their children to schools otherwise funded by religious institutions (Todd, 2000). These are extraordinary developments. Rather than simply allowing religious institutions to exist and to supplement government services, the new mandates actively support parts of their religious missions. The new mandates even allow religious institutions to supplant government-provided services, since legislatures and the public perceive some services provided by private, religious institutions as more effective than those provided through strictly secular, governmental efforts. Thus, for example, early critics of voucher programs argued (and lost the argument) that vouchers may compel students either to attend sectarian schools or to forfeit their chance for a better education, since vouchers would leave public schools underfunded and increasingly deficient (Weishaar, 1994). How and whether religious groups will be able to expand their social service delivery by taking advantage of federal and state monies remains to be seen. It appears indisputable, however, that the general trend toward privatizing government services increasingly views religious institutions as viable alternatives to be treated like other private or non-for-profit service organizations.

Critical Implications of Shifting Mandates

A succinct review of complex jurisprudence necessarily glosses over important controversies. The danger of playing down controversies seems especially worth emphasizing in the contexts of laws that regulate religion. Members of the Supreme Court themselves and commentators often report feeling frustrated by what many of them view as inherently contradictory principles on the subject, a frustration that has led some to accept the impossibility of devising a coherent theory of the government's regulation of religion and to pronounce efforts to formulate such theories "foreordained failures" (Smith, 1995, p. 45). In our analysis, though, we avoid much of the controversy and replace it with a consideration of the significant implications that arise from existing trends.

The first critical implication of the current regulation of religion involves the focus on protection from state coercion and state promotion of religious beliefs and practices. For example, a key justification for the movement from separation to religious accommodation in school cases is the legitimacy of providing government funds in a way that allows individuals to make private choices (in contrast to following government mandates) that result in the promotion of religious views and/or benefits to religious institutions. Also included under the "private actor" rationale is judicial receptivity toward incidental government aid that flows to religious schools because of individual parents' private decisions about where and how to educate their children. The Supreme Court has distinguished aid that takes the form of an impermissible "direct subsidy" from aid that represents a permissible "transfer," much as a state employee may donate part of his or her salary to a church. The revitalized child benefit doctrine, coupled with the view that aid can legitimately flow to religious entities as a result of personal choices, ensures greater government assistance to religious schools and increased governmental support for religious expression. It also means that current laws must focus on eliminating coercion from governmental actions. This focus will become increasingly significant, as we see in the next chapter, because it will allow the Supreme Court to support the constitutionality of private decisions to allow expressions of religious ideology in public schools on the rationale that such expression does not come from the government. The Court now interprets many instances of religious activity in public settings as "private" and therefore permissible expressions.

The second key implication involves the failure to focus on active support from the state in a nondiscriminatory manner. The Supreme Court has recognized that the government cannot condition benefits on the relinquishment of constitutional rights (*Rust v. Sullivan*, 1991), but it has not gone so far as to hold that the government must subsidize the exercise of such rights. If it did, then parents could assert not only a right to select private education for their children but also an entitlement to the same services for those children that the government provides for public school students. Requiring evenhanded treatment of sectarian and secular enterprises would seem to be the next logical step. The nondiscrimination and private-actor rationales support the claim that, in allocating benefits, the government must treat religious organizations the same as it treats secular entities. Although the claim may seem rather unreasonable or even absurd, the Supreme Court itself offered the possibility. The Court's 1995 *Rosenberger* decision, for example, seemed to support such nondiscrimination. As the Court noted in its plurality: "We have held that the guarantee of neutrality is respected, not offended, when government, following neutral criteria and evenhanded policies, extends benefits to recipients whose ideologies and viewpoints, including religious ones, are broad and diverse" (*Rosenberger v. Rector & Visitors of the University of Virginia*, 1995, p. 839). Expanding the nondiscrimination principle to protect all religious claims against differential treatment, perhaps paralleling protections afforded on the basis of gender or race (see Mawdsley, 1998), does not seem to constitute much of a doctrinal leap, especially in light of the past two decades' jurisprudential developments. Establishment Clause jurisprudence ultimately may adopt a more relaxed interpretation of restrictions on government action, but the result may vary considerably from what many theorists have sought or imagined. "Neutrality" most likely will remain an important value of the religion clauses, but unqualified adherence to this goal would allow—indeed, arguably, demand—state advancement and support of religion far beyond the traditionally more narrow jurisprudence delimiting the bounds of church-state separation.

The third key point involves the extent to which shifts and expansions matter. The recent expansion of permissible governmental accommodations of religion under the Establishment Clause portends significant implications for church-state relations in schools where the accommodation concept originated. Accommodations made to respect Free Exercise rights in public education (e.g., exemptions from specific school observances, activities, and assignments) may affect mainly the individuals treated differently and do not alter secular school programs or activities. The stakes appear considerably higher with accommodations permitted under the Establishment Clause. For example, accommodations in terms of student-initiated devotionals in a public school touch the entire student body. Moreover, accommodations that encourage parents to select parochial schools for their children (e.g., through voucher plans) potentially affect all students. Likewise, many may be tempted to dismiss as insignificant the controversy over student-initiated devotional activities in public education and incidental aid to parochial schools. A brief student-led prayer in a public school graduation ceremony may not highly influence students' religious beliefs, and the provision of remedial services in parochial schools may not present a serious threat to the advancement of secular schools. Yet, accommodating religious groups by allowing such sectarian observances in public schools or aid to parochial schools eventually may contribute to an increasingly assertive religious establishment in education. Most proselytizing religious activities may come to seem permissible (or at least something to be accommodated) in public education if the Supreme Court ultimately interprets the Establishment Clause as not barring devotionals as long as students make the decision to conduct them and as not barring religious meetings as long as representatives of religious sects work as private individuals, rather than in their official capacities, when functioning within the public school. Similarly, most types of government aid to religious schools may gain protection under a new umbrella if the federal judiciary continues to uphold government aid that flows to religious schools because of parents' decisions to select parochial education for their children, on the basis that the funds merely follow the child. As we already have seen and will see more concretely in the next chapter, these scenarios are not as outrageous as they may seem and actually approximate quite closely current legal trends and realities.

In addition to being significant, the changing legislative and jurisprudential trajectories may be difficult to undo, given their remarkable momentum. Indeed, one can hope that the expansion of participation by religious institutions in public programs, permitted by the change in standards in government funding cases, will produce salutary consequences. In the broadest sense, these unforeseen benefits might go beyond efficient delivery of government benefits to include a broadly enriching peace and cooperation among diverse social traditions, faith-based and otherwise. However, perfect religious equality, pluralism, and toleration seem inconsistent with the drive to know "ultimate truth" that lies at the heart of many faiths, and whether such a peace can be attained is debatable, even if it is not currently being debated. Nevertheless, such a hope remains a part of constitutional faith, expressed through the gloss on the religion clauses and the overarching and consistent concern with civil peace as a primary end of the constitutional experiment with religious freedom. If anything, the momentum at least suggests a need for an equally aggressive effort to guide policies toward positive outcomes.

A last point worth highlighting involves the manner in which the Court determines whether a religion qualifies as a religion for the purposes of receiving protection. The cases we have discussed tend to skirt the issue; the Court seemingly has opted for an inclusive, expansive interpretation of what constitutes religion. The proposition has considerable merit. The Court, for example, does not require membership in an organized religious group (*United States v. Seeger*, 1965), or adherence to tenets, beliefs, or teachings supported by organized religious bodies (*Frazee v. Illinois Department of Employment Security*, 1989) in defining who is religious. Indeed, in *Welsh v. United States* (1970, p. 340), a case that put an end to a series of cases involving conscientious objectors to federal conscription, the Court held that "if an individual deeply and sincerely holds beliefs that are purely ethical or moral in source and content but that nevertheless impose upon him a duty of conscience . . . those beliefs certainly occupy in the life of that individual 'a place parallel to that filled by . . . God' in traditionally religious persons." It is difficult to imagine a broader approach to determining what constitutes religious beliefs for the purpose of attaching constitutional significance. The need for an inclusive approach is understandable when we consider the fullness of the constitutional clauses that regulate religion; the prospect that churches and states might launch legal actions to determine what has religious meaning touches the very core of the constitutional guarantee against the establishment of religion. As a result, the Court aptly has noted that the judiciary must avoid examining the tenets of religious bodies because these inquiries impermissibly entangle the government in religious affairs (*Employment Division, Oregon Department of Human Resources v. Smith*, 1990). Given this need, the Supreme Court generally uses the term "religion" unreflectively, as though it were completely self-defining (see Pealver, 1997).

Although it leaves room for considerable discretion and properly admits that injustices might follow efforts to evaluate the "religiousness" of beliefs and practices, the Court's approach to defining religion for the purposes of protection from state interference is cause for concern. Most notably, the Court allows infringement of religious beliefs when they violate broader social norms. For example, the Court allows infringement on religious rights when they contradict established ideals—monogamous marriage (as when the Court upheld laws prohibiting polygamy, *Reynolds v. United States*, 1878) and established laws providing for child protection (as when the Court upheld laws against child labor, *Prince v. Massachusetts*, 1944). The Court does protect religious beliefs when they reflect dominant social values (as when it allowed the Amish to remove their adolescents from public schools on the ground that the Amish religious community offers proper civic values, *Wisconsin v. Yoder*, 1972). The three case examples may not seem especially troublesome in that they reveal a need to abide by dominant social values, but it does seem peculiar that religious protections seem to offer less protection to minority religions (see Wybraniec & Finke, 2001). To be sure, the cases may have been rightly decided and may reflect the deep roots of our jurisprudential values. But, precisely which values are protected increasingly becomes contentious. Many leading scholars now note that the current Court has become more accommodationist toward the "major" or "mainstream" religions, namely the dominant Judeo-Christian faiths, and substantially less protective of "smaller" or "nonconforming" religious groups (Choper, 2000). The shift in Establishment Clause jurisprudence, the rise and fall of strict church-state separation, has seemed to mirror the rise and fall of anti-Catholicism, since many religious groups now share the Catholic Church's political and social agenda (e.g., regarding religious schools and family life) (see Berg, 2001). Such protection seems inevitably linked to the need to operate within the confines of existing jurisprudential values that are rooted in and influenced by the dominant religious faiths (e.g., those that protect beliefs and truths that emerge from unfettered, individual choice rather than from state or church mandates; see, e.g., Craycraft, 1999). Yet, even if the courts could look beyond their own religious beliefs and dictates, they nevertheless must rule within established legal boundaries. In fact, shifting jurisprudence may account for the observed trend toward less strict separation between church and state, which may be seen as the product of greater judicial restraint on the part of the present Supreme Court with respect to individual rights—an increased willingness by the prevailing majority to defer to the elected branches of government and a greater reluctance to reject government action in favor of personal liberty. Regardless of the reasons for the trend, the inconsistencies become increasingly problematic in light of emerging legislative directions. As we have seen, modern legislation and jurisprudence increasingly allows the state to provide support for faith-based institutions; how the legal system will determine whether less dominant religious groups can receive similar support remains to be determined. In the end, it seems clear that tradition has rendered the Court reluctant to examine religious tenets and institutions, except when they are not those that are socially dominant—an exception that leads commentators to conclude that the legal system remains much more responsive to some religions than to others (see Palmer & Hardman, 1999).

Conclusion

How society regulates religion is the cornerstone of our effort to understand how society approaches the role of religion in adolescent life. This is especially significant given the recent increase in legislation that directly involves religious institutions and in Supreme Court cases that involve religious issues. Understanding the opportunities and boundaries, as we have seen, requires a firm understanding of constitutional mandates. Regrettably, the mandates are far from clear. On its face, the Constitution's Free Exercise Clause prevents the government from banning religion, while the Establishment Clause prevents the government from requiring anyone to practice a particular religion. Given the pervasive reach of governmental regulations in modern life, efforts to adhere to these principles result in complex jurisprudential doctrine. To further complicate matters, the intentions of the drafters of the U.S. Constitution are not clear and do not point unequivocally to any indisputable single way to address the First Amendment's guarantees. The divergent initial positions of the Founding Fathers are reflected in the divergent views of the current Supreme Court justices and of Americans on religion's place in civil society.

These divergent views often lead many to describe the Supreme Court's religion case law as confused, inconsistent, and incoherent. Jurisprudence admittedly remains incoherent if one searches for a reasoned elaboration of principles grounded in constitutional text or tradition, or if one expects a sustained, systematic effort to discern and develop the underlying values of the religion provisions in a reasonably consistent, narrow manner. The Supreme Court renders split decisions in almost all important issues relating to freedom of religion and separation of church and state. The Court often draws nuanced lines on the basis of the particular facts offered in each case. This approach means that the current development of the law provides more general themes than categorical principles and fails to provide the doctrinal stability expected from the Supreme Court. This area of law, however, is not as incomprehensible as many claim. Much can be gained by a close examination of what the Court has found worth addressing and worth ignoring.

A close look reveals significant doctrinal shifts that, in and of themselves, reveal important jurisprudential stances. First, the shifts in jurisprudence reveal the fundamental point that interpretations of the clauses can and do in fact change. Second, the shifts reveal that attempting to define the constitutional parameters for the conflicts among citizens, religious entities, and the government does not yield any clear, absolute rules. The Constitution casts the religious clauses in absolute terms; the two clauses, if expanded to a logical extreme, clash with each other. Third, the divergent initial positions of the Founding Fathers are reflected in the divergent positions of judges and justices today, leading to a lack of a single, dominant voice from the courts. Fourth, creating a coherent jurisprudence from the clauses remains exceedingly challenging, given the remarkable extent to which public life implicates religious beliefs and organizations. Fifth, the Court necessarily must balance numerous interests and efforts to protect religion. Last, despite consistent litigation asking the Court to balance the competing interests and to offer guidance, it has yet to address many important issues. In what it has addressed, however, its rulings reveal much substance.

The Supreme Court has increasingly allowed governmental interference even in the face of Free Exercise challenges. The Court has developed a remarkably low governmental standard for evaluating the permissibility of infringements upon Free Exercise rights. Generally, laws that incidentally burden religious practice are permissible if they are neutral laws of general applicability. Despite this apparently permissive approach to regulating religion, however, when laws are not meant to be neutral or of general applicability and if they concomitantly burden other important rights, the Court will impose more demanding obligations on the state to justify its laws and will require that they be narrowly tailored in their intrusion on religious practices. As we have seen, the Court especially looks favorably upon Free Exercise claims when governments can accommodate those claims without excessively burdening themselves. Despite these considerations, it seems that laws do not reach this high level of scrutiny and that religious rights do not remain as inviolate as many would believe.

Establishment jurisprudence reveals an equally impressive amount of settled doctrine and some notable voids. The Supreme Court has generally accepted that government actions can entangle religious interests. Leading cases generally permit the neutral provision of public funds to religious institutions that provide secular services; when the services are not entirely secular, the benefits may still reach sectarian institutions. In its analyses of public funding of services provided by religious institutions, the Court has focused on individual beneficiaries and is more likely to permit disbursement when it can frame the provisions as allowing individuals to decide whether to use the public funds for private, sectarian ends. In addition, the Court is generally uninterested in empirical investigations to determine the extent of establishment. The Court seemingly presumes regularity in the actions of publicly funded employees in religious settings and emphasizes that law must be decided first and foremost on constitutional, rather than political or statistical, grounds. It remains to be determined how this lack of concern for empirical fact will manifest itself in other areas of regulations that deal with religious matters.

Complexities abound as we try to understand and regulate religion's role in public and private life. When we look at how the legal system regulates adolescents' religiosity and religious environments, the complexities become even more striking. The general rules that guide constitutional jurisprudence still hold, but they take a different shape because they must consider (but often fail to do so) adolescents' abilities and vulnerabilities. We now turn to those complexities.

5

Regulating Adolescents' Religious Environments

Adolescents occupy a peculiar place in law and in policy-making. What constitutes adolescence for the purposes of legal regulation varies, the law typically classifies adolescents either as children or as adults. The legal system's response to adolescents' religious beliefs, development, and environments is similarly dichotomous. As already described in chapter 1, sometimes the legal system presents adolescents as especially vulnerable to religious ideas and to normative social pressures (*Lee v. Weisman*, 1992; *Santa Fe School District v. Doe*, 2000), while at other times the system emphasizes adolescents' maturity and ability to resist religious ideas and social pressures (as in *Board of Education of Westside Community Schools v. Mergens*, 1990). In other contexts, adolescents' abilities are not even considered worth mentioning; the legal system either views adolescents' vulnerabilities as adequately protected by their parents (*Good News Club v. Milford Central School*, 2001) or frames adolescents' rights as clashes between parental rights and governmental obligations (*Wisconsin v. Yoder*, 1972). In yet other contexts, the state's higher interests may be at stake, and those interests may be so compelling that there is no inquiry into parents' right to direct their children's upbringing or of adolescents' abilities to resist the effects of religious proselytizing or to decipher the state's role in supporting religious beliefs (see *Bowen v. Kendrick*, 1988). Last, the legal system remains silent in the vast majority of decisions that directly impact adolescents' religious development and environment; as a result, it often bestows on parents and communities plenary authority to direct adolescents' religious development and environment (*Pierce v. Society of Sisters*, 1925).

Although we may lament the lack of an overarching approach to regulating adolescents' religious environments, an understanding of the legal regulation of adolescents necessarily requires a close examination of the rights of adolescents in different contexts and across different relationships. This chapter details the current understanding of jurisprudence that affects adolescents' religious development and religious environments. The previous chapter already has established that the law in fact highly regulates religion; it remains for us to explore how the law balances the state's interests in protecting religious liberties and the religious rights of parents, communities, service providers, and adolescents themselves.

Parents' Religious Rights

The nature of parents' religious rights mirrors, and even buttresses, their broad parental rights. The law pervasively grants parents enormous power to control their children's upbringing and even assumes that parents take that responsibility seriously. The law charges parents with their children's basic care and with the duty to protect them from harm. As a result, parents enjoy the highest protection against laws that may interfere with their ability to raise their children as they see fit. That authority includes parents' right to make decisions on adolescents' behalf on matters broadly ranging from nutrition, medical treatment, and residence to the choice of friends and reading materials. Although the state firmly recognizes parental rights, it places important limitations on the exercise of those rights. Given that society eventually may bear the burden of parental failures, the state subjects parental authority to governmental supervision. As a result, and most notably, if parents fail to provide adequate care, the state can intervene to protect their children's welfare. Much of the policy debate in this area of law, then, focuses on the allocation of authority between parents and the state. Understanding how the legal system envisions parental rights and delineates its boundaries helps clarify the magnitude and the nature of parents' religious rights.

Parents' right to direct the upbringing of their children was established in a now famous trilogy of cases that balanced the role of parents and states in fostering children's development. In the first case, *Meyer v. Nebraska* (1923), the Supreme Court established that the Due Process Clause of the Fourteenth Amendment protects parents' broad authority to raise children as they see fit. The Court unequivocally held that parents could provide for their children's education without reasonable interference. In *Meyer*, a teacher had violated a statute that prohibited the teaching of a foreign language to any student who had not passed eighth grade. After having read a German passage to a ten-year-old pupil, the teacher was convicted and fined. The Supreme Court reversed the conviction. Rather than focusing on the rights of teachers, however, the Court held that the statute violated the rights of *parents* to choose a suitable curriculum for their children. The Court reasoned that parents should be able to make choices for their children without state intrusion and that parents typically act in their children's best interests. The Court closed by observing, "That the state may do much, go very far, indeed, in order to improve the quality of its citizens, physically, mentally, and morally, is clear; but the individual [parent] has certain fundamental rights which must be respected" (*Meyer v. Nebraska*, 1923, p. 401). Although this often repeated dictum continues to resurface in support of parental rights, the resulting legal rule clearly places limits on those rights. Despite the high regard with which the state views parents, it still possesses the ultimate authority to control parental actions so long as the state respects certain fundamental parental rights. Exactly what those rights are, and the circumstances under which they may be infringed, have been the subject of important refinements and continued challenges.

The second case, *Pierce v. Society of Sisters* (1925), further elaborated on the rights of parents and the reasons states can intervene. In this case, the Court delimited the nature of the state and parents' obligations. The issue at hand involved a Catholic school's challenge to an Oregon statute that compelled public school attendance and prohibited other forms of education. The Court noted that the state may retain an important interest in deciding which educational facilities meet the state's high standards. However, the Court concluded that the statute went too far in mandating that only public education was suitable. The Court found that the statute imposed too heavy a burden on the rights of parents to direct the education of their children. The majority insisted that, notwithstanding the "power of the state reasonably to regulate all schools" (*Pierce v. Society of Sisters*, 1925, p. 534), the statute nevertheless unreasonably interfered with the "liberty of parents and guardians to direct the upbringing and education of children under their control." The Court then continued to provide the basic rationale for that rule, noting that:

> The fundamental theory of liberty upon which all governments in this Union repose excludes any general power of the State to standardize its children by forcing them to accept instruction from public teachers only. The child is not the mere creature of the State; those who nurture him and direct his destiny have the right, coupled with the high duty, to recognize and prepare him for additional obligations. (*Pierce v. Society of Sisters*, 1925, pp. 534–535)

The Court, then, emphasized that parents share the state's interests in the development of children, that those interests center around the need to prepare individuals to become productive citizens, and that there exists no standardized manner through which individuals learn to fulfill their social obligations.

The third central case, *Wisconsin v. Yoder* (1972), forcefully upheld *Meyer, Pierce*, and intervening cases as it delineated who could control children's upbringing. In *Yoder*, as has been discussed, Amish parents of fourteen- and fifteen-year-old children were prosecuted and convicted for violating Wisconsin's compulsory school attendance law, which required children to attend public or private school until age sixteen. The parents claimed that the mandatory school attendance law infringed upon their First Amendment rights because "their children's attendance at high school . . . was contrary to the Amish religion and way of life" (*Wisconsin v. Yoder*, 1972, p. 209). By invalidating the legislation that forced Amish parents to send their children to schools past the eighth grade, the Court took the opportunity to severely limit the state's power to circumscribe parental rights to raise their children according to their religious beliefs. The Court explained that a "state's interest in universal education, however highly we rank it, is not totally free from a balancing process when it impinges on fundamental rights and interests, such as those specifically protected by the Free Exercise Clause of the First Amendment" (*Wisconsin v. Yoder*, 1972, p. 214). Borrowing language from *Pierce* and from *Meyer*, the Court reasoned that the role of parents in raising their children is "firmly rooted in American tradition" and "beyond debate" (*Wisconsin v. Yoder*, 1972, p. 232). The parents' right to the free exercise of religion outweighed the state's interest in ensuring education for adolescents. Highly determinative in the balancing was the need to protect the rights of parents during their children's transition to Amish adulthood. The Court noted that it was precisely because of the critical period of adolescence that parents' religious rights in this case needed to supercede the state's interests. Finding that adolescence was a critical period in the development of Amish adolescents and that regular education would increase the chances that Amish adolescents would leave their religion and way of life, the Court held that compulsory public school attendance for older Amish adolescents (but not for younger children) violated Amish parents' constitutional rights to direct their children's religious, moral, and educational upbringing.

Although the *Yoder* case serves as the foundation for parents' religious rights claims, it also is instructive for the concomitant limits it places on parental rights. The *Yoder* Court limited its implicit preference for parental religious freedom to cases where the state's interest did not involve protecting children from harm. The Court noted that *Yoder* did not raise issues of "harm to the physical or mental health of the child . . . " (*Wisconsin v. Yoder*, 1972, p. 230). Although removing adolescents from public schools might limit their opportunities and affect their choice whether to leave their Amish communities, the Court explained that "the power of the parent, even when linked to a free exercise claim, may be subject to limitation . . . if it appears that parental decisions will jeopardize the health or safety of the child, or have a potential for significant social burdens" (*Wisconsin v. Yoder*, 1972, pp. 233–234). Finding that the teens' remaining faithful to the Amish community would neither harm Amish adolescents nor burden society, the Court implicitly endorsed a harm standard as a measure of a compelling interest that allows the state to infringe upon even fundamental rights.

The extent to which the Court in *Yoder* accepted that parents' religious rights could be infringed highlights the idea that parents may retain a liberty interest in the custody and management of their children but that that liberty comes with a corresponding duty to provide appropriate care. Although *Yoder* stood for favoring parents' religious claims, it left undisturbed the general rule that parents must maintain a threshold level of care, even though reaching that level may require acting contrary to a deeply held religious belief. That rule had been firmly established several years earlier, in *Prince v. Massachusetts* (1944). The legal controversy in *Prince* started when Sarah Prince, the custodian and aunt of a nine-year-old girl, sought to teach her the practices of her religion and allowed her to help sell religious literature on street corners. Prince continued to practice her faith, despite warnings by a state official—a local policeman—that she was violating child protection laws.

As a result of her persistence, a state court convicted her of violating Massachusetts's child labor laws. Prince countered that, as a devout Jehovah's Witness, she had to follow her faith's principle that children have a religious duty to work before the age of eighteen. Noting that "neither rights of religion nor rights of parenthood are beyond limitation" (*Prince v. Massachusetts*, 1944, p. 166), the Supreme Court held that enforcement of the state child labor laws had not violated Prince's Free Exercise Clause right to direct the child's religious upbringing. In recognizing the state's preeminent right, the Court wrote: "A democratic society rests, for its continuance, upon the healthy, well-rounded growth of young people into full maturity as citizens, with all that implies" (*Prince v. Massachusetts*, 1944, p. 168). The crippling effects of child employment served as an adequate rationale for Massachusetts to enforce its child labor laws by prosecuting Prince. In vivid language, the Court concluded: "Parents may be free to become martyrs themselves. But it does not follow they are free, in identical circumstances, to make martyrs of their children before they have reached the age of full and legal discretion when they can make that choice for themselves" (*Prince v. Massachusetts*, 1944, p. 170). Under this line of reasoning, then, the state may act to preserve human life, ensure a productive and self-perpetuating society, and, at the same time, conserve the state's resources. The Supreme Court has recognized the state's preeminent right to protect children from serious and irreversible harm; even when faced with Free Exercise Clause challenges, the Court gives weight to an interest in the child's future freedoms at the expense of the current claims for both parent and child.

Despite continued public debate about the unjust attacks on the rights of parents, cases that directly involve parents' rights strongly affirm (and actually have helped to create and buttress) an unwavering tradition in jurisprudence on the regulation of adolescents. That tradition privileges the rights of parents and assumes that they will act in their child's best interests; it provides them with broad discretion to raise their children as they see fit. That tradition is so entrenched and reaches such high levels of regard that infringement can be justified only if it is required by a compelling state interest. As a result, even though parents' religious rights claims may be neither absolute nor immune from state intervention, few challenges succeed. Regardless of the eventual outcome of any challenges, however, the rights and interests of the adolescents involved do not figure prominently when pitted directly against parental rights *except* as a means for the state to devise a more general rule and to declare society's superior interest over the rights of both parents and their children.

Religious Communities' Rights

Although parental rights tend to be conceptualized in terms of individual rights, they actually involve community interests. This is particularly the case with the religious rights of parents, for those rights frequently include the right to bring up children in a particular religious community. Although analyses generally ignore community rights, these rights do enjoy considerable protection, especially when coupled with a Free Exercise claim. A particular religious community's rights, then, presents another important dimension of the manner in which the legal system regulates adolescents' religious experiences. Though community concerns present endlessly varied issues, three cases capture the most important jurisprudential concerns addressed so far by the Supreme Court. The cases involve well-established and stable communities with beliefs and practices that become problematic to the extent that they either are beyond the pale of mainstream norms or otherwise interfere with important social programs.

The first leading case to address a religious community's rights and efforts to meet adolescents' needs was a case that we already have recognized as a leading parental rights case. That case, *Wisconsin v. Yoder* (1972), focused on parental rights but did so in the name of protecting the Amish communal group's autonomy. One of *Yoder*'s lessons is that child rearing constitutes a vital community issue, especially for highly distinctive communities. As we have seen, the parents in *Yoder* defended their rights on the ground that compulsory education beyond the eighth grade violated their right of free exercise of religion. Parents who sent their children to high school, they argued, exposed themselves to the risk of "the censure of the church community, . . . [and] also endanger[ed] their own salvation and that of their children" (*Wisconsin v. Yoder*, 1972, pp. 208–209). The Amish community accepted education at the elementary levels as necessary in order for them "to read the Bible, to be good farmers and citizens, and to be able to deal with non-Amish people when necessary" (*Wisconsin v. Yoder*, 1972, p. 212), but it objected to education beyond that level because it taught values "in marked variance with Amish values and the Amish way of life" and exposed children to a "'worldly' influence in conflict with their beliefs" (*Wisconsin v. Yoder*, 1972, p. 211). These claims acquire particularly powerful legal and practical significance, the plaintiffs argued, because following the law's secular mandates in this instance would jeopardize the salvation of children, parents, and the entire community.

Although analyses of the case focus on the claims of parents, these claims spoke for communal interests. Given that the Court viewed the Amish community's objection to formal education beyond the eighth grade as firmly grounded in its central religious concepts, the Court's Free Exercise analysis focused on the need to protect the religious interests of the community. After emphasizing the sincerity and depth of the Amish belief that secular education posed a threat to both the community and, potentially, to individuals' salvation, and after stressing the fact that this belief was an integral part of Amish religion, the Court found the state's interest in compulsory higher education important but not sufficiently compelling to override the Amish claim:

> Whatever their idiosyncrasies as seen by the majority, this record strongly shows that the Amish community has been a highly successful social unit within our society, even if apart from the conventional 'mainstream.' . . . We must not forget that in the Middle Ages important values of the civilization of the Western World were preserved by members of religious orders who isolated themselves from all worldly influences against great obstacles. . . . There is no specific evidence of the loss of Amish adherents by attrition, nor is there any showing that upon leaving the Amish community Amish children, with their practical agricultural training and habits of industry and self-reliance, would become burdens on society because of educational shortcomings. (*Wisconsin v. Yoder*, 1972, pp. 221–224)

The Court's concern clearly centered on the clash between the competing demands of the community and the broader society and the extent to which the smaller community could fulfill its obligations toward the larger social group.

The *Yoder* Court balanced communal and individual interests and found that potential communal clashes essentially override individual interests. The threat to the Amish people's communal values were neither incidental nor accidental. The Amish feared that exposure to outside contacts and ideas gained by public education beyond a certain level would contaminate their communal culture. The Court considered the three-hundred-year heritage of the faith and the importance of community and lifestyle to its perpetuation. As the Court put it, secondary schooling exposes "Amish children to worldly influences in terms of attitudes, goals and values contrary to beliefs, and by substantially interfering with the religious development of the Amish faith community at the crucial adolescent stage of development, contravenes the basic religious tenets and practice of the Amish faith, both as to the parent and the child" (*Wisconsin v. Yoder*, 1972, p. 218). From the state's perspective, on the other hand, that exposure was precisely the point of compulsory education, which it viewed as necessary for fully functioning citizenship. The Court properly observed that compulsory education serves two related but distinct social purposes: it enables children both to function as citizens of the society and to make mature, informed individual choices. Although these purposes sometimes are explained in terms of training in basic skills, education actually is a form of socialization in the culture and values of one's society (Levesque, 2002). However, framing issues in terms of community clashes ignores the interests of individual Amish adolescents and the fundamental goal of compulsory education—to enable children to make better future choices about their lives in general. Even taking a more communal approach, however, one could posit that the broader society's interests include equipping children to make future choices free from the authority of a localized community. The concern about "unharnessing" children from the authority of the community aims to empower them to make such informed future choices. Although couched in arguments of individual rights, the Court's reasoning fundamentally ignores those rights in favor of community concerns. The Court's action to foreclose the risk that Amish adolescents might choose to leave the community, then, could well be seen to support the community's overwhelming power to foreclose individual choice, rather than, as the Court reasoned, its success in molding productive citizens who just happen to be Amish. The *Yoder* decision assumes that Amish adolescents' religious identity essentially has been determined, that their faith and that of the Amish community are one.

The second and more recent case to address community interests on the basis of religious claims, *Board of Education of Kiryas Joel Village School District v. Grumet* (1994), essentially presents the converse of *Yoder*. It examines the extent to which regulations may directly and affirmatively advance a religious community's peculiar interests. The case involved the village of Kiryas Joel, which New York created as a religious enclave of Satmar Hasidim, a Jewish sect whose distinctive beliefs and practices require virtual separation from others. The village was created under the usual provisions of New York law, which permits residents to petition their town government to form a village within its boundaries. To overcome opposition by nonsectarian residents of the town of Monroe, the community drew the boundaries for the new village so as to include only the land owned by Satmars. In effect, the Satmars were permitted to secede from the secular community to form a religious community. The legal challenge that developed in *Kiryas Joel* did not involve the state's role in establishing the enclave. Rather than challenging the incorporation of the separate religious community, the challenge involved the subsequent creation of a special school district designed to accommodate handicapped Satmar children. These children required special education services unavailable in the private religious schools attended by Satmars but available in the secular schools of the Monroe school district. However, when the Satmar children attended the public schools, they suffered bitter emotional distress from being placed in an environment vastly different from their religious community. After most of the Satmar parents withdrew their handicapped children from the public schools, the state responded by creating a special school district to accommodate the handicapped Satmar children's educational and religious needs.

Despite the state's laudable attempt to address the Satmar children's needs, the Supreme Court invalidated the special school district. The Court did not object to the special district because it provided public support for religious education; the services were secular and not notably different from those provided to all state residents. Instead, the Court objected to the state's accommodation of the Satmars' religious needs by creating a school district for them alone. The Court viewed the creation of a state-sponsored special school district for sectarian purposes as tantamount to a law delegating "civic authority to 'the Grand Rebbe'" and a clear violation of the Establishment Clause of the First Amendment.

The third leading case does not explicitly deal with communal religious claims but does reveal another dimension in the manner in which the legal system addresses group claims, including religious claims, in the context of the rights of children. That case, *Mississippi Band of Choctaw Indians v. Holyfield* (1989), illustrates the Supreme Court's responses to legislative attempts to gain special protection for cultural minorities, in this instance indigenous native peoples protected by the Indian Child

Welfare Act (1978). As understood by the Court, the Act was prompted by the wholesale removal of Indian children from Indian homes through adoption by or foster care placement in non-Indian families, which accounted for approximately 90 percent of placements involving Indian children. Although the Act was a response to concerns about harm accruing to the parents and their children as a result of the separations, the Act centered on the harm to the Indian tribes and their culture. Children placed with non-Indian families were denied exposure to Indian culture and traditions; the tribes were thus denied "the only real means for the transmission of the tribal heritage" and were undercut in their "ability to continue as self-governing communities" (*Mississippi Band of Choctaw Indians v. Holyfield*, 1989, p. 34). In order to "protect the rights of the Indian child as an Indian and the rights of the Indian community and tribe in retaining its children in its society" (*Mississippi Band of Choctaw Indians v. Holyfield*, 1989, p. 37), Congress provided the tribes with nearly complete authority over child custody. Tribal courts were given exclusive jurisdiction over custody proceedings involving Indian children who resided or were domiciled within the tribe's reservation. In the case of children not domiciled on the reservation, tribal courts were granted presumptive jurisdictional preference over state courts: on petition of the parent or the tribe, custody jurisdiction would be transferred to the tribe absent good cause, an objection by either parent, or declination of jurisdiction by the tribal court. The reach of the tribal courts, then, could essentially lead anywhere.

The case that challenged the Act's long reach, *Holyfield*, involved an interpretation of the scope of the exclusive jurisdiction provisions. *Holyfield* specifically addressed whether two babies born two hundred miles off the reservation to parents who were both residents and domiciliaries of the Choctaw Reservation were, by virtue of their parents' domicile, themselves residents of the reservation. The state supreme court held that they were not; it emphasized the mother's considerable efforts to give birth off the reservation and the fact that the parents had arranged for adoption by a non-Indian couple (with whom the children had resided for three years by the time the case reached the Supreme Court). The Court found these considerations irrelevant and ruled that the statute intended that the child's domicile be that of the parents, regardless of the parents' preferences regarding placement in a non-Indian home. The Court emphasized that the Indian Child Welfare Act (1978) sought primarily to protect the tribe's interest, an objective that superseded contrary parental claims.

Without doubt, the Act establishes the tribe's right to preserve its culture and community over and against both the rights of the natural parents and the prerogatives of the larger society, the state. Most striking about the Act and the Supreme Court's approval is that both place the welfare of the children in the hands of the Indian community solely for the sake of the community. The move seems quite radical in its apparent defiance of the parents' rights (here the parents clearly had objected) and in its disregard for the state's own prerogatives in adoption policy. Yet, by actually placing child-rearing rights in the hands of communities, *Holyfield* actually takes only a modest step beyond *Yoder*. The Court in *Yoder* had skirted the question of how adolescents' own interests in schooling would be taken into account as it subordinated those interests to the interests of the parents and of the Amish community. In both cases, then, the local community norms trumped those of the broader society and ignored the rights of the children involved.

Although *Holyfield* and *Yoder* have important similarities, there are also key differences. We can easily move beyond the fact that *Holyfield* deals with American Indian claims and sovereignty issues. If the legislature and the Court had had a higher regard for the rights of individual parents or, even more radical, the individual rights of children, finding for them would not have been that unimaginable. Recall, for example, that, in *Lyng v. Northwest Indian Cemetery Protective Association* (1988), the Court did not find constitutional harm when Native Americans challenged the Forest Service's decision to construct a logging road through sacred lands. The federal government has the prerogative to take precedence over claims that its actions infringe on religious and other liberties that the federal government itself has bestowed on groups (see Brown, 1999). For our purposes, however, it is important to observe the ways in which the distinctions between *Holyfield* and *Yoder* highlight the extent of community rights. The Amish in *Yoder* did not seek plenary legal jurisdiction over their adult members; Amish parents, representing their community and their children, asked only that the state leave them alone. The Amish community did not challenge the rights of children from other communities to go to public schools. Yet, in principle and in effect, this is the power *Holyfield* granted the tribe. The Court required children who had become part of one community and family to return to the control of a community foreign to them and to those they had not viewed as their parents. From this perspective, there is only a subtle distinction between these two cases. The parents in *Holyfield* were not irrevocably bound to the community; the law presumably allowed the parents to exercise their preference by forfeiting their domicile in the reservation. Rather than doing so, the parents sought to retain the benefits they enjoyed as residents on the reservation without submitting to all of the community's norms. The Indian Child Welfare Act (1978) required parents simply to choose between living by the community's norms and leaving the community. Given that the *parents* did not want to leave the community, the civil society encompassing the community could intervene, but only to enforce the community's right to self-governance. Just as in *Yoder*, the Court in *Holyfield* acted to ensure that the community was left alone.

Taken together, the three cases reveal the conflict between the need to respect different conceptions of the "good" and the need to promote conditions of individual choice. In essence, the tension involves the extent to which civil society can protect what "right-thinking people" know or believe and what they can reasonably demand that others know or believe. In determining how to balance these concerns, the cases are consistent. In *Yoder*, the claims of the community trumped those of the broader society, generally because the community claim mirrored the broader society's most ambitious aims in producing citizens. In *Kiryas Joel*, the claims were trumped because the community asked more than to be left alone; respecting its claims required affirmative assistance to the community singled out for specific governmental support. In *Holyfield*, the group's interests survived the challenge; the Court ruled that, as long as individuals want to be bound by particular communities, they should follow its rules. The rule that emerges, then, supports the view that when a religious community seeks protection, the Court will protect the religious community's interest in being left alone but will not offer it affirmative assistance. Where the rights and needs of adolescents fit into these patterns remains to be determined, in these cases, the potential individual interests of adolescents were not separated from the conflict among the parents, the communities, and the state. In fact, in all three cases, we do not know much about what the "children" wanted, even more important, the cases did not require that we know their preferences.

Service Providers' Rights

The great power bestowed upon parents and localized communities finds its mirror opposite in the current articulation of religiously affiliated service providers' rights. In general, the state may severely constrict interactions between professionals and adolescent clients. The limitations essentially derive from the law's assumptions about adolescents' immaturity and incompetence and from the general rule that parents determine children's access to social services. The former generally results in limiting access to services, unless (1) the service is not very intrusive or an emergency requires intervention, (2) the service is intrusive but individual adolescents evidence a level of maturity and competence that justifies allowing them to determine access to services, or (3) the public's interest, such as public health and safety, is so great that not allowing access would be detrimental to either the adolescent or society (see Levesque, 2000a). The last rule, with some protections against parents' actions that would not be in their children's best interests, generally means that parents can both require their adolescents to receive services and opt out of services for them (Levesque, 2000a). These legal rules provide the appropriate context from which to understand service providers' rights to follow their own religious principles and firmly held convictions, for they highlight the remarkable extent to which the state may limit adolescents' access to services.

The manner in which the legal system regulates service providers' interactions with adolescent clients directly bears upon the extent to which the religious beliefs of service providers may impact the services adolescents receive. Although a complex set of laws regulates communications and actions between service providers and their clients, three areas of regulation determine the relationships' boundaries from the providers' perspectives. The first concern is the extent to which the government can distort interactions, including those that would be influenced by religious beliefs, between service providers and adolescent clients. The second issue is the extent to which the government can support religious providers in their provision of services that necessarily are influenced by religious beliefs. The third major issue is the extent to which individual and institutional service providers may *refuse* to provide services because of profoundly held religious beliefs.

The law that regulates the first issue—how service providers can conduct interactions that would be proper and legal except for the government's desire to encourage certain outcomes—imposes restrictions on service provider speech. The state may severely limit discussions that occur between clients and professionals in order to foster what the state sees as the best outcome among competing legally permissible ones. Two recent cases help clarify how the state can direct interactions; both cases deal with a morally contentious and religiously divisive issue that directly affects adolescents—family planning.

Rust v. Sullivan (1991) asked the Supreme Court to determine the constitutionality of the interpretation of Title X by the Department of Health and Human Services (HHS). This legislation partly authorized HHS to administer grants and enter into contracts to establish and operate family planning projects. The regulation at issue involved a panoply of restrictions upon the conduct and speech of Title X grantees, including physicians who worked in publicly funded family planning clinics. The speech-related regulations prohibited physicians from providing "counseling concerning the use of abortion as a method of family planning" and from providing referrals to women seeking abortions (*Rust v. Sullivan*, 1991, p. 193). In response to a specific request for a referral to an abortion provider, the regulations suggested that physicians state that "the project does not consider abortion an appropriate method of family planning and therefore does not counsel or refer for abortion" (*Rust v. Sullivan*, 1991, p. 193).

The physicians who challenged the regulations argued that the regulations interfered with the First Amendment rights of Title X health-care providers and Title X patients. They alleged that the regulations (1) impermissibly discriminated among grant applicants on the basis of their views on abortion and (2) impermissibly conditioned the receipt of Title X funds upon the surrender of a fundamental right to freedom of expression. In addressing the doctors' arguments, the Court's majority did not begin by determining the level of protection appropriate for doctor-patient speech. The majority opinion did not include a First Amendment analysis of the regulations from the standpoint of either doctors' speech or patients' right to receive medical information. Instead, the Court's analysis of the speech-related restrictions was rooted in its interpretation of the doctrine of unconstitutional conditions. Under the doctrine of unconstitutional conditions, "government may not condition the receipt of its benefits upon the nonassertion of constitutional rights" (Tribe, 1988, p. 681). Relying on this analysis, the Court determined that the HHS regulations were a constitutional exercise of the government's power to fund some activities and not others. According to the Court, conditioning the receipt of federal funds on remaining silent about abortion simply conforms with the government's power to make funding choices in accordance with its preference for childbirth over abortion. With respect to the impact of the regulations upon physicians' speech, the Court explained that:

> a doctor employed by the project may be prohibited in the course of his project duties from counseling abortion or referring for abortion. This is not the case of the Government "suppressing a dangerous idea," but of a prohibition on a project grantee or its employees from engaging in activities outside of its scope. (*Rust v. Sullivan*, 1991, pp. 193–194)

Thus, the Court endorsed the proposition that government may, to promote its reasonably held viewpoint, censor the speech of publicly funded speakers, including physicians.

Because the regulations at issue in *Rust* applied only to publicly funded physicians, and because the case was decided under the doctrine of unconstitutional conditions, the *Rust* opinion left unclear whether the Court would permit the government to impose requirements on the content of doctor-patient speech that occurred in private settings. In *Planned Parenthood v. Casey* (1992), the Court jettisoned this implied limit on the government's power to regulate the content of physicians' speech. *Casey* involved the constitutionality of a Pennsylvania statute that regulated abortions. In addition to imposing limitations on physicians' conduct, the statute's "informed consent" provision required that physicians tell every patient seeking an abortion about the health risks of abortion and childbirth and the probable gestational age of her fetus. It also mandated that physicians tell patients about the availability of printed materials that described the fetus and listed agencies that offered alternatives to abortion; stated that the child's father was liable for financial assistance to support the child (even if he had offered to pay for the abortion); and stated that medical assistance might be available for prenatal care, childbirth, and neonatal care.

Unlike the restrictions at issue in *Rust*, the provisions in *Casey* did not silence physician speech. Like the HHS regulations, however, the Pennsylvania statute's speech-related provisions directly regulated the content of physician-patient discourse for the purpose of persuading patients to elect a governmentally preferred course of action. Thus, the Court in *Casey* again faced the question of whether the First Amendment prohibited the government from imposing viewpoint-based restrictions on doctor-patient speech. The Court's free-speech analysis was based on whether the restrictions "unduly burden" a woman's privacy right to choose an abortion and concluded that the speech-related provisions did not violate the First Amendment and, thus, that the government could use doctor-patient conversation as an instrument for expressing a viewpoint or for persuading patients to opt for a governmentally preferred course of action. *Casey*, then, moves beyond regulating speech that occurs in a publicly funded setting to allow the government to impose content regulations in private settings. While the plurality acknowledged that the challenged regulations implicated physicians' speech rights, they summarily dismissed this concern, stating that advising patients is merely a "part of the practice of medicine, subject to reasonable licensing and regulation by the State" (*Planned Parenthood v. Casey*, 1992, p. 2824). Under *Casey*, then, the authority to license and regulate the practice of medicine empowers states to compel publicly and privately financed physicians to make certain statements to patients as long as those statements further a legitimate state interest and are not false or misleading. This approach essentially locates the governments' power to regulate the content of doctor-patient speech in the states' police power to license and regulate physicians.

Rust and *Casey* stand for the proposition that the state may limit the contents of service provider-client discussions, but the government's power to compel speech has its limits. In *Casey*, the Court did not vest the state with an absolute power to dictate the content of physicians' statements to patients by compelling physician speech. Regulations are unconstitutional if they require physicians to make statements that are false or misleading. Under *Rust*, restrictions on publicly funded physicians are unconstitutional if patients have an "all-encompassing" relationship with their doctor that gives rise to a reasonable expectation of receiving comprehensive medical advice. Likewise, the government may not impose viewpoint-based restrictions that do not afford publicly financed physicians the opportunity to distance themselves from the state's message. Given the types of regulations presented in the cases, however, the law clearly may go far in regulating interactions between service providers and clients. *Rust* and *Casey* reveal that the Supreme Court adheres to the view that the First Amendment does not prohibit the government from attempting to influence patients' medical decision making by regulating the content of doctor-patient discourse, whether it occurs in a private or a public setting.

Service providers who interact with adolescent clients also have rights, which raises the issue of whether the government may support religiously affiliated service providers in their efforts to provide services biased by religious beliefs. The Supreme Court has spoken to this issue in a leading case that focuses on how the government may support the provision of limited services and include religious providers to further governmental ends. That case, *Bowen v. Kendrick* (1988), also involves family planning matters and adolescents. At issue in *Bowen* was the *Adolescent Family Life Act* ("AFLA") (1982, 1988), which supplies federal funds to both governmental and independent-sector nonprofit organizations that do research or provide services (counseling and education) for adolescents in the areas of premarital sexual relations and teenage pregnancy. The Act sought to support these programs by promoting the use of other family members, religious and charitable organizations, voluntary associations, and other groups. Rather than simply including religious organizations as possible grant recipients, however, the statute specifically calls for the active participation of religious organizations in providing the specified services and does not limit their use of granted funds to secular purposes. In addition, AFLA increases the likelihood that funds will go to religious organizations, given that the Act stresses premarital abstinence, prohibits granting money for "family planning services," and grants funds to programs that "do not provide abortions or abortion counseling or referral" or encourage abortion (*Bowen v. Kendrick*, 1988, pp. 596–597).

Because religious organizations were included among the recipients of the grants, a group of taxpayers challenged AFLA, arguing that funding religious organizations violated the Establishment Clause. The Supreme Court upheld the statute on the grounds that the statute did not violate the Clause "on its face"; the Court remanded for reconsideration of the "as-applied" challenge in instances in which grant recipients were pervasively sectarian organizations. In its analysis, the Court applied the *Lemon* test's three major concerns, reviewed in chapter 4. Regarding *Lemon*'s first prong, that the legislation have a secular purpose, the Court was highly deferential to the legislature. The Court agreed, along with both parties involved, that, as a whole, furthering religious concerns was not the sole motivation behind the Act and that AFLA had a valid secular purpose—to prevent teenage pregnancy and premarital sex, both of which cause economic and social injury. *Lemon*'s second prong requires the Court to determine whether the principal or primary effect of the law impermissibly advances religion. The Court identified nothing inherently or specifically religious about the activities or social services provided by the grantees to adolescents with questions about premarital sexuality.

Moreover, simply because AFLA expressly required religious organizations to be considered among the available grantees and demanded that the role of religion be taken into account by secular grantees, those provisions did not have the effect of endorsing a religious view of how to solve the problem under consideration. As to grantee eligibility, the Court interpreted AFLA as religion-blind because Congress required that all organizations, secular and religious, be considered on an equal footing. Further, the legislation did not violate the Establishment Clause merely because there was an overlap between certain religious beliefs and the moral values urged by AFLA. Critical to the result was the majority's refusal to hold that faith-based teenage counseling centers were necessarily pervasively sectarian. Although the federal assistance took the form of a direct cash grant, the First Amendment was not necessarily offended as long as the grantee was not pervasively sectarian. Since the Court found no reason to believe that any significant portion of the funds necessarily would flow to "pervasively sectarian" institutions, it perceived no risk that the organizations would use the funds to advance their religious mission. The third prong of *Lemon* requires the Court to consider whether the statute in question fosters an excessive administrative entanglement between religious officials and the offices of government. The Court noted that HHS appropriately could (and must) monitor AFLA grantees to prevent the misappropriation of public funds. HHS did not require faith-based grantees to follow any federal guidelines concerning the content of the advice given to teenagers or otherwise modify their programs. HHS also did not address nondiscrimination requirements as they pertain to the served beneficiaries. Since the Court did not view the religious grantees as necessarily pervasively sectarian, it logically concluded that the federal agency's limited oversight could not be deemed excessively entangling. In deciding the case, the Court spoke in sweeping terms as it sanctioned governmental aid on the basis of equality in the rules for awarding grants. The Court spiritedly declared that "religious institutions need not be quarantined from public benefits that are neutrally available to all" and that "this Court has never held that religious institutions are disabled by the First Amendment from participating in publicly sponsored social welfare programs" (*Bowen v. Kendrick*, 1988, p. 609). The Court's vibrant language firmly ensures religious individuals and institutions an important (if not primary) place among publicly funded service providers.

The last central issue involving service provider's rights concerns the extent to which individual and institutional service providers may refuse to provide services because they violate the provider's profoundly held religious beliefs. Although it would seem that this would be a settled area of law, the law actually leaves the rights of service providers largely undetermined and underdeveloped. The degree to which service providers can follow their religious beliefs and refuse to take part in objectionable services remains somewhat unclear. The most established area of law involves physicians' religious convictions in the context of whether they can refuse to participate in abortion, sterilization, or the use of abortifacient or to carry out directives to withdraw feeding tubes. Although no Supreme Court cases address this area of law, states have enacted many forms of protections, largely in response to the Court's highly charged abortion rights cases. Generally, these statutory protections address medical issues and protect providers from employers that would retaliate against professionals who refuse to perform specific procedures (Wardle, 1993). Relatedly, a complex set of laws regulates the extent to which institutions can avoid providing services that conflict with religious principles espoused by the religious institutions that provide financial support for the service provision. Generally, religiously affiliated institutions may be required either to provide offending services or to make alternative arrangements if they receive state or federal assistance; otherwise, they remain private institutions that may determine the nature of their service provisions (see White, 1999).

Although quite unclear in many regards, the rights accorded service providers clearly illustrate the extent to which the legal system regulates the nature of the services adolescents receive. We already have seen that *Rust* and *Casey* ruled on the important lesson that service providers may not be allowed to provide services they deem appropriate and that they may be required to express viewpoints that they consider inappropriate. In addition, medical providers seemingly are protected from being required to perform procedures they find objectionable. It may well be that other providers will gain similar rights, but this remains to be determined. These issues clearly affect adolescents' environments. Yet, these concerns, especially the refusal of service providers to participate in services they personally find objectionable, have not been addressed by the law or by those seeking to establish the nature of adolescents' legal rights. The legal system has yet to address fully the extent to which adolescents can be protected by the religious rights of service providers and, equally important, the extent to which the religious rights of service providers factor in services when the interactions of adolescents and service providers are not the subject of close regulation. The law typically assumes that individuals can determine for themselves whether the offered services are biased and that they have access to other services if they choose not to receive services they see as biased. How adolescents can realistically access services and when they have an independent right to do so remain debatable and have not been addressed in analyses of adolescents' rights, to which we now turn.

Adolescents' Rights

As we already have seen, understanding the nature of adolescents' rights means understanding the rights of others and the extent to which those rights impact adolescents' experiences and environments. The central theme in adolescents' rights jurisprudence is the principle that parents control the upbringing of their children. It is therefore not surprising to find that the legal system generally has not recognized that adolescents have the right to direct their own religious upbringing and environment (see *Wisconsin v. Yoder*, 1972). The Supreme Court has noted, however, that adolescents may have religious rights (see, e.g., *Prince v. Massachusetts*, 1944, p. 165). Although these two approaches to adolescents' religious rights seemingly conflict, the Court has not gone so far as to hold that adolescents have constitutional religious rights separate from their parents' right to religious liberty. As a result, the extent to which adolescents' own rights are recognized when they conflict with those of their parents remains unclear. There is some basis for the claim that adolescents retain constitutional rights inside their parents' homes vis-à-vis their parents, and there is even more basis for the claim that adolescents have religious rights outside their family environment. Our analysis necessarily must consider both of these contexts, as well as the extent to which adolescents' religious claims may diverge from those of their parents.

The U.S. Supreme Court certainly has not foreclosed the possibility that adolescents may have recognizable religious rights separate from those of their parents. Despite its deference to parental rights and obligations, the Court repeatedly has held that adolescents possess constitutional rights. *Prince v. Massachusetts* (1944), the leading case in allowing state restrictions on a guardian's free exercise right to control her ward's upbringing, arose in the context of the child's own free exercise claim. In strikingly broad dicta, the Court noted that "the state has a wide range of power for limiting parental freedom and authority affecting the child's welfare; and that includes, to some extent, matters of conscience and religious conviction" (*Prince v. Massachusetts*, 1944, p. 167). The ultimate result of *Prince*, however, belies that proclamation. Even though the minor knew that her proselytizing efforts were necessary to avoid her "condemnation 'to everlasting destruction at Armageddon'" (*Prince v. Massachusetts*, 1944, p. 170), the Court nevertheless restricted her ability to act on her unequivocally articulated beliefs, even when they clearly complemented those of her legal guardian. In light of the other cases we have reviewed that place the control of adolescents' religious rights squarely in the hands of parents, especially *Pierce v. Society of Sisters* (1925) and *Wisconsin v. Yoder* (1972), Supreme Court jurisprudence cannot be read as clearly recognizing that children have religious claims distinct from their parents' religious rights.

Given that constitutional case law does not stand for the proposition that adolescents may not have an independent right to exercise their religious dictates and that cases involving adolescents' religious rights skirt this concern, it seems reasonable to turn to the Court's other analyses of adolescents' First Amendment rights to discern the potential scope of adolescents' religious rights. The high-water mark of this area of the law, at least in terms of efforts to recognize and protect adolescents' independent First Amendment rights, came in the form of a challenge to the rights of three teenagers to wear black armbands to protest the Vietnam war. In that case, *Tinker v. Des Moines Independent Community School District* (1969, p. 511), the Court, for the first time, stated that "students in school as well as out of school are 'persons' under our Constitution" and that they possess "fundamental rights which the State must respect." Although *Tinker* still stands and ushered in a period of high hopes, it did not provide the expected, and even broadly assumed, breakthrough in adolescents' rights. The cases that followed curtailed *Tinker*'s apparently broad recognition of adolescents' First Amendment rights (see *Hazelwood School District v. Kuhlmeier*, 1988), because of the legitimate need to shield adolescents from themselves and to protect broader social interests (see *Ginsberg v. New York*, 1968). Even more problematic for efforts to champion adolescents' own religious rights is the remarkable extent to which, in even the leading cases that boldly proclaim adolescents' rights, adolescents' interests and views often are essentially indistinguishable from those of their parents (see *West Virginia State Board of Education v. Barnette*, 1943; *Tinker v. Des Moines Independent Community School*

District, 1969). Indeed, the decisions that purportedly proclaim adolescents' own First Amendment rights may be read as supporting parents' right to impart their values to their children and to encourage their children to express those views and values against state claims aimed at limiting this parental right. Thus, there appears to be a fundamental rule that minors may have First Amendment rights, but those rights are much less broad in scope than those of adults, especially those of their parents and other caretakers (for a review, see Levesque, 2000a, 2002).

Cases outside First Amendment jurisprudence also may offer insight. As with the history of First Amendment cases that apparently recognize adolescents' rights, only to be followed by other cases that limit them, these cases also mitigate against claims that adolescents possess religious rights independent of their parents. In cases where assertions of adolescents' constitutional rights directly or indirectly conflict with those of their parents, adolescents commonly have failed to secure separate constitutional protection as the courts adopt a stance deferential to parents. This approach rests on the law's concept of the family, in which parents are presumed to possess what the child lacks in maturity, experience, and capacity for judgment. The approach also rests on the recognition and assumption that "natural bonds of affection lead parents to act in the best interests of their children" (see *Parham v. J.R.*, 1979, p. 602). According to this approach, although the interests of the adolescent and parent may conflict, the family relationship fosters a natural accommodation of each to the other. The natural accommodation helps account for the high interest in protecting family relationships made even more evident in the leading parental rights case, *Pierce v. Society of Sisters* (1925). As it struck down the Oregon statute forbidding sectarian education, the Supreme Court highlighted both the "right of parents to choose schools where their children will receive appropriate mental and religious training" and also "the right of the child to influence the parent's choice of school" (*Pierce v. Society of Sisters*, 1925, p. 532). Parents and children have the right to choose between public schools and available sectarian alternatives; strife within the family may exist, but the family is nevertheless presumed to operate on mutual self-interest. Intrafamily resolutions of potential conflicts are deemed preferable to any judicially imposed balancing of the conflicting interests, an approach that raises the specter of state standardization of its children and that seems inimical to both the letter and the spirit of the Constitution (see *Meyer v. Nebraska*, 1923, pp. 399–403; *Pierce v. Society of Sisters*, 1925, p. 535).

We have seen, however, that there are clear exceptions to the conditions in which parents may raise their children with minimal state interference. The state is not without control over parental discretion when that discretion may jeopardize the children's physical or mental health. The most far-reaching, and controversial, exception to the parent-deferential approach to adolescents' constitutional rights within their families is the Court's analysis of a minor's right to obtain an abortion. That recognition called for the establishment of a state-operated mechanism to facilitate the exercise of minors' constitutional rights independent of their parents' support or even knowledge (e.g., the judicial bypass that allows sufficiently mature adolescents to decide whether to obtain abortions without parental consent or notification; see *Bellotti v. Baird*, 1979; *Planned Parenthood v. Casey*, 1992; Levesque, 2000a). Unlike the First Amendment cases we have reviewed, however, these cases rely on the right to privacy, and it remains to be seen whether the analysis controls other First Amendment cases.

Commentators who have examined the religious rights of children note that the exception to parental discretion based on the constitutional principle of privacy may support an argument for recognizing children's free exercise rights (see Steinberg, 1995; Buss, 1999). These commentators, however, have not actually made the argument, they have contented themselves with noting that it can be made. Presumably, the right-to-privacy cases could help support adolescents' religious rights claims to the extent that the Constitution explicitly protects the free exercise of religion, it seems at best disingenuous to place a higher value on an individual's right to make a decision regarding abortion than on the right to worship as one pleases.

Although it may seem jurisprudentially reasonable and feasible to graft the Court's reasoning in cases involving adolescents' abortions (privacy) onto cases involving religious rights, the argument actually faces important challenges. At least as envisioned by the Court, the abortion decision remains unique. The Court seemingly has decided that the abortion decision requires greater autonomy from one's parents and has viewed these privacy decisions as sui generis, rather than as extending to other constitutional rights, given that the decision is irreversible, urgent, and more grave (see *Bellotti v. Baird*, 1979, p. 639). Even if it could be argued that decisions about having an abortion and practicing a given religion are similar and must be similarly respected, it is important to note that it remains to be determined whether even the right-to-privacy cases respect adolescents' rights (see Levesque, 2000a). Arguments about adolescents' rights to privacy caution against championing adolescents' religious rights through parallel mechanisms (such as judicial bypasses); use of currently envisioned mechanisms would not necessarily result in fostering adolescents' exercise of their religious rights, even though these rights would be formally recognized. The practical operation of adolescents' religious autonomy would enlist the state (generally in the guise of the court system) in determining whether the religious decision at issue was best for the individual adolescent. This role raises rather challenging issues, such as the competence of courts to understand adolescents' needs, the potentially massive intrusion of the courts in family decision making, the high risk that harm would result from efforts to determine adolescents' religious beliefs, and the assumption that adolescents lack the experience, perspective, and judgment to influence important decisions that affect them. Given these challenges and the inability of courts and commentators to address them satisfactorily, it is not surprising to find that the legal system generally has not addressed adolescents' right to practice their own religious beliefs, even when those beliefs do or might conflict with those of their parents.

As we also have seen, although parents retain powerful rights and although society exhibits a tendency to distrust state involvement, the state does not always stay out of familial affairs and decisions that impact adolescents' religious beliefs. Although intact families may provide no occasion for the state to intrude and inquire as to an adolescent's preferences, when families fracture or are at risk for fracturing the state may intervene. This intervention typically occurs in two contexts. The first context involves the parents' request for state intervention during separation or divorce. In those instances, judicial proceedings customarily elicit and give due weight to adolescents' views (Drobac, 1998). These inquiries may be most relevant in deciding the minor's best interests and in respecting the development of his or her religious identity, but they do not amount to granting independent rights to the adolescent. Courts generally ignore teens' viewpoints in cases where the parents agree, and courts have not addressed whether children have separately cognizable rights of religious exercise.

The second context for state intervention involves risks of harm, or potential harm, to the adolescent or to societal interests. In these instances, the state may limit parental rights in favor of the state's own obligation to protect the adolescent or otherwise to recognize the adolescent's rights.

When the state protects adolescents in this manner, however, adolescents again do not necessarily control their own religious rights. Rather, the state controls adolescents' right to protection and generally determines the nature of that protection. The state, through its officials, essentially becomes a substitute parent to guard against parental failure (this was the result, for example, in abortion cases where the pregnancy may be assumed already to have fractured the family, as evidenced by the minor's refusal to consult with her parents). Even inquiries that do infringe on parental discretion, then, do not necessarily mean that the legal system recognizes adolescents' independent rights.

As expected, the adolescents' rights we have just highlighted may not apply where parental authority is reduced and state control is increased. As we have seen, the nature of adolescents' rights varies in different situations, especially in contexts that may not involve direct parental control and that may involve the rights of other adults. Adolescents may gain greater control over their rights when they are deemed mature minors or when other circumstances (such as public health or the need to preserve democratic principles) warrant granting them greater control. Adolescents may be deemed to possess rights that relate to their religiosity, most notably in contexts that are outside of direct parental control.

Although the legal system typically does not directly address the rights of adolescents and seemingly prefers to address the rights of those who can exert control over them, recent jurisprudence has explored the nature of adolescents' rights outside familial contexts. That jurisprudence typically involves educational rights and public school environments. In fact, this seems to be the only context in which the Supreme Court addresses adolescents' own religious rights. Although it remains unclear how the conception of adolescents' rights in these contexts may transfer to other contexts, jurisprudence in this area is instructive, since school settings offer a peculiar context that necessarily balances the rights of parents, communities, those who implement legal rules (school boards and teachers), peers, and adolescents themselves. The balancing in this context seems quite significant, given the vast power schools wield to socialize students and the vast power the law grants parents and communities to socialize their children to function within their own environments. Thus, the lessons learned from this area of law tell us much about the nature of adolescents' rights and how they are balanced with other people's rights and with broader social interests. In educational contexts, two related situations that directly involve adolescents warrant close analysis.

The first situation involves the permissibility of private, nondisruptive religious exercises by students. Like other jurisprudence involving religion, Supreme Court jurisprudence on students' voluntary participation in devotional activities has changed dramatically in recent years. From the 1960s well into the 1980s, for example, the Supreme Court interpreted the Establishment Clause as barring daily prayer and Bible reading in public schools. Most notably, in *Wallace v. Jaffree* (1985), the Court concluded that an Alabama statute that permitted a moment of silence "for meditation or voluntary prayer" was unconstitutional because it indicated the state's intent that prayer be considered the favored practice (*Wallace v. Jaffree*, 1985, pp. 40, 60–61). The Court found it irrelevant that participation was voluntary; the simple fact that such devotionals were sponsored by public schools abridged the First Amendment. Given the Court's ultimate ruling, it is not surprising to find that the case fostered the general presumption that the Establishment Clause barred religious speech from governmental forums. Although the *Wallace* decision might at first appear to be an effort to remove all varieties of religion from schools, its real target was official observances or sponsorship of religion, not the private free exercise of religion in the schools. The Court plainly made this point when it emphasized that "the legislative intent to return prayer to public schools is, of course, quite different from merely protecting every student's right to engage in voluntary prayer during an appropriate moment of silence during the school day" (*Wallace v. Jaffree*, 1985, p. 59). The Court left open the possibility that a moment of silence law could be valid if it was more neutrally worded and did not carry the heavy legislative history baggage attached to the Alabama statute. The Court's ruling was much more measured than many believed; not surprisingly, the Court soon refined the nature of permissible devotional activities and increasingly allowed religious practices in public settings, even including, as we explore later, moments of silence for prayer in public schools.

Refinements in the Supreme Court's approach to devotional experiences came most obviously in *Lee v. Weisman* (1992). In *Lee*, the Court faced the question of whether a prayer given by a clerical member as a part of an official school graduation ceremony violated the religion clauses of the First Amendment. The prayer was actually given at a middle school graduation ceremony, although the Court held that the point was not moot because the objecting student was also likely to face a similar invocation and benediction at her high school graduation ceremony.

The challenged activities in this case started when the school principal invited a rabbi to deliver prayers at the graduation exercises. Knowing that religious invocations could offend some individuals, the principal provided the rabbi with a pamphlet entitled "Guidelines for Civic Occasions" and advised the rabbi to offer a nonsectarian prayer. The rabbi complied and gave two prayers that lasted no longer than a minute each, including a "respectful" moment of silence both before and after each prayer.

Despite the principal's and the rabbi's efforts to remain nonsectarian, the Court held that even this short, nonsectarian, nonproselytizing prayer violated the Establishment Clause because it coerced students into participating in a state-sponsored religious exercise. The majority unequivocally asserted its position:

> the sole question presented is whether a religious exercise may be conducted at a graduation ceremony in circumstances where, as we have found, young graduates who object are induced to conform. No holding by this Court suggests that a school can persuade or compel a student to participate in a religious exercise. That is being done here, and it is forbidden by the Establishment Clause of the First Amendment. (*Lee v. Weisman*, 1992, p. 599)

The *Lee* court noted that the "dominant facts" that "marked and controlled" this decision were that the state officials controlled the prayers given at the graduation ceremony and that the attendance of students who objected to the prayer was essentially mandatory (*Lee v. Weisman*, 1992, p. 586).

Lee is particularly noteworthy for the various jurisprudential doctrines the Court harnessed to limit devotional activities in schools. The first factor in the Court's decision concerned whether the prayer was state sponsored. The Court found that the principal's determination that a prayer would be said was the equivalent of a state mandate that a religious exercise be held at the graduation ceremony. The principal also decided who would give the prayer. In addition, the Court observed that the principal controlled and directed the content of the prayers by furnishing the invited rabbi with guidelines. The Court stated that "'it is no part of the business of government to compose official prayers for any group of the American people to recite as a part of a religious program carried on by government,' and that is what the officials attempted to do" (*Lee v.*

Weisman, 1992, p. 588). Another factor in the Court's decision involved the mandatory nature of attendance at the graduation ceremony. Although attendance was not required and students could receive their diplomas without attending the ceremony, the Court observed that graduation is one of life's most important events. Missing graduation would mean the "forfeiture of those intangible benefits which have motivated the student through youth and all her high school years" (*Lee v. Weisman*, 1992, p. 595). Noting that the Constitution forbids the state from exacting religious conformity from students as a price for attending graduation, the Court rejected the school board's argument that students voluntarily attend graduation exercises and can miss them if they disagree with the recitation of an invocation and benediction. Finally, the Court considered the unique coercion present in public schools, which places primary and secondary school children at particular risk. In its analysis, the Court noted that students were under subtle psychologically coercive pressure and had no alternative that would have allowed them to avoid an appearance of participation in the prayer. The Court determined that this type of coercion clearly violated the Establishment Clause. Included in the controlling opinion is a caveat that "what to most believers may seem nothing more than a reasonable request that the nonbeliever respect their religious practices, in a school context may appear to the nonbeliever . . . to be an attempt to employ the machinery of the State to enforce a religious orthodoxy" (*Lee v. Weisman*, 1992, p. 592). The result, reinforced by an emphasis on the significance of "private" peer group pressure on possible dissenters, makes a constitutional violation possible even when the state itself is not the direct agent of the religious exercise. State inaction is just as important as state action if that inaction can reasonably be perceived as an endorsement of religious beliefs. Although the concept of private action remains controversial in this context, *Lee* at least stands for the principle that the government cannot facilitate private prayer in circumstances that force dissenters to assent to that which they do not believe.

The holding in *Lee* and the value of some of its supporting doctrinal analyses have become quite evident in and settled by the more recent case involving school prayer, *Santa Fe Independent School District v. Doe* (2000). *Santa Fe* revolved around the permissibility of student-led prayer at public high school football games. After suit was filed against the practice of having the "student council chaplain" deliver a prayer before home football games, the school district enacted a policy that remitted to student elections the question of student-led prayer at commencement and sporting events (*Santa Fe Independent School District v. Doe*, 2000, pp. 2272–2273). First, the students voted on whether the event should be preceded by an "invocation and/or message . . . to solemnize the event, to promote good sportsmanship and student safety, and to establish the appropriate environment for the competition" (*Santa Fe Independent School District v. Doe*, 2000, p. 2273). If the students so voted, they would then vote again to select which student would deliver the message. The only resulting legal issue concerned whether the student speech should be viewed as private or whether it should be attributed to the state. There was little doubt that the sectarian religious messages that had been uttered immediately prior to the games for many years would, if spoken by agents or employees of the school, have violated the Establishment Clause. This is significant, given that all justices agreed that the Establishment Clause forbade school-sponsored prayer and that prayers by students in their private capacity were protected by the Free Speech and the Free Exercise Clauses.

Disagreement in *Santa Fe* arose only because the prayers were school-sponsored. Although students led the prayer prior to high school football games, the Court found that the prayer contravened the Establishment Clause because it constituted public speech (in this case, by being school sponsored), not private speech. The Court had little problem disposing of the private-public question by ruling against the school district. Relying on an elaborate history of the district's official involvement in decisions to support prayer exercises at commencement and other school functions, the Court found numerous doctrinal impediments to the school's policy. Building upon and expanding *Lee*, the Court held that the prayer policy involved state-supported coercion of public school children to acquiesce in prayer. The Court gave little credence to the district's arguments that no students were coerced by such a policy. The Court noted that players, cheerleaders, band members, and students who wished to participate in important school events were required to attend and found that none of these students should be forced to choose between avoiding a religious exercise and attending an event of importance to the school and community. Invoking the now-entrenched theory of endorsement, the Court held that the district's policy endorsed prayer at football games because the text of the policy explicitly mentioned "invocation" and "solemnization" as among the purposes of the pregame message and because the trappings of a high school football game would make any prayer over the public address system appear to an objective listener "as stamped with [the] school's seal of approval" (*Santa Fe Independent School District v. Doe*, 2000, p. 2278). The reluctance to find for the district became even more obvious when the Court referenced *Lemon*, which many commentators and even some justices had thought long dismissed, to find that the school's policy, understood in local, historical context, lacked a secular purpose and therefore constituted an unconstitutional establishment of religion. Relatedly, the doctrine of political entanglement, also played down in many decisions for more than two decades, reappeared in a new and exceedingly persuasive context. The Court's opinion reasoned that the elections authorized by the district's policy "entrust[ed] the inherently nongovernmental subject of religion to a majoritarian vote" (*Santa Fe Independent School District v. Doe*, 2000, p. 2283) and therefore violated the Constitution, regardless of the outcome of the vote. An election mechanism in which all understand that the choice focuses on the possibility of prayer "encourages divisiveness along religious lines" (*Santa Fe Independent School District v. Doe*, 2000, p. 2283); the Constitution bars the state from instigating such divisiveness. The student election, said the Court, effectively silenced the views of the minority, put the minority at the mercy of the majority, and did not protect diverse student speech. In fact, far from making the prayer permissible, the election served to intensify the offense of the minority. Viewed in this manner, the Court's resolution of the disputes in *Santa Fe* seems quite reasonable; it simply confirmed that a group plebiscite cannot obliterate an individual's liberty to decide whether and how to participate in a religious occasion. After *Santa Fe*, secular issues with religious overtones inevitably and appropriately remain to be resolved by the state's political processes, but political processes may not resolve issues of religious observance in public institutions. *Santa Fe*, then, solidified the notion that schools must not encourage prayer and that students are free to exercise their religious beliefs so long as they do so under their own volition and do not offend other students' rights.

Santa Fe portends change; the Court's recent reluctance to address the issues involved indicates that it may pause to let the lower courts address similar issues first. Most notably, the Court has yet to rule on a slew of cases that permit moments of silence that seem strikingly analogous to the moment of silence statute the Court held unconstitutional in *Wallace v. Jaffree* (1985). In *Brown v. Gilmore* (2001), the Court refused to review a Virginia policy that, like many other state laws, allows a moment of silence for meditation, reflection, or prayer. The policy explicitly states that its main purpose is to guarantee students' right to the free exercise of religion and to provide the least possible pressure on students to engage in or refrain from religious observation on school grounds. Yet, the *Santa Fe* Court found the Santa Fe district's policy facially invalid because the policy unquestionably had the "purpose and . . . perception of encouraging the delivery of prayer" (*Santa Fe Independent School District v. Doe*, 2000, p. 317). Since the statutes indicate that one of their purposes is to guarantee the students' right to the free exercise of religion, they ostensibly create the constitutionally impermissible perception that the schools encourage prayer. Given the Court's reluctance in *Santa Fe* to give credence to the statutory purpose of the statute, *Santa Fe* renders questionable the rationale for Virgina's statute: providing the least possible pressure on students to engage in or refrain from religious observation on school grounds. Despite the attempts to structure such policies so that they can survive constitutional scrutiny by offering only pure "moments of silence," the constitutionality of such laws remains uncertain, and the Supreme Court has yet to address the claim.

The second situation involving adolescents' religious rights in educational settings deals with whether and how a publicly funded open forum can allow students to participate in religious exercises. This issue first appeared in *Widmar v. Vincent* (1981), which asked the Court to determine whether state actors can prohibit private speakers from using university facilities for religious speech. In this leading case, the University of Missouri at Kansas City enacted a new regulation that prohibited an evangelical Christian group of students, which previously had been permitted to use the university's classrooms, from holding meetings that included prayer, hymn singing, Bible study, and discussion of religious experiences. Although the university permitted a large variety of organizations to use school facilities for meetings, the new rule prohibited use "for purposes of religious worship or religious teaching" (*Widmar v. Vincent*, 1981, p. 265). The Court began by noting that the university's provision of facilities to registered student organizations created an open forum; such a forum required the state university to justify its exclusion of religious clubs under the applicable constitutional norms. The norms at issue involved the Free Speech Clause, which the Supreme Court understands as prohibiting the singling out of religious views from other private expression for differential treatment; such treatment conflicts with the Free Speech Clause because it constitutes unconstitutional discrimination based on a particular point of view. By prohibiting religious clubs from using its facilities, the university imposed a content-based exclusion that could be sustained only by a showing that the exclusion was necessary to serve a compelling state interest and that it was narrowly drawn. The university asserted compliance with the Establishment Clause as its compelling interest. While agreeing that compliance with the Establishment Clause establishes a compelling interest, the Court did not find the "equal access" policy necessarily at odds with the Establishment Clause. In concluding that an open-forum policy that was nondiscriminatory toward religious speech would provide religion with only incidental benefits, the Court held that an open forum does not confer "any imprimatur of state approval on religious sects or practices" (*Widmar v. Vincent*, 1981, p. 274). The Court noted that, by providing access to a range of student groups, public institutions of higher education advance a secular purpose and do not excessively entangle the state with religion. By focusing its analysis on the expressive aspect of the students' devotional activities, the Court found that the university's ban on religious meetings violated students' free speech rights. The Court also was persuaded by the fact that the open forum was available to religious as well as nonreligious speakers; the Establishment Clause does not prohibit the extension of general benefits to religious groups. Although the university could establish reasonable restrictions on the time, place, and manner of meetings (or simply deny access to all groups), the Court found that the university must provide access to a forum created for student expression on state-supported college campuses. The Court, then, required state-funded, public institutions of higher education to act neutrally toward religious groups that sought the same accommodations offered to secular groups.

This area of law gained immediate significance because it could readily be expanded to cover primary and secondary school settings. At its core, *Widmar* involved an important exception to prohibitions of religious expression and protected students' rights to express their religious beliefs in an active group manner; it even potentially could be used to justify proselytizing. But, *Widmar*'s analysis did not address whether the Establishment Clause prohibited similar student-initiated devotional meetings held on public secondary school premises. Unlike university students, high school students could be viewed as a more captive and thus more vulnerable audience that might understand the meetings as representing the school's policies. Congress directly challenged that concern when it adopted the Equal Access Act (EAA) (1984, 1994). The Act requires "equal access" policies in public secondary schools under certain circumstances and in compliance with specific guidelines. The EAA provided that:

> It shall be unlawful for any public secondary school which receives Federal financial assistance and which has a limited open forum to deny equal access or a fair opportunity to, or discriminate against, any students who wish to conduct a meeting within that limited open forum on the basis of the religious, political, philosophical, or other content of the speech at such meetings. (Equal Access Act, 1984, 1994, § 4071(a))

In addition to requiring equal access policies, the EAA delineated certain criteria for what constitutes fair opportunity. They include the following: that meetings be student initiated and voluntary; that the school, government, or its agents or employees not be meeting sponsors; and that agents or employees of the school or government attend religious meetings only in a "nonparticipatory capacity" (Equal Access Act, 1984, 1994, § 4071(c)(1)-(c)(3)). The EAA further prohibited the states and their political subdivisions from requiring student participation in religious activities and from influencing the content or form of any religious activities. By enacting these provisions, Congress sought to distinguish clearly between activities initiated and controlled by students and those initiated and controlled by the school or government, simply because the Constitution permits the former and prohibits the latter.

The EAA apparently extended *Widmar* in ways the Court had not addressed, and its constitutionality was challenged in an important case, *Board of Education of Westside Community Schools v. Mergens* (1990). That case started when Westside High School denied Bridget Mergens's request that her Bible club be allowed to use school facilities; she promptly brought suit, claiming that the school's decision violated the Equal Access Act, as well as the First Amendment. In its analysis, the Court recognized that, in passing the EAA, Congress had extended the holding in *Widmar* to public secondary schools. The Court proceeded to interpret the meaning of the phrase "noncurriculum-related student group" to determine whether Westside High School was required to provide equal access to religious clubs (*Board of Education of Westside Community Schools v. Mergens*, 1990, pp. 234, 237). While a majority of the Court agreed with regard to the statutory interpretation, the Court was divided on the rationale needed to uphold the EAA as constitutional.

The plurality opinion examined *Widmar* and agreed with its conclusion that opening university facilities to religious groups sends a message of "neutrality rather than endorsement" (*Board of Education of Westside Community Schools v. Mergens*, 1990, p. 248). Following that logic, the plurality concluded that a school does not endorse student speech when that speech is allowed on a nondiscriminatory basis. The plurality did express concern that school actions might be perceived as representing the government's support of religion in the "eyes of impressionable youngsters" (*Board of Education of Westside Community Schools v. Mergens*, 1990, p. 250). However, it also observed "a crucial difference between government speech endorsing religion, which the Establishment Clause forbids, and private speech endorsing religion, which the Free Speech and Free Exercise Clauses protect" (*Board of Education of Westside Community Schools v. Mergens*, 1990, p. 250). The Court noted that the statute provided important safeguards; it prohibited active participation by school officials and limited meetings to "noninstructional time," all of which resulted, according to the Court, in little, if any, risk of state endorsement or coercion (*Board of Education of Westside Community Schools v. Mergens*, 1990, p. 251). By noting that the Act prohibits sponsors from participating in meetings, the Court dismissed concerns about faculty sponsors and other accouterments of secondary school life. The Court further reasoned that, "[t]o the extent a school makes clear that its recognition of respondents' proposed club is not an endorsement of the views of the club's participants, . . . students will reasonably understand that the school's official recognition of the club evinces neutrality toward, rather than endorsement of, religious speech" (*Board of Education of Westside Community Schools v. Mergens*, 1990, p. 251). Whether schools actually make their positions clear in practice even though they can do so in theory was not the Court's central concern. From the Court's view, secondary school students are mature enough to recognize that the school is not endorsing student speech that it merely permits. Given this line of reasoning, the Court rejected an Establishment Clause challenge to the EAA and affirmed secondary school students' rights to meet and to share religious beliefs on school grounds. The Court declared that, "even if a public secondary school allows only one 'noncurriculum-related student group' to meet, the Act's obligations are triggered and the school may not deny other clubs, on the basis of the content of their speech, equal access to meet on school premises during noninstructional time" (*Board of Education of Westside Community Schools v. Mergens*, 1990, p. 236).

The recent cases addressing the use of public schools by student-initiated groups reflect the judicial sentiment that religious speech and other private expression deserve equal treatment in public schools. This is distinct from, but often confused with, public school access by student-initiated groups during noninstructional time, which is governed by the EAA. When public schools are made available for community use, the school functions as a proprietor and does not supervise the activities as it does with student groups. In 1993, the Supreme Court held, in *Lamb's Chapel v. Center Moriches Union Free School District* (1993), that if officials allow groups to use the public school to present particular topics (i.e., family values, child rearing) from a secular perspective, they cannot bar meetings on these topics that include religious perspectives. In essence, school districts run afoul of the Free Speech Clause if they enforce policies on the use of their facilities that discriminate against a religious group's message. In *Lamb's Chapel,* the Court examined the school district's rules, which permitted use of school property by nonstudents for social, civic, or recreational uses but not for religious purposes. On the basis of those rules, the school district had denied the request of Lamb's Chapel to use school facilities to show a film series that offered a religious perspective on family life and child rearing. The Court found that control over the use of public property that is not a public forum can be based on subject matter but that the control must be reasonable in light of the forum's purpose and must be viewpoint neutral. A speaker cannot be excluded from a nonpublic forum because of the view that the speaker would espouse if the speaker's subject is one that would otherwise be includable in the forum. While such an exclusion is normally unconstitutional, the discrimination can be saved by an Establishment Clause defense. The school district's denial could be tolerated, that is, if permitting the use of the facilities by the group in question would be an establishment of religion. Here, however, the Court rejected fears of an Establishment Clause violation. The Court assumed that there was no real danger that the community believed that the school district was endorsing a general or a particular religion.

The reasoning in *Lamb's Chapel* has been extended to student groups, even those that want to use school premises not only for discussion of religious subjects but also for worship services in which adolescents (and children as young as six) are called upon to commit themselves to Christianity and to proselytize other students. In a recent case exploring those issues, *Good News Club v. Milford Central School* (2001), a student religious organization led by adults from a local church had been denied access to a public elementary school for an afterschool meeting place because the school's policy excluded groups engaged in religious instruction. Relying on *Lamb's Chapel*, the Court reasoned that the exclusion amounted to viewpoint discrimination and violated the group's free speech rights. The Court held that, since school officials had opened the door for outside groups to meet after school, school officials could not deny use of those facilities to groups expressing a "religious viewpoint." Unlike previous decisions that focused on fears that students might feel coerced to attend, the Court reasoned that parents must give permission to their children to participate in afterschool activities, which protects students from unwarranted pressures and helps clarify that schools do not endorse the meetings. Thus, the Court consistently has held that the Establishment Clause allows religious speakers to use public school facilities when there is no appearance of government endorsement of religion. So long as the government actor affords religious speakers who are not government actors the same opportunity to use government facilities as is given to secular speakers, there is no danger that a perception of government endorsement will exist. Absent such a perception, the actions do not violate the Establishment Clause.

Generally, then, adolescents possess no religious freedom rights inside their homes unless their parents choose to grant them such privileges. When outside the direct control of their parents, however, the rights of adolescents are beginning to mirror those of adults. Indeed, recent developments in adolescents' rights reflect the fall of the "wall of separation" metaphor, and litigation involving adolescents seems to be setting numerous leading legal precedents. The extreme focus on maintaining a firm wall of separation between "church and state" appears to have been replaced by the concept of nondiscrimination or equal treatment of religious and other personal speech. Without doubt, there remains much uncertainty, but the Court has announced some general rules. A state does not establish religion merely by granting access to facilities to students who are promoting religious beliefs as long as it grants a similar access to those who promote secular ideas. In fact, schools are required to treat students' religious and nonreligious speech equally once they have opened the forum to any student speech. However, a state cannot convey a preference for religious speech in the guise of allowing students to exercise their rights to free speech. The Court now concludes that private (personal) religious and nonreligious expression should be treated similarly under the Free Speech Clause, an approach that gives great weight to the religion clauses' First Amendment companion.

Critical Implications of Legal Mandates

We now have seen multiple features of the laws that regulate adolescents' experiences. Conceptions of adolescents' own rights relating to their religious experiences and environments remain pervasively unexplored to the extent that adolescents' rights generally do not factor greatly into decisions that seek to balance the rights of others. Although there may not be one, overarching jurisprudential or policy direction, five points seem especially worth emphasizing.

First, as with other areas of regulation involving adolescents, parents retain the plenary right to direct their children's upbringing. The significance is so great that, when community rights play into the balancing, their role tends to be quite stealthy and is framed in the language of parental rights and the extent to which broader communities can override those rights. As a result, the legal system frames the rights of adolescents as it addresses the rights of parents (or the state acting as *parens patriae*). For instance, *Meyer* directly protects the right of parents to control their child's upbringing; it does not protect a child's right to receive an education in a foreign language. Similarly, *Pierce* recognizes parents' right to direct their child's education by choosing parochial schools rather than private schools; it does not recognize the child's right to a private education. In *Kiryas Joel*, the Court denied the rights of parents to create a special school district to protect their religious beliefs and framed the opinion in terms of parents' (and, through them, the community's) concerns for their children and the right of civil society not to actively support religious sects, rather than in terms of children's rights to education. *Prince* authorizes the state, as *parens patriae*, to enact child labor statutes and to infringe on parental rights; it does not extend specific constitutional rights to children. Thus, it appears that when a conflict arises between adolescents' rights and those of their parents, adolescents' rights gain protection when there is a threat of physical harm, and that protection comes in the form of governmental action as *parens patriae*, not in the form of the right to act independently of one's parents.

Second, when addressing religious concerns, the legal system has tended to individualize communal experiences. Playing down the communal aspects of religiousness results in peculiar outcomes. First, it essentially rejects consideration of the free exercise interests of members of religions to which the idea of a worshiping community is central. This was the essential result in *Kiryas Joel*. In that case, the Court found it inappropriate to prevent a religious sect's children from associating with their neighbors and appropriate to allow contact with those who are different, even though it may result in "panic, fear and trauma" and even though two-thirds of the full-time students at the school in question were Hasidic children from outside the village (*Board of Education of Kiryas Joel Village School District v. Grumet*, 1994, pp. 694, 701–702). Second, the approach fails to consider the interests of individuals who see themselves as claimed by religious commitments they have not chosen. These individuals do not experience religious commitment as a result of a free marketplace of ideas, a metaphor that dominates jurisprudential efforts to treat religious beliefs and institutions in a neutral manner and to view religious beliefs as something that can be ignored. Prime examples of these considerations arise when religious convictions are ignored and assumed to be amendable to change, as evidenced in the aunt's convictions in *Prince* and the law's failure to accommodate. Third, and especially in terms of adolescents, it allows for a focus on the rights of parents. By focusing on parents, the legal system overlooks the impact of legal rules on the parents' adolescent children or on groups of adolescents. The *Good News Club* case illustrates well what may result from such a narrow focus. The environment produced by religious groups may leave some students feeling excluded from religious exercises and feeling coerced to participate, but those risks tend to be underestimated by the courts because they individualize the community aspects of religion, assuming that parents (not the adolescents themselves) can resist coercion by other adolescents.

Third, the current approach seemingly plays down the omnipresent power of the state. The legal system's approach to issues of coercion is illustrative. The Establishment Clause unequivocally forbids the state from coercing or compelling persons by direct legal sanction to conform to religious sentiments sanctioned by the state. But government has many different ways to enforce its preferences short of direct coercion. Government is ever-present in almost every aspect of our cultural and personal lives, and its influence may be felt in subtle ways. Government may indirectly coerce conformity to its religious messages by making nonconformity more difficult or costly, regardless of whether it carries any threat of legal penalty. Expanding the concept of coercion beyond direct or legal means provides a reasonable and necessary check on the power of the modern regulatory state. The Court's current approach to coercion, however, is limited as it determines what constitutes an indirect coercive threat. For example, in *Lee v. Weisman* (1992), the Court responded to a classic example of indirect coercion. In that case, the student could have avoided participating in prayers offered at her graduation ceremony by choosing not to attend; by avoiding graduation, she would not have been offended, nor would she have been forced to participate in religious exercises or say what her conscience would not allow. The approach embraced by the majority in *Lee* suggests that social pressure to engage in religious activity may give rise to an Establishment Clause violation *if* the government has created and retains sufficient control over the setting in which that pressure occurs. When the state creates the occasion, controls the program, and is the sole or principal religious speaker, it is easier to attribute to the state's coercive process any pressures to conform to that message that occur. The government violates the coercion principle when it forces individuals to choose between participating in or protesting a religious exercise and forfeiting some right, benefit, or privilege. This approach remains potentially limiting. So long as graduation prayers confront dissenters with the choice of participating unwillingly in a religious activity, openly dissenting, or electing to miss their graduation exercise, it makes little difference whether student volunteers or school officials deliver the prayers; participation in such prayers can hardly be considered voluntary. Rather than reducing the effects of peer pressure, the school's policy actually sanctions it, allowing students to use the machinery of the state to impose their approved religious exercise on nonadherents. The approach parallels the manner in which the Court approached Establishment Clause challenges to afterschool religious club meetings, as affirmed in *Good News Club*. It would be disingenuous to ignore the pervasively coercive environment of public secondary schools. It also is disingenuous to think that, in environments as controlled and authoritarian as public high schools, officially sanctioned student activities involve teachers in only a neutral, nonparticipatory, custodial fashion. How the legal system approaches the regulation of adolescents' environments, then, plays down the state's omnipresent role in adolescents' lives.

Fourth, the legal system does not necessarily address the rights of adolescents when adolescents are outside the control of their parents. This is most explicit in analyses of service providers' rights. Restrictions on the speech of publicly funded physicians, for example, pose a threat to patients' interests. Compelling physician speech, like silencing it, implicates both doctors' right to speak and patients' right to receive information. The Court characterized and decided *Rust* and *Casey* as though they involved a bipartite conflict between physicians' right to speak and states' right to regulate professionals, rather than a tripartite conflict among physicians' speech rights, government's power to regulate professionals, and patients' right to receive information. The Court's neglect of these audience-based concerns led it to formulate standards for judging the constitutionality of content restrictions on doctor-patient speech that do not necessarily protect patients. For example, restrictions on publicly funded physicians are unconstitutional only if patients have an "all-encompassing" relationship with their doctor that gives rise to a reasonable expectation of receiving comprehensive medical advice. As a result, publicly funded specialists can be forbidden to speak to patients about specific medical procedures. Further, under *Bowen v. Kendrick* (1988), teachers and counselors may be prohibited from providing full information and from engaging in open dialogue with students, despite their professional responsibility to do so. Government may not impose viewpoint-based restrictions that do not afford publicly financed physicians the opportunity to distance themselves from the state's message. Although physicians may distance themselves by offering disclaimers, they are not likely to do so if they agree with the government's position about a particular treatment. If a doctor's silence is compelled, that doctor's patients will be deprived of governmentally disfavored information without notice that they are receiving only partial information and advice. In the case of compelled speech, these patients will hear a biased message intended to steer them toward the government's preferred course of treatment, with no knowledge that the message stems from state opinion rather than from their physician's best medical judgment.

Last, the law fails to consider adolescents' peculiar circumstances. Adolescents' potential susceptibility to coercion is a clear example of this, as we have seen, but equally illustrative is the manner in which the legal system views adolescents' impressionability and their ability to perceive (or to misperceive) government endorsement of religion or of religiously influenced practices. The circumstances the Supreme Court let stand in *Bowen v. Kendrick* (1988) are illustrative. In *Bowen*, the Court did not find an unacceptable risk that the no-establishment principle was violated, even though the ultimate beneficiaries of services were adolescents in circumstances that could challenge their ability to decipher the biases behind the offered services. The Court therefore let stand the use of government funds to pay religious organizations to teach and counsel adolescents on a highly sensitive subject of considerable religious significance, even on church or parochial school premises and without any requirement that religious symbols be removed from the sites. The impact of adolescents' impressionablity under these circumstances appears worth considering, especially because the risk of imposing a government orthodoxy is particularly great. Pregnant teenagers seeking information from AFLA programs, for example, may be in crisis, both economically and psychologically, and, given the general paucity of adolescent health services, may be both unable to go elsewhere for unbiased and complete counseling and unaware that the counseling they are receiving is one-sided. When AFLA programs take place in public schools, the arguments against government orthodoxy seem even stronger because of the compulsory character of school attendance and because of the role of public educators in preparing students for full citizenship participation in a democratic society. Against a background of deference to the school board's authority to define the curriculum, the Supreme Court has recognized the free speech rights of students in the public school setting to speak, to "opt out" of speaking, and to receive uncensored ideas (Levesque, 2002). When the education and health programs take place in sectarian, non-school settings, their impact may even be greater; locating AFLA programs away from public schools removes the traditional political checks that might ordinarily be provided by the reactions of school boards, teachers, taxpayers, and state education departments. AFLA grantees are not likely to protest against the program restrictions, because they have already been preselected (they by definition accept the anti-abortion orientation required of grant applicants) and already censor their speech because of their own preexisting missions; the subject matter of AFLA closely aligns with the fundamental tenets of some religious faiths. Under such circumstances, the government's speech will not be neutralized by political or institutional forces but instead will align with sectarian and ideological interests. Indeed, the reasoning behind Charitable Choice is that faith-based organizations have a better track record than the government in solving adolescents' problems precisely because of their religious orientations (see Hamilton, 2001). Given that reality and the very aim of AFLA, the law formally assumes that adolescents are not impressionable, an assumption that renders the law permissible, while hoping that they are not (and that they will accede to the law's effort to have them act as the government, and some dominant religious beliefs, would have them act).

Conclusion

Given the wide diversity in approaches to adolescents' place in law and society, it is not surprising to find that legal responses to adolescents' religiosity and religious environments are a mixed bag. No simple calculus determines the nature of religious rights as they affect adolescents and how those rights may be balanced with the rights of others. When the legal system does address adolescents' own rights, the analyses remain limited, and the extent to which the rights are tailored to adolescents remains debatable. Although the current laws seem comprehensive in their regulation of adolescents' religious development and environments, a close look reveals that they have important limitations. Even when the needs of adolescents are thought to be addressed, other values are at stake.

The lack of simple rules makes inevitable the need to take a close look at adolescents' place in society and law. To a large extent, adolescents continue to be the objects of regulations rather than their subjects. For example, the parental presumption focuses only on extremes: only real or potential harm to the child or the burden that a parent's actions threaten to place on society leads the legal system to recognize that adolescents have rights or potential interests different from those of their parents. Likewise, communities can exert important control over adolescents' religious identities; they have so much power that the leading cases that deal with communal claims do not even bother to address adolescents' rights. We have seen that some cases that deal with service provision do address adolescents' rights, but those discussions remain far from complete. To complicate matters, when adolescents' rights are addressed, even the most liberal policies remain quite limited. The legal system has yet to confront its extensive regulation of religious concerns, its lack of appropriate responses to private actions, and its pervasive failure to even consider adolescents' peculiar needs.

PART IV

Conclusion

6 Not by Faith Alone

Religion and matters of faith played powerful roles in everyone's lives, including the lives of nonbelievers, long before modern social sciences turned their attention to religious institutions and their impact. Indeed, the social sciences often have ignored the manner in which religious institutions and beliefs affect society's response to individual and social needs. Although considerable attention is now being paid to the subject, the resulting research remains narrow. None has focused on how religions' roles and new policy developments relating to religion affect adolescents' development and their place in civil society.

The previous chapters suggest that four factors account for the pervasive failure to examine more systematically the impact of religion on adolescents and to explore more deliberately how social policies regulate adolescents' religious environments. The first is the misperception that the legal system formally excludes consideration of religion, that there exists a firm "separation of church and state." As we have seen, the common use of the phrase does not match its relevance to the task of legal interpretation or to the law's impact on religion. Absolute separation between the state and religious institutions and beliefs is neither a goal nor a possibility. There is a long history of attempts to remove religion from public policy, even though, as we have seen, religious values play powerful but highly stealthy roles in policy-making. Despite this, religion continues to be viewed as a private matter of personal belief, with little practical relevance to policy, largely because considering religious issues in public policy-making would suggest the imposition by the government of religious beliefs. This belief assumes the possibility of a clean and clear distinction between the religious and the secular and constitutes the second reason religion is ignored by policymakers: the distinction between public and private may be recognized by nonreligious individuals but is likely to arouse objections among religious individuals, who are likely to find faith relevant to all aspects of their lives. The assumption that people can and should separate their religious beliefs from their social and cultural identities ignores the inseparability of religious (and antireligious) beliefs from people's worldviews. The third factor is the popular perception and social tradition that affirms parents' right to control, guide, and even determine both adolescents' religious orientations and their religious environments. Although parents can and often do play powerful roles in adolescents' lives, perceptions of that power frequently overestimate parental influences. Indeed, our close look at religious environments, the impact of religious beliefs, and the development of religious orientations suggests many reasons to question the extent to which parents directly control adolescents' religious experiences and developments. This third factor leads to the fourth—the wide power the law bestows upon parents to raise their children as they see fit and the prevailing tendency to treat adolescent minors as children under the law. As we have seen, parents do retain wide latitude, but the legal system, to a degree that varies with the context, allows adolescents to exercise their own religious rights and protects those rights from state-sponsored encroachment. Even where parents retain great control, the law allows the state to intervene to protect the rights of adolescents or to address potential conflicts that may arise when the rights of parents interfere with the broad interests of civil society, local communities, or others who interact with adolescents. Equally important, the law impacts religions, which, in turn, impact parents and adolescents—and the influence on adolescents does not necessarily flow directly through parents, as seen by the regulation of religious communities and adolescents' peers. To conclude, then, that religion is of little significance to policies that affect adolescents requires us to ignore the fundamental social, developmental, and legal realities that impact adolescents' religious development, their religious environments, and the demands of living in a civil, religiously pluralistic society.

To suggest that there are significant misperceptions about adolescents' place in religion, law, and policy, and to challenge accepted beliefs about parents' control of their children's religious development, is undoubtedly controversial. Yet, that these misperceptions exist undergirds the importance of evaluating the role of religion in adolescent development and policies that affect adolescents. Several features of the historical moment, coupled with these misperceptions, render urgent a close look at intersections among religion, adolescent development, families, communities, and the law.

A Critical Juncture

Adolescents today face changed and often troubled realities in their families, communities, and peer relationships. Adolescents live in distressed families and communities that limit opportunities. An increasing number of families struggle to nurture and care for their children and even to provide basic physical necessities (Montemayor, Adams, & Gullotta, 2000). Likewise, adolescents painfully experience a loss of effective community ties and supports. Many locales no longer offer the important sense of place and connection to others that long has characterized neighborhoods and local institutions such as schools (see Weissbourd, 2000). In addition, large numbers of adolescents grow up without opportunities to ponder transcendent meanings, and religious communities often fail to address adolescents' needs (Roof, 1994). Evidence of the resulting failure to support adolescent development abounds, particularly in rates of victimization, school failure, and unmet mental health needs among adolescents (see Levesque, 2002). Importantly, even if adolescents are able to grow up in stable families and in communities that have strongly established social and religious groups, these groups, even with the best of intentions, may in fact harm adolescent development or foster biases against individuals or members of other groups and, by doing so, place a burden on broader society. Adolescents, families, and communities cannot ignore adolescents from their own and other communities; adolescents face many communal challenges even as communal institutions themselves face challenges.

In addition to changes in social relationships, shifts in governments' role in providing for basic human needs makes necessary a reconsideration of religion's role in adolescence. Even as communities and private community groups confront major social challenges, they are increasingly assuming responsibility for providing necessary social services. Although religious institutions always have played a dominant role in the provision of social services, the pace of this shift is indicative of a lessening sense of public responsibility and accountability in this area (see, e.g., Bane, Coffin, & Theimann, 2000). In addition to proving potentially problematic for adolescents, this change may prove problematic for the very fabric of society. The push for using public dollars to pay for private and religious schools, for example, challenges the shared context that connects private and public lives. Current reforms encourage use of public funds to educate adolescents *outside* the common, public system. This focus dismantles the very idea of a common school committed to the inculcation of a shared American tradition and capable of integrating students from a variety of backgrounds (Levesque, 2002). The policy shift runs deep. The recent injection of private religious options into the delivery of welfare services becomes particularly worrisome in view of the rampant discrimination revealed by numerous evaluations of child welfare services (Levesque, 2000b). Discrimination in the provision of social services is difficult to eradicate, and the risk of discrimination may increase as we increasingly rely on private religious groups to provide services (see Minow, 1999).

There have been revolutionary developments in the laws and policies that regulate religion, family life, and adolescents. We have seen how the legal system increasingly entangles religion with matters of state. For example, the legal system has become increasingly receptive to claims that allocations of public funds for social services must not discriminate against religious groups that also offer those services. Likewise, we have seen how the legal system now appears ready to intervene in family life and to protect adolescents from religious behaviors. Recent changes in medical neglect laws have reversed past efforts to protect the religious rights of parents. Although parents and religious communities still retain the right to raise children according to their religious dictates, the states appear increasingly willing to intervene and protect children (see, e.g., Palmer & Hardman, 1999). Last, we have seen that the legal system increasingly recognizes that adolescents are persons worthy of constitutional protection in their own right. For example, the legal system now bestows on adolescents procedural rights when they become involved in the juvenile justice system, a recognition fueled by the due process revolution of the 1960s (see Levesque, 1997). The imprints of these legal developments are most prominent in the public sphere. The legal system protects adolescents' expression of religious views and prevents state officials from supporting environments that would coerce adolescents into supporting or professing certain religious beliefs (*Santa Fe Independent School District v. Doe*, 2000). All of these developments in the regulation of adolescents' private and public lives reflect a recognition of the need to protect adolescents from harm and to prepare them for responsible citizenship. These developments, however, remain far from systematic, coherent, and as liberative and as pernicious as alleged by those who champion a more aggressive turn to parental rights (see Hafen & Hafen, 1996). Many incongruities mark the legal systems' responses to adolescents, and developments in law and policy do not necessarily respond well to the current understanding of adolescent development, adolescents' place in society, and the social world they inhabit.

The extent to which the legal system fails to respond to adolescents' peculiar needs, and thus those of society, is yet another reason to reconsider the legal regulation of religion as it affects adolescents. The legal system has had considerable trouble responding to the current understanding of adolescent development. Much of the failure is understandable in light of the nature of the law; law cannot respond to or address one issue without considering that issue's impact on others. Given the extent to which adolescents must rely on others to meet their needs, it is necessary to consider other peoples' rights and responsibilities as one considers the jurisprudence that affects teens. However, the legal system cannot achieve greater efficiency and ensure justice if it responds to misperceptions about adolescents, rather than to reality. The traditional view of adolescents is that they are passive recipients of social beliefs and traditions, including religious ones. Although that view appropriately highlights the relative power of social forces and their dominance over adolescents, it fails to note adolescents' active participation in their own development. As we already have seen and will see more clearly, research on how adolescents idiosyncratically evoke, interpret, and respond to their social environments continues to revolutionize the study of their development (see Levesque, 2000a, 2002). This new understanding of the adolescent period reinforces the need to examine the legal regulation of adolescents' religious development.

Given the context-dependent peculiarities of the legal regulation of adolescents, the legal system should welcome an improved empirical understanding of adolescence to ensure that its provisions closely correspond to adolescents' realities. Yet, political or moral expedience often substitutes for policy based on evidence from the social sciences. As we have seen, many appropriately lament the legal system's failure to consider the realities of adolescents' experiences and attribute the failure to the lack of an overarching doctrine, a lack that results in inconsistency, confusion, incoherence, and an apparent inability to recognize adolescents' peculiar needs. To a large extent, the diversity of views of and responses to adolescents' place in society reflects the complexity of adolescence, making essential attention to specific contexts. The lack of an overarching doctrine and the need to accept context-dependent approaches, however, need not necessarily translate into inadequate, ill-thought-out, problematic, or even harmful approaches to the regulation of adolescents and their social contexts. Given the existence of important social science findings related to adolescents, we find peculiar the legal system's continued hesitancy about considering and benefiting from them.

The misperceptions, diversities, failures, and complexities we have discussed do much more than suggest a need for the current evaluation. They complicate our effort, though they do not render it impossible or futile. The previous chapters suggest that a sharper focus on adolescents' basic rights and a firmer social science understanding of adolescence could help us achieve the desired outcomes of secure, healthy, and responsible adolescents. This focus, importantly, responds to the complexities of adolescent life. Thus, it does not unnecessarily limit the firmly established rights of parents and communities to influence their children's religious environments and development. Instead, it clarifies and supports those rights and rejects the popular view that envisions and assumes the inevitability of a clash between adolescents' rights and those of their communities—for, whether or not adolescents, parents, or the broader society recognize it, adolescents are community members. This point deserves emphasis, since those who adhere to a firm parental rights model may lose sight of the fact that others too have rights that must be considered.

Although the following analysis considers extending innovations in adolescents' rights, these expansions are far from radical and are not beyond the realm of possibility. Likewise, although the analysis suggests taking an alternative direction to achieve a traditional end, the proposals are consistent with numerous legal mandates. These suggestions are supported by existing legal structures that are largely undeveloped or simply ignored. The proposal, then, in some ways is quite simple: we can better develop legal and policy mandates, respect adolescents' basic rights, reflect the realities of adolescent life, and foster healthy individual and social development while addressing adolescents' religious needs and their religious environments. To achieve those ends, we propose the following as one possible course of action.

Steps Forward: Law, Policy, and the Developmental Sciences

The legal system undoubtedly can benefit from insights generated by research in the social sciences. The diversity of approaches to addressing different contexts, and the obvious attachment of religious beliefs and religious environments to deep moral and political issues, suggests that empirical accounts may not contribute greatly to the legal regulation of religion that affects adolescents' experiences. Yet, regardless of the political and religious stances taken by policymakers and those who interpret and implement existing laws, they must address adolescents' peculiar place in society and balance adolescents' rights against those of their parents, communities, and civil society. The hope that adolescents, parents, and society can benefit from findings in the developmental sciences as they relate to adolescents and religion lies in policymakers' ability to integrate these findings into the values that guide jurisprudence and public policy. Regulating adolescents' behavior and environments in ways consistent with social science suggestions necessarily requires a close examination of the rights of adolescents in different contexts and across different relationships.

The discussions in the preceding chapters suggest some basic principles that can guide policies and reforms that seek to balance more effectively the rights of adolescents, parents, communities, and the broader society. The principles champion respect for adolescents and ensure young peoples' membership in a society that takes their interests seriously and that fosters the growth of citizens who also take their own and societal interests seriously. The following discussion examines these principles, clarifies where they diverge from current law and policy, evaluates available social science suggestions regarding the appropriateness of the principles, and responds to potential challenges to a serious consideration of their content.

Two caveats arise as we move from a more descriptive to a more proscriptive analysis. First, the highlighted principles are meant to serve as guiding tools. None has particular talismanic qualities; all the principles deserve close consideration, and their content obviously must be weighted differently according to contextual requirements and to the peculiar exigencies that arise from the need to respect other people's interests and those of civil society. Second, the principles all interrelate and even overlap. The models certainly could be much more conceptually tidy, their essence shaped to maximize their independence, but such contorted models and efforts would not necessarily offer much in terms of practical utility. The conflictual interactions among the worlds of law, religion, and adolescence present complex and, when in need of rebalancing, highly context-dependent outcomes. Bearing in mind these realities, we now revisit our previous discussions to detail possible directions in the law's approach to adolescents' religiosity and religious environments.

Recognizing Adolescents' Competencies

Traditional jurisprudence in adolescents' rights generally bestows the control of adolescents' rights upon parents. For example, parents generally control the context of family life. Where parents do not have control, the state intervenes to act as parents otherwise would to control adolescents' rights. It may therefore be difficult to imagine that this system, which also bestows upon adolescents control over their own rights, substantially differs from traditional orientations. Yet, modern jurisprudence has accorded adolescents increasing control over their own rights by allowing them to share in the experience of autonomy rights enjoyed by adults. Despite the greater attention to adolescents' rights in recent decades, this trend has not eliminated all paternalism from the legal treatment of adolescents. Importantly, even when innovative developments in adolescents' rights do seek to recognize adolescents as subjects of their own rights, those developments firmly attach to traditional jurisprudence and continue to recognize the fundamental reasons for giving parental rights much of their valence and the government so much power—to protect adolescents and to produce socially responsible citizens (see Levesque, 1997). In terms of law, then, taking adolescents' competencies seriously might mean beginning to bestow the control of adolescents' rights directly upon adolescents themselves, rather than continuing to confer them upon parents or the state.

The law's attitude toward the question of whether some adolescents' decisions should be treated no differently from adults' decisions largely has been based on assumptions regarding minors' decision-making competence. As we have seen in chapter 5, laws that deal with adolescents' competencies and their place in the law often markedly vary, depending on the context. The law sometimes must explicitly or implicitly judge the competence of minors to make significant decisions about the course of their lives. To the extent that minors exhibit competence as great as that displayed by adults, it seems to follow that their decisions should be respected, at least no less than those of similarly situated adults. Respecting adolescents' decisions, as with adult decisions, can mean depriving the adolescent of the beneficial consequences of those decisions on the grounds that those consequences are outweighed by negative consequences either to the decision maker or to others. Respecting an adolescent's decision also might mean allowing the teen to experience negative consequences. Given the risks involved, the determination of what constitutes appropriate competency undoubtedly is quite tricky. The social sciences have much to offer as we strive to formulate general rules in this area.

An increasingly large body of work has empirically examined the quality of adolescent decision making and its similarity to or difference from that of adults (see, e.g., Woolard, Reppucci, & Redding, 1996). Although early work in this area received considerable criticism for being politically biased, vague, and based on imperfect assumptions, the work nevertheless yielded significant conclusions and trends. The vast majority of studies on the cognitive elements of decision making have supported the conclusion that the decision-making competence of older adolescents does not differ greatly from that of adults (Scott, Reppucci, & Woolard, 1995). That research has now been complemented by studies that index a wide range of factors that might lead adolescents to outcomes significantly different from those that adults would experience when making significant life choices. Existing literature on these noncognitive factors tends to support the notion that the differences between older adolescents and younger adolescents are more prominent than are differences between older adolescents and adults (Steinberg & Cauffman, 1996; Scott, Reppucci, & Woolard, 1995). Most notably, studies that focus on the ability to make decisions independently, that is, free of parental or peer group influence, show that, while peer group pressure clearly increases during early adolescence, both peer pressure and parental influence diminish during the high school years. For other factors, such as impulsivity and sense of personal responsibility, there are also greater differences between early and late adolescence than between late adolescence and adulthood. Thus, even when a particular context demands more than cognitive skills to make sound decisions, adolescents may not be particularly deficient in those skills, or so incompetent as to influence the way others treat them.

Decisions about religious matters are no exception to this rule. In terms of adolescents' religious environments, religious development, and religious consciences, existing research reveals many complexities and—although much still remains to be determined—important conclusions. Parents greatly influence adolescents' religious development and religious beliefs. This influence is particularly powerful with regard to early experiences that establish the working models individuals use to approach their social worlds. As we also have seen, however, by adolescence, children's religious beliefs are impacted by factors besides parental influence. Available religious institutions, community beliefs, peer groups, and social environments such as schools and social service providers also impact adolescents' religious development, attitudes, and behaviors. In addition, religious development is an active process, a process of exploration within the available social environments. Indeed, the most recent research reveals that the most healthy forms of religious identity emerge through adolescents' active exploration of alternatives and subsequent decisions regarding their own religious orientations. Likewise, it becomes increasingly clear that to understand the social influences on adolescents, one must recognize the many recursive influences on those who interact with them, such as the impact of the community and peers on parents.

These findings suggest that matters of faith may not involve issues of competence as traditionally described by available research. Matters of faith seemingly depend on others for their development, and decisions regarding matters of faith clearly depend on others to the extent that religion involves communal experiences. Likewise, decisions on matters in which faith plays a part (such as medical decision making) also are complex, with religious issues weighing differently in decisions depending on the adolescents' faith and the nature of the religious influences around them. Adolescents, then, actively make decisions about the significance of religion in their lives; religious beliefs impact their other decisions; and others are involved in the development and exercise of their faith. To a large extent, we may conclude that adolescents seem competent enough to have faith in a religious sense and to hold religious beliefs and attitudes worth respecting. That conclusion is significant: adolescents can be deemed to hold religious convictions.

It remains to be determined whether the law can benefit from these findings. The research suggests the need to at least consider ways that the legal system can better address adolescents' religious needs as it grants adolescents greater control over their own rights and as adolescents respond to religious environments and contexts that may be influenced by others' and their own religious beliefs. A close look at the current laws that regulate adolescents' religious environments and religious development reveals that providing greater respect for adolescents' abilities to make decisions regarding matters of faith is both feasible and far from radical. In fact, it is much less extremist and less radical than the traditional trends that have viewed adolescents as possessions over which parents and others must exert power and control. Two examples demonstrate that the approach remains consistent with less extreme traditional laws *and* that it respects traditional concerns.

Providing adolescents with greater autonomy to exercise their religious rights requires accepting the state's control of adolescents, a reality that is not surprising given that the state also ultimately controls adults. That limitation, however, still leaves immense leeway. Our approach proposes that greater scrutiny be given to efforts that deny adolescents the rights others enjoy. That is, it simply demands that adolescents be treated with the respect given other groups and that they not suffer from invidious discrimination. For example, innovative policies already ensure that adolescents have exit options, for example, in the form of judicial relief or social service delivery when their rights would otherwise be violated by parents or others acting in ways that control adolescents. We have seen that the highly charged example of parents' rights to control their children's reproductive freedoms provides the starkest example of the many ways the law skirts parental rights to provide adolescents with ways to exercise their own rights when they conflict with those of their parents (see also Levesque, 2000a). We have seen other examples, in the context of community service requirements in public schools, that have been challenged on students' own free exercise claims, with the result that schools must provide alternative programs. Schools even have included religious harassment in their new policies to combat school violence. All of this has been spurred by recent Supreme Court decisions that recognize the need for schools to address civil rights violations against adolescents (see Levesque, 2002). Likewise, we have seen that courts have required alternative services when delinquent adolescents would otherwise be required to participate in treatment programs that require the expression of religious beliefs. All of these developments directly protect adolescents' religious rights and shield them from environments that infringe on those rights. Importantly, adolescents have the competence to hold religious convictions, and such convictions are worthy of respect.

Although these laws may remain controversial, they respect traditional concerns to the extent that they take seriously society's interest in assisting victimized and vulnerable adolescents. This societal interest challenges efforts to justify denying adolescents access to resources and benefits simply because they are not seen as adults. The extent to which adolescents can receive greater protection from victimization because of their own religious orientations reveals how the legal system can respond in a manner that takes adolescents' needs more seriously. Traditionally, when harm occurred in relationships with families and service providers, adolescents did not receive assistance because they were not recognized as legal actors and essentially were not recognized as having been harmed. Our approach's plausibility and feasibility finds clear expression in the manner in which some states provide adolescents with direct access to courts and services and the manner in which schools now must enact policy responses to religiously motivated harassment and conditions that fail to respect adolescents' religious rights.

Although the legal system increasingly recognizes the peculiar legal needs of adolescents and their potential competencies, the recognition that adolescents possess significant competencies has not eliminated continued challenges to the idea of conferring on them greater access to legal mechanisms meant to protect them from harm and to foster their healthy social development. The challenges may seem particularly great, but they reveal the complexities that jurisprudence and litigation must address even when the rights of adults are at issue. Even these recent developments accept the realities of adolescents' special incapacities and their roles in families and communities. This means that adolescents' increased control over their own rights as they gain rights does not translate into society's abandoning them to their own rights. Like adults, adolescents remain social beings with social responsibilities. The next issue becomes, then, how to determine when, and in what contexts, adolescents should be allowed to express and experience religious environments and beliefs as they see fit.

Supporting Adolescents' Dynamic Self-Determination

Recognizing that adolescents may be competent enough to hold religious beliefs worth protecting and that they may be deemed legally competent to control their rights in some contexts underscores only part of what the legal system may need to consider if we are to take adolescents' religious rights seriously. Taking their rights seriously would mean that the legal system would respect their right to self-determination and structure their social environment in ways that foster self-determination simply because it remains unlikely, and inappropriate in most contexts, to bestow upon adolescents total control of their rights. Their rights and obligations must be balanced against those of others. In envisioning how to achieve this balance, the need to enhance dynamic self-determination is key in guiding the development of policies to regulate adolescents' religious development and environment.

The central tenet of dynamic self-determination directly focuses on the manner in which the legal system bestows and frames rights. The standard suggests that adolescents must gain more control over their own decisions as they develop and that they should enjoy autonomous rights consistent with their evolving capacities. Most fundamentally, the principle demands that adolescents be allowed space within their personal relationships, families, religions, and cultures to find their own mode of *individual* fulfillment. As conceived, self-determination aims to position adolescents to develop their own perceptions of their well-being as they enter adulthood, rather than foreclose on the potential for such development. The principle also involves ensuring that the legal system structures adolescents' environments so that they can recognize violations against their religious rights; the legal system can play a powerful role in raising adolescents, parents, and broader society's awareness. Despite the focus on individual fulfillment, the principle recognizes the social nature of development and seeks to involve adolescents in community membership and to foster responsible citizenship. A close look at social science evidence and the law reveals that efforts to achieve an appropriate balance are already occurring *and* that adolescents, families, and society would benefit from legal environments that recognize and respond to adolescents' need for self-determination.

Much controversy clearly would accrue to any suggestion that adolescents would benefit from being allowed to explore, question, and resolve their own matters of faith. Even more resistance would arise if the proposal were to suggest that parents and communities would benefit from such self-determination. Yet, existing research clearly favors those proposals, rather than authoritarian, external impositions of religious beliefs and practices. Active self-determination in supportive environments may be difficult to imagine after years of thinking of religion as something imposed according to traditional beliefs, but the reality that emerges from research is that traditions become most relevant to individuals whose beliefs result from internal pressure, rather than from external sources. This is especially true in our complex, civil society, which lacks clear religious rituals for adolescents that would cement respect for religious principles and traditional hierarchies (as are present in more traditional, violent rites of passage, for example). In authoritative environments, adolescents actually become more religious, in the sense that they search for meaning in life, evaluate the importance of faith in their lives, commit themselves to community causes, and become more accepting of others' religious beliefs and practices. We also have seen that adolescents who approach religious issues in this manner have the healthiest social and psychological outcomes. Those who fear that adolescents may become irreligious and shirk responsibilities, then, may feel confident that the opposite seems to occur. Those who fear that adolescents will become more religious may be the ones with more grounds for concern, but their concern loses legitimacy to the extent that our legal system formally seeks to protect matters of faith and protect individuals' religious choices.

As with other legal developments discussed earlier, existing legal systems already attempt to foster self-determination. The use of the maturity standard illustrates the feasibility of this approach. The standard, for example, has been used to determine access to medical services, including access to abortion, medical testing, and treatment. Following their initial reluctance to interfere with the judicially recognized sacred duty of parents to raise their children, the courts have since the late 1970s taken a more active role in ordering medical care for children, religious objections by parents notwithstanding. In some instances, the courts have been willing to order life-saving and even life-enhancing medical treatment, no matter how remote the threat of death or how dangerous the procedure, so long as the medical treatment offers the hope of improving the child's health. Through these developments, the courts have infringed on the rights of parents, including their religious rights. Thus, as adolescents' rights have come to be seen as fundamental, and adolescents have been deemed able to determine the outcomes they themselves want, jurisprudence and legislation has increasingly allowed adolescents access to services, courts, and extraprocedural protections.

Although our legal systems increasingly recognize adolescents' right to self-determination, the principle is still applied narrowly and in limited contexts. For example, current conceptions of the maturity standard generally do not allow for direct intervention into family dynamics and redirection of familial dynamics consistent with adolescents' levels of maturity. Nor has the principle typically been used to regulate adolescents' educational environments and their access to such environments. In those instances, clear lines dictate how groups of adolescents gain rights. Few jurisdictions expansively accept the need to better recognize adolescents' right to self-determination as part of the effort to respect adolescents' individual rights.

Given the limited attention that is formally paid to adolescents' self-determination, it is fruitful to examine instances where the legal system distinguishes less between adolescents and adults and more between adolescents and civic concerns. Such instances highlight the critical point that respecting adolescents' need for self-determination does not mean that adolescents' rights necessarily trump those of their parents and their communities, but it does mean that adolescents should be treated as individuals able to determine their beliefs and to suffer the consequences when those beliefs lead to harm to themselves or to others. The Supreme Court's response to the recent slew of statutes to stem hate crimes, including crimes resulting from religious hate, reveals the limits of self-determination. Interestingly, both of the leading cases that set the standards involved adolescents and, equally interestingly, the fact that adolescents were involved simply did not matter to the resolution of the constitutional issues. The first case, *R.A.V. v. City of St. Paul* (1992), involved a group of juveniles who shaped broken chair legs into a cross and burned it inside the fenced yard of an African American couple who lived across the street from the juvenile who appealed his conviction. The city chose to charge the juveniles under the St. Paul Bias-Motivated Crime Ordinance, which provided that:

> Whoever places on public or private property a symbol, object, appellation, characterization or graffiti, including, but not limited to, a burning cross or Nazi swastika, which one knows or has reason to know arouses anger, alarm or resentment in others on the basis of race, color, creed, religion or gender commits disorderly conduct and shall be guilty of a misdemeanor. (*R.A.V. v. City of St. Paul*, 1992, p. 2541)

Although recognizing that such statutes try to address hate-motivated conduct, the Supreme Court nevertheless found that they abridge First Amendment rights by regulating the content of expressions. In other words, the ordinance invalidly attempted to regulate expression on the basis of favoritism, or hostility to a certain message. The legislature's power to proscribe "fighting words" does not extend to the proscription of "fighting words" that embrace a particular message disfavored by lawmakers (*R.A.V. v. City of St. Paul*, 1992, pp. 2544–2545). The Court conceded that the city did have a compelling interest in protecting the physical and emotional well-being of citizens who traditionally have been harassed, but it also noted that the only reason for the city to impose a content-based proscription on expression was to express "their disfavor with certain types of expression" (*R.A.V. v. City of St. Paul*, 1992, pp. 2549–2550). The Court found the legislature's efforts unconstitutional, holding that nonverbal expressive activity can be banned because of its action but not because of the idea it expresses. This meant that, while the defendant in *R.A.V.* could have been prosecuted for violating state statutes that prohibit criminal damage to property, threats of terror, and arson, he could not be prosecuted under an ordinance that punishes the burning of crosses and other hateful expressions.

The Supreme Court construed hate action differently in the next opportunity it had to address the issue, in *Wisconsin v. Mitchell* (1993). In that case, as we have seen, the Court unanimously supported the state's penalty-enhancement statute for hate crimes. In *Mitchell*, an African American adolescent received an enhanced penalty for a criminal assault on a white teenager who was targeted after the adolescent had watched a scene from *Mississippi Burning* in which a gang of white thugs beat up an African American child while he was praying. In his ruling, Justice Rehnquist distinguished *R.A.V.* from *Mitchell*, noting that the former case concerned the content of speech, while the latter concerned "conduct unprotected by the First Amendment" (*Wisconsin v. Mitchell*, 1993, p. 2196). The distinguishing features of these cases are their violations of what have been referred to as "hate speech statutes," as opposed to "hate crimes statutes." The Court noted that "bias-inspired conduct" is "more likely to provoke retaliatory crimes, inflict distinct emotional harms on their victims, and incite community unrest"—all valid state concerns (*Wisconsin v. Mitchell*, 1993, p. 2201). Accordingly, Wisconsin's penalty-enhancement legislation passed the constitutional balancing test: the interest of the state in redressing these harms outweighed its obligation to protect the beliefs or biases of the defendant.

The two cases delimit the contours and the significance of self-determination. *R.A.V.* stands for the freedom to express one's thoughts and to act on them within the context and confines of content-based expressive conduct, while *Mitchell* stands for the state's freedom to regulate proscribed conduct and to impose more severe sentences for crimes motivated by thoughts and beliefs. The distinction between the cases is important because the perpetrators of hate crimes cannot shield themselves with First Amendment speech protections when they commit crimes in order to express their hatred for a group. The holding in *Mitchell* punishes more severely those who succumb to these protected predilections. The approach is significant; it allows individuals to express themselves, and it assumes that they can determine their actions, regardless of whether they are adolescents. The approach also is significant for one of the key jurisprudential principles on which it stands—faith in the marketplace of ideas. While it could be argued that *R.A.V.* supports belief in the marketplace of ideas that nurture hate crimes, the Court addressed that concern rather pointedly. A society dedicated to a free exchange of ideas from diverse communities has the responsibility, if not the obligation, to confront bias-motivated hate messages, and, equally important, a community that confronts hate messages will ensure that destructive messages do not win out in the marketplace of ideas, including adolescents' marketplace of ideas. Society has an obligation to ensure support for responsible self-determination, even and especially for adolescents.

Although these cases directly concerned adolescents and highlight the significance granted to the market of ideas in order to support adolescent development, other settings also take self-determination in a marketplace of ideas seriously enough to allow infringement on religious rights. Especially notable are the cases that regulate school settings that allow the exclusion of religious claims in order to further interaction with prescribed curricula deemed necessary for adolescents' healthy development and the proselytizing of some students by other students. For example, a common religious freedom claim involves religious parents whose children are required to hear or read objectionable material in the public school curriculum. Religious claimants have asserted a right to "opt out" of classroom assignments that allegedly promote "secular humanism," most obviously in the form of sex education programs that allegedly push promiscuity and physical education classes that expose students to allegedly immodest dress. The parents and organizations that have brought these challenges typically lose most of these claims (Levesque, 2002). The courts often couch their reasoning in the need to grant school officials the flexibility necessary to respond to the public's interest in fostering responsible citizens. The proselytizing cases also reflect the recognized need for adolescents to interact with unlike peers. In those cases, the courts evince a greater willingness to make age-based distinctions about the appropriateness of unlike-peer interactions. Indeed, in those contexts, courts are willing to show less deference to parents' desire to protect their children from interactions with unlike peers as those children mature. These cases reflect a greater level of attention to adolescents' changing needs and roles. They note adolescents' decreasing vulnerability to confusion or trauma resulting from ideological discussions among peers and the increased importance of these interactions as adolescents approach adulthood. These cases also highlight much more than the manner in which the legal system recognizes adolescents' rights differently in different contexts; they highlight the need to ensure that adolescents participate in communities.

Considering the Centrality of Adolescents' Participation

The next principle challenges the traditional view that others should control adolescents' decisions and may ignore adolescents' potential role in decisions that affect them. The effort seeks to foster adolescents' participation in decision making and to develop decision-making skills that enable adolescents to become responsible members of society. Although adolescents are usually presumed to be dependent and unable to participate actively in determining their own development, social science evidence emphasizes the need to promote adolescent development by fostering adolescents' participation in determining their own outcomes. Research consistently indicates that healthy adolescent development results from social and familial contexts that support autonomy, provide structure, and supply warm and involved socializing agents (Levesque, 2002). These contexts offer informational feedback and decision-making rationales and administer consistent consequences as they encourage self-initiation, minimize use of controls, and adopt adolescents' perspectives. The healthy outcome of autonomy allows adolescents to act on their own volition while still relying on others for support, unlike the frequently fused and confused concept of independence, which implies detachment and minimal reliance on others.

As we have seen in chapter 3, similar dynamics influence adolescents' religious development and their religious environments. Many studies that assess the values and parenting behaviors of specific religious groups conclude that religiosity engenders authoritative parenting. In turn, these behaviors help account for many positive adolescent outcomes. Importantly, authoritative parenting apparently helps to diminish the negative effects of some parental behaviors deemed harmful by many but encouraged by some religious communities and beliefs, such as corporal punishment. Likewise, certain styles of parenting and family environments clearly contribute to many of the negative outcomes associated with adolescents' religious behaviors, such as becoming involved in hate groups and delinquent cults. Development largely involves the highly active process of negotiating the development of one's personal set of values in the context of one's important relationships. Although the process continues over the entire life course, the need for engagement seems particularly central to the adolescent period. Adolescents evince a high need to belong, a need for attachment to others and to communities, that makes significant the context-driven nature of religious engagement. As we have seen, the nature of that engagement changes as age significantly structures the manner in which young people negotiate their environment, the most obvious change being the developmental process of separation from parental influence, coupled with the increase in the role of peers who help rechannel the nature and saliency of religious concerns. Understanding the role of religion in adolescents' experiences, then, requires a close look at how they are allowed to engage in the community, peer relationships, and families in which they find themselves.

The law increasingly allows for adolescents' participation in decisions about their own lives, although still in a limited manner. Respect for the principle of participation most frequently involves efforts to provide adequate information in interventions and to foster inclusion into society. For example, the law already mandates the principle in educational provisions, through its focus on inclusion, its respect for cultural and religious differences, and its recognition of students' rights to free speech. Likewise, laws provide adolescents in other settings with opportunities to express and explore options, for example, by requiring that adolescents receive counseling before medical services or materials can be provided. In addition, and as we have seen, the law provides some adolescents with access to genuine avenues of complaint in situations in which they feel mistreated, ignored, or abused. In sum, the basic principle reflected in due process—the right to be heard and recognized and to participate in governmental processes that impact one's rights—increasingly applies to adolescents and different parts of their lives that do not directly involve courts.

Despite this recognition, legal contexts that involve religious matters do not encourage adolescents' participation as much as research suggests they could. Two examples are illustrative. The first example limits discussion largely on the basis of religious grounds, while the second excludes discussion of religious issues on the basis of adolescents' presumed inability to participate and learn. Both approaches reveal not only the limitations of approaches that ignore adolescents' need to participate in matters that affect them but also the positive outcomes that would result if the law were changed to foster adolescents' participation.

The first example involves recent limitations placed on the nature of discussions between service providers (including school counselors) and adolescents seeking reproductive services and receiving sexuality education. As we have seen, the Supreme Court has approved policies that seek to stifle informed participation. The Court assumes that patients will respond to physician expressions of state policies rationally and critically and that they will be neither unduly influenced nor confused. According to the Court's vision, as a matter of course patients will simply question physicians who fail to distance themselves from a governmentally dictated message to determine whether their advice or their silence expresses their own professional judgment or results from state policy. Likewise, the Court accepts that patients will take at face value disclaimers by physicians who verbally distance themselves from state-imposed messages and will be unaffected by the apparent dissonance between the views of the doctor and those of the state. The Court's assumptions about how patients will respond to and be affected by messages conveyed by physicians about state policy conflict with the findings of research that details the dynamics of doctor-patient communication (see Maynard, 1991). As that research shows, the purpose and structure of the doctor-patient relationship vest physicians with immense authority and power in the eyes of patients. Physicians' authority derives from their superior knowledge, education, and high social and economic status. Patients' disempowered position leads to passivity, a reluctance to question or challenge physicians, and an inability to take control of the topics of discussion; patients suspend their critical faculties and defer to physicians' opinions (Beisecker & Beisecker, 1990). Findings from doctor-patient interactions strongly indicate that the Supreme Court overestimates the extent to which patients' critical rationality and inquisitiveness will neutralize the coercive effect of government messages delivered by physicians. Patients are not likely to challenge physicians' governmentally dictated silence or their biased presentation of medical options. Rather, patients are likely to give great weight to physicians' expressions of state preferences, not because they are persuaded by the messages but merely because physicians deliver the messages. The asymmetrical, highly emotional quality of doctor-patient interaction makes it probable that physician expressions of the state's viewpoint will intimidate patients, particularly those who are nonwhite, poor, elderly, female, or adolescent, and lead patients to respond with confusion and deference. Although adolescents (like adults) may be unable to interact the way the Supreme Court assumes they can, we already have seen that adolescents do have a level of competence that could allow them to sort out differences in information. As we have seen, evidence supports the contention that adolescents are not so incapable of complex thought that the state needs to preselect information for their welfare. Importantly, adolescents seem more likely to have the skills needed to evaluate information than those the Supreme Court has recognized in mature teenagers, such as their ability to make decisions regarding abortion and contraception without state or parental input. The available evidence suggests, then, that adolescents could benefit from being more fully included in decision making in areas that affect them.

The second example involves the central argument that underlies the refusal to provide religious groups equal access to elementary and high schools. The argument tends to be that access advances religion because adolescents are impressionable and might misinterpret the school's involvement in the religious meetings. To the extent that adolescents' impressionability remains relevant, properly designed policies can help eliminate their reasonable misperceptions of endorsement. Most notably, if a school's actions might lead to reasonable misperceptions, then the school's missions and reason for existence require that it attempt to correct it. This view suggests, in fact, that the equal access controversy in public schools presents an educational opportunity. Rather than exclude religion, schools might be better advised to teach adolescents the differences among endorsement, hostility, and neutrality. In addition, and as we will return to, schools could educate students on the fundamental principles of the First Amendment, on the difference between private and public action, and on the reasons to tolerate divergent views. It is the public school's role to educate students and to inculcate values of inclusion and tolerance, not to foster the values of exclusion and intolerance. By responding to equal access by student religious groups, adolescents can learn to respect the values and beliefs held by others without endorsing those values. Beyond teaching, officials can take other precautions to minimize adolescents' reasonable misperceptions of endorsement. School officials could specifically disclaim any government endorsement or disapproval of religion. Schools can inform students that groups that meet after school are gathering because they want to meet and not because of any state prescription. In fact, no psychological evidence indicates that a clear message stating that the school neither endorses nor disapproves of the groups that meet on school property after school hours would not suffice to clarify any reasonable misperceptions. Adolescents have sufficient maturity to understand policies that help ensure that school officials' words and deeds neither endorse nor disparage religion. Through improved education and clear disclaimers, equal-access policies can reduce the possibility of reasonable misunderstanding. These possible school responses are far from far-fetched; they are precisely what seems to be required by the recent Supreme Court decision *Good News Club v. Milford Central School* (2001), which permits religious groups to meet on school grounds once certain protections are in place. In that case, the Court did not give unfettered and absolute free access to religious groups; the Court required that steps be taken to ensure neutrality toward religion. In fact, the Court reasoned that religious groups must be offered equal treatment for fear that not doing so would breed hostility toward religious beliefs.

Taking adolescents' participation seriously in the context of schools raises the uncomfortable issue of religious education in public schools. We already have seen that the Court prohibits religious instruction in public education and that the legal system has long been widely understood as forbidding public schools from fostering or inculcating religious belief (see *Lemon v. Kurtzman*, 1971). Yet the Court also has a long history of clearly allowing public schools to require the study of religion as part of a well-rounded curriculum (see *Abington Township v. Schempp*, 1963). In a deliberate effort to clarify the apparent confusion regarding the place of religious instruction and education in public schools, the federal government has taken the unusual step of issuing guidelines to help school districts design their own policies relating to religious expression, inform teachers and principals of their responsibilities and the rights of their students regarding religious expression, and provide parents with accurate information about their children's right to religious expression. The government has done so through its guidelines, *Religious Expression in Public Schools* (United States Department of Education, 1995). The admittedly brief guidelines simply clarify that, although public schools may not provide religious instruction, schools may provide religous education to the extent that they can teach about religion, including study of the Bible or other scripture (in the context of the history of religion, comparative religion, or the role of religion in history); consider religious influences on art, music, literature, and social studies; and even teach about the religious aspects of religious holidays. In addition, the guidelines emphasize that schools may actively teach civic values and virtues even though those qualities also may be held by most religions. Although rather simple distillations of complex jurisprudential controversies, these guidelines and related efforts nevertheless help refute the myth that religious concerns must be excluded from public schools and convey the reality that schools must address religious issues as part of the educational process.

Although the educational system can, and actually may need to, address religious matters, more systematic efforts to foster religious education inevitably encounter many difficulties. At least two challenges seem to arise that highlight the need to consider how adolescents themselves must participate in finding a proper place for religion in educational settings. The first challenge, and the most likely reason for not offering religious education in a more systematic fashion, involves the inevitability of opposition from some participants whose religious views challenge the notions of openness, dialogue, impartiality, and tolerance that generally serve as a primary rationale for religious education. For example, students who abide by an epistemology based on the authority of revelation may approach religion in ways that challenge knowledge (and actions) based on personal experience, reasoning, and the need to accommodate others' values. Although this fear may have much merit, we have seen that our legal system already has adopted a pragmatic solution to this concern as it officially requires at least the government-sponsored approaches to religious education to acknowledge that others can hold different beliefs equally sincerely or choose different ways of life. Like our constitutional system, this approach to religious education adopts as its core purpose the promotion of understanding of and respect for people whose cultures and beliefs are different from one's own and the promotion of positive attitudes toward living in a pluralistic society. These values serve, for example, as the basis for moral and character education movements in public schools, which have been enshrined in many state and federal laws as the foundation for education and for living in our civil society (for a review, see Levesque, 2002). Indeed, it is precisely because we live in a pluralistic society with individuals who hold extreme views that these approaches gain significance.

The second, and obviously more controversial, challenge involves efforts to promote the religious and spiritual development of pupils. This challenge is problematic in that such attempts, at least overt ones, are in fact seen as impermissible and as affronts to the Constitution. Yet, schools already profoundly influence adolescents' spiritual development (or religious development, broadly defined), even though they may not do so directly; any educational process that shapes minds inescapably conveys a spirituality (see, e.g., Rodger, 2000). Given the paucity of research that examines religious education in public schools, the extent to which schools inevitably address spiritual or religious issues suggests that efforts to address those concerns more directly may benefit from parallel proposals submitted as necessary to address issues of cultural diversity and moral education (see Levesque, 2002) and to approach highly controversial subjects, such as sexuality education (see Levesque, 2000c). The latter analyses reveal that policies that address the broad array of religious concerns would benefit from efforts that seek to (1) recognize the potential centrality of religious faith and practices in the lives of students; (2) affirm the worth of all students, regardless of their religious or secular beliefs; (3) provide students with opportunities to develop skills necessary to gain an understanding of different ways of life and religious perspectives; (4) protect students' right to practice their religious beliefs in a manner consistent with the demands of civil society; and (5) guard the rights of students to be free from religious intrusion, harassment, or verbal or physical assault if they choose to exercise their freedom from religion. These efforts obviously constitute only a starting point; they essentially offer a minimalist approach that needs much fine-tuning, evaluation, and development before it can address more fully the religious and spiritual needs of adolescents in schools.

Although the law may broadly structure policies that encourage schools to enact and to follow our basic guidelines, the reality is that religious education in public schools remains an emergent field that requires us to advance in uncertain territory. Yet, that territory is far from uncharted. We may draw on our understanding of educational processes and religious development. In addition, policies developed and implemented in other countries, most notably Britain and Canada, may provide useful starting points (see Leicester, Modgil, & Modgil, 2000). Foreign educationalists have shown an extraordinary increase in interest in spirituality and religiosity (see Carr, 1996; Gay, 2000). They have designed a standardized curriculum and sought to address how teachers, even with the best intentions, may lead students to adopt intolerant views. These efforts that address both overt and covert religious perspectives are an important step. At a minimum, they provide evidence that we can more systematically approach religious issues in school settings, since everyone in our society must confront the conceptual indeterminacy and competing views of what should constitute this form of education. They also underscore the need to recognize the role adolescents themselves must play in fostering environments that respect their religious rights and those of others.

The social sciences recognize the central need for adolescents to participate in their social environments, especially those wittingly or unwittingly infused with a religious belief. The legal system, however, has only started to capitalize upon those findings. As we have seen, the participation principle operates at several levels and is buttressed by numerous rights. It most notably includes adolescents' rights to receive adequate information, to be heard in matters that concern them, and to exercise freedom of expression. It also includes the principles we already have discussed, such as those that encourage policies that allow adolescents to share in decision-making processes and to gradually control the outcomes of important decisions as they achieve legal independence. All of these principles and rights will help adolescents gain respect for and develop their own religious beliefs.

Affirming the Significance of Adolescents' Best Interests

The legal system would be much improved if it were to recognize adolescents' competencies, foster their dynamic self-determination, and support their participation in matters affecting them. Those principles, however, do not provide a standard by which to determine the extent to which adolescents' religious rights would be protected. The frequently adopted "best interests" standard emerges as the clearest standard that already is accepted as the fundamental legal starting point to determine how we should treat adolescents.

Reconsidering the types of social environment that would protect adolescents' religious rights necessarily involves considering what legally constitutes adolescents' best interests. Both traditional and innovative efforts follow the rule that those who control adolescent development and adolescent's rights should act in adolescents' best interests. Without that fundamental rule, the power granted to others loses its legitimacy. Yet, if it is taken seriously, it may lead to results that seem peculiar and unexpected and that hold institutions and individuals legally accountable when they otherwise generally escape accountability. This is particularly critical in that, as we have seen in chapter 4, the standard holds the promise to address private as well as public actions, one of the key limitations of jurisprudence and legislation relating to religion. Despite that promise, it remains difficult to realize that both private and public actors must take adolescents' interests seriously. In fact, that difficulty helps account for many spirited attacks on the best interest standard that charge that it remains too indeterminate and too subject to decision-maker biases. Yet, the standard still constitutes a key concept in child welfare law. In fact, a close look at existing laws and at the supporting rationales for those laws reveals that the legal system actually allows for holding individuals accountable when they do not act in adolescents' best interests and that institutions already aim to foster accountability that would lead individuals to act in the best interests of adolescents. Understanding how these rationales could impact adolescents' religious environments and the laws regulating them, then, first requires a quick look at those laws and their supporting rationales. Understanding these laws, and how they could work with the previously enumerated principles, could help further respect adolescents' religious rights.

Numerous laws that regulate adolescent life, and those who interact with adolescents, already focus on adolescents' best interests. First, parents already must act in their children's best interests. For example, the best interests standard serves as the basic assumption that guides legal responses to family life. It provides the major rationale for allowing parents to control adolescents' educational experiences and prevents state intervention in families not viewed as abusive. In addition, the law mandates that parents act in their adolescents' best interests, assumes that adolescents' best interests guide parent-child relationships, and views that situation as best for society. Second, despite the existence of communities that foster beliefs and practices that may harm adolescents, several legal rules require communities to address factors that lead to clear developmental deficits and to act in adolescents' best interests. Most notably, several efforts direct resources to adolescents: considerable allocations of what has been called "welfare" continues to be directed toward adolescents and children. In addition, laws indirectly reallocate access to resources through education. Third, peers also must act to foster adolescents' best interests. Admittedly, developments in law still do not reflect the powerful role peers play in adolescent development and in efforts to foster healthy lifestyles. However, several avenues may be taken to ensure that adolescents' actions are more consistent with their peers' interests. At a minimum, efforts could aim to prevent direct harms perpetrated by peers. Efforts also could encourage adolescents to treat each other with more respect and concern for one another's interests and ensure that people act in others' interests as well as their own. Although these mandates may seem far-fetched, legal mandates already exist to develop peer relations in these ways. The two institutions that provide the environment in which peers interact and that react to those interactions—schools and juvenile justice systems—already aim to encourage and foster healthy peer interactions (see Levesque, 2000a, 2002). Last, adolescents themselves are an often ignored control and influence on other adolescents. Healthy adolescents act in their own best interests. This is not to negate adolescents' profound role as social actors but to emphasize that they actively contribute to their own development and that they have responsibilities to themselves and others. Again, laws already interfere in and regulate adolescents' lives for their own good, as well as protect them from potential harm. In fact, the dual purpose of the law remains the general rationale that allows legal systems to treat adolescents and children differently from adults. Laws that restrict adolescents' freedoms do so to ensure that adolescents' best interests are considered and acted upon when practicable.

Despite the slew of legal rules and principles, it still remains difficult and disingenuous to argue that society and individuals actually follow the best interest mandates. The failure is especially true in the context of religious environments that affect adolescents. For example, states do not act on adolescents' best interests when they allow or encourage discrimination in education, medical neglect, problem family dynamics, or even family violence, all of which have been linked to a focus on protecting religious rights. Likewise, society sometimes stands idly by and, out of respect for religious beliefs, tolerates beliefs that may foster intolerance, prejudice, and hate activity—even though the legal system, as we have seen, hopes, but the law does not require, that society will help foster appropriate responses to problematic manifestations of religiosity. When states deny the basic interests of adolescents in those circumstances, they pursue the best interests of parents, communities, and institutional actors who deny adolescents' rights, rights that adolescents would enjoy but for their status as adolescents. As the state allows these conditions to continue, however, its actions become increasingly difficult to justify on either traditional or modern jurisprudential grounds. If anything, then, the focus on adolescents' best interests is far less peculiar than the failure to put adolescents' interests first when those interests arguably should take precedence in a civil society that aims to protect the rights of all its citizens.

Society, law, and policy typically frame the harms adolescents suffer in these circumstances simply as collateral costs in a system geared toward having private citizens (parents) be responsible for the upbringing of children and as the result of society's inability to address all harms individuals suffer. Yet, the legal system could respond more effectively to the harms we now recognize as linked to respect for religious beliefs. How the legal system addresses the harms that arise from religion's role in education and social service delivery is a clear example of the ability of the legal system to respond simply by harnessing already established legal mandates. Recall that the legal system currently tends to allow parents to use vouchers in support of sending adolescents to schools that provide instruction consonant with parental beliefs; similarly, religious groups now receive governmental support to provide services without changing their religious environments and practices. Although such efforts may not affront adolescents' rights, the real possibility exists that some schools may foster religious and other forms of prejudice and that some adolescents may not receive services they would receive if they had access to secular service providers. That is, these policies raise the possibility that those entrusted with the care of adolescents may not respond appropriately to adolescents' own best interests. The legal system, however, already allows for reconsidering educational and social service delivery while still respecting adolescents' religious rights, their parents' rights, and civil society's basic foundational principles.

Greater protection for adolescents' best interests can come in many ways consistent with our democratic ideals (see Minow, 1999), but programs that consider adolescents' roles in civil society would at least benefit from considering the following possibilities, which are already permitted by our legal system. First, no private school or program should be eligible for support through public dollars or vouchers if its mission or practices exclude individuals on the basis of religion, race, ethnicity, or nationality. Second, schools that receive public funds or vouchers must engage in educational programming to address the legacies of intergroup hatred and conflict and to promote tolerance and respect across religious, racial, and ethnic groups. Such programs should include not only in-class curricula but also common activities that directly connect students with peers from other schools with a different religious affiliation, or none. Third, to permit effective comparisons of their operations and outcomes, all participating programs and schools must supply comparable information to watchdog and public groups. These requirements offer minimal protection against segregated, self-confirming enclaves and the erosion of the shared public space necessary for a complex, vibrant, and pluralistic society. Last, neither Charitable Choice nor school voucher programs should be allowed to eliminate a genuinely accessible and attractive secular option for adolescents who do not want to deal with a religious provider of services or schooling—an option that requires providing adolescents and their parents with the type of information necessary for making informed choices. Overall, the approach reflects the fundamental purpose of the public policy of this constitutional order, which involves the need to prepare people to become liberal democratic citizens. That constitutional mandate means promoting an overall order in which the values of bonding, sect-like associations are, on the whole, subordinate to the virtues of bridging associations. As much as many may argue against the rights of adolescents or champion the rights of parents to determine their children's religious environment, that fundamental civic purpose remains the foundation on which our society stands.

Matters under familial control clearly offer a more fundamental challenge. The current focus on the best interest of the child places very high obstacles in the path of efforts to intervene directly in family life: adolescents must suffer, or be at risk for suffering, a recognized form of harm (such as child abuse). That the focus on harm exists, though, is not the end of our discussion. We already have seen how commitment to pluralism, religious freedom, and parental rights long has provided parents with the authority to opt out of the public schools. Yet, we also have seen that the power to do so is not absolute. Even when parents opt for home schooling, for example, they still are subject to state regulations. When the rights of adolescents are deemed fundamental, the state has an obligation to protect those rights. Likewise, removing adolescents from schools can be done only in the presence of alternatives; this was emphasized in the leading parental rights case that recognized the power of parents to remove their adolescent children from school, *Wisconsin v. Yoder* (1972). In addition, it is important to keep in mind that to think of state intervention as direct involvement in families is to take an unnecessarily narrow view of state intervention in family life and to miss what may constitute the vast majority of laws that do regulate families. We have seen, for example, that changing schools and peer relations means impacting families, as does changing the nature of social service delivery. Recognizing and responding to the powerful social forces outside families—especially religious institutions—does not mean that policies must sever the powerful bonds between parents and adolescents. Rather, recognizing the role of social forces outside families simply means that they must be considered if we are to ensure that adolescents have options when their relationships fail to operate in their best interests and to help them determine what they deem potentially so important and central to their humanity.

Conclusion

Religious institutions and beliefs shape us and the future of our democracy. As it has throughout the history of our nation, religion plays a key role in determining the nature of communities, family dynamics, interpersonal relationships, and even social policies. As we have seen, despite efforts to exclude religion from the public (and sometimes private) sphere, religious beliefs and groups play central but often enshrouded roles in policy-making and in shaping individuals' cultural and personal identities. We continue to witness rapid changes in the legal foundation of adolescents' religious environments, and evidence suggests that we have reached a critical juncture. The rapid changes in the regulation of religion and the impact of those changes on adolescents lead us to conclude that we cannot hope to address adolescents' and society's needs without more structured responses to adolescents' place in the law, that we cannot rely on the received faith that adolescents' interests are addressed adequately as we address the rights and needs of those who control adolescents' social environments. Although it may seem irreverent and even antireligious to suggest that we cannot rely on past faiths, it serves to highlight the law's now wide-ranging regulation of religion. Given the law's impressive reach, we cannot cling to traditional paradigms and assumptions about the inherently private nature of religious faith that champion efforts to leave religion alone and subject to the will of those who traditionally have controlled adolescents. We cannot do so simply because the legal system structures our private and more public religious environments. Because of the law's role and the civic ideals imbued in law, the legal system requires close scrutiny to ensure that it furthers freedoms, rather than stifles them.

A close look at how the legal system approaches adolescents' religiosity and religious environments reveals many inadequacies. The failure to address more directly adolescents' needs, abilities, and peculiar place in families, communities, and the law becomes quite suspect when we consider two points. First, issues relating to religious freedom clearly involve rights that deserve protection of the highest order. Although we have seen that recent Supreme Court cases do not grant religious rights absolute power over other, potentially competing rights and societal interests, religious rights nevertheless remain highly regarded and respected. Second, the Supreme Court recognizes that the legal system must respect religious rights even when the rights holders are adolescents. The Court continues to extend important due process rights to adolescents, affirms fundamental (but highly controversial) privacy rights, and even recognizes adolescents' rights involving freedom of conscience. These extensions of adolescents' rights, however, have yet to take full shape and have been explored only in limited contexts. In fact, we have seen that much of the analyses of adolescents' rights involves analyses of other groups' rights.

The need to address adolescents' religious rights more directly makes critical the type of analyses just offered. Available social science evidence and innovative laws suggest a need to rethink adolescents' rights and offer paths to greater legal recognition. Equally important, existing laws already provide mechanisms consistent with new conceptions of adolescents' rights, and the existence of several innovative efforts to address adolescents' rights suggests that such efforts are far from futile. As we have seen, effective efforts genuinely involve adolescents in the planning and implementation of programs and laws and recognize that groups and communities channel and constrain behaviors, even those as profoundly intimate as religious activity. Thus, it increasingly makes sense to ensure the fundamental democratic principle that all people have a right to self-determination, to participate in society in a democratic manner that appreciates individuals' basic interests, and to exert more control over the rights that would benefit them. All of these ingredients have been identified as important to positive adaptations to civil society.

Although the social sciences may suggest and the law may support and even require the development of the types of principles presented, we also have seen that much controversy would attach to such efforts. This is unsurprising; the development of adolescents' rights always has engendered much controversy. We also have seen that recognizing the right to explore ultimate questions with an unfettered conscience may prove a difficult challenge for a society that places much weight on the rights of parents to determine their children's religious upbringing. Rather than ignoring objections to efforts to grant greater respect to adolescents, our analyses reveal the need to accept those challenges and to respond to them properly. The legal system can help recognize issues and structure responses. That the legal system can guide such intervention in adolescents' lives as it regulates religion is just as critical for reform and inquiry as are efforts to protect adolescents from overt violations and discriminations that are now routine matters for the child welfare system, as well as the educational and the juvenile justice systems. In fact, laws already exist that may be harnessed to help foster healthy and supportive adolescent development, including adolescents' religious identity and the impact of that identity on their social dispositions.

A close look at what the law can help accomplish also highlights its limits. Given the needs of civil society, of a society poising itself to foster and respect religious freedom, the law cannot embark wholeheartedly on many of the suggestions offered by those who seek to harness governmental resources to involve adolescents in religious activities. The legal system cannot do so even though the social sciences support the claim that religiosity is linked to an impressive array of positive outcomes for society and even for adolescents themselves. Although religious beliefs are linked to positive outcomes, we have seen that the links are not as direct as current discourse suggests. In fact, adolescents' responses to religious concerns—adolescents' search for meaning in their highly controlled environments—includes many potentially negative outcomes. The existence of either possible outcome may reveal limitations on the law's impact, but it also suggests the central role of law—how the law can and must structure adolescents' environments.

Understanding the potential role of law has meant that our investigation necessarily has ventured beyond current constitutional, jurisprudential, and legislative mandates. It has meant, first, understanding the role of religion in adolescent life. The investigation revealed that much depends on adolescents' internal and external resources. No single psychological, social, or cultural factor explains adolescents' religious dispositions (including antireligious dispositions) and the impact of those dispositions on adolescents' developmental outcomes. Rather, a constellation of forces work together to achieve positive outcomes and healthy development (or their opposite). Drawing on available social science research that has explored the nature of adolescence and the possible influence of religious concerns on adolescent development, we have offered legal principles to help better guide responses to adolescents' peculiar place in society and law. Considering these principles in our efforts to respond to adolescents can help ensure that we recognize both adolescents' abilities and their peculiar social condition in a manner that will help ensure greater respect for adolescents' own religious needs.

It may be naive to suggest that greater recognition of adolescents' individual needs and concerns may lead to positive outcomes, not the least of which is allowing adolescents to negotiate their own paths to meaningful lives. Parents, not society, traditionally have carried the burden of socializing and preparing adolescents for adulthood. But, time, society, families, and the law have changed. There is a pressing need to move beyond simple causal models in order to understand fully both adolescents' religious development and the impact of religious development on adolescents' developmental outcomes. Religion infuses adolescents' environments, and it is that infusion that impacts religiosity and developmental outcomes. The current social science understanding of those environments led to our development of key guiding principles, which find remarkable parallels in those that our legal system views as foundational to participation in modern civil society. Although religious choices, and the provision of environments to help recognize and foster choices, surprisingly have not been part of the discourse on adolescents' rights and youth policy, we have seen that they can play a role in fostering a more mature, appropriate, and just development of adolescents' rights. We have seen that jurisprudential and legislative developments, coupled with rapidly changing social conditions that affect adolescents, place us at a critical juncture that we can no longer ignore. Achieving the goals of our civil, pluralistic society requires a recognition that all people deserve and have real opportunities to develop and follow their convictions in a society that understands and responds to the potential of those convictions in everyone's lives.

References

Abington School District v. Schemmp, 374 U.S. 203 (1963).

Adolescent Family Life Act, 42 U.S.C. § 300z *et seq.* (1982 & Supp. 1988).

Agostini v. Felton, 521 U.S. 203 (1997).

Aguilar v. Felton, 473 U.S. 402 (1985).

Ahn, Helen Noh (1994). Cultural diversity and the definition of child abuse. In Richard Barth, Jill Duerr Berrick, & Neil Gilbert (Eds.), *Child Welfare Research Review* (pp. 28–55). New York: Columbia University Press.

Ainsworth, Mary D. S. (1989). Attachment beyond infancy. *American Psychologist*, 44, 709–716.

Alley, Robert S. (1996). *Without a Prayer: Religious Expression in Public Schools*. New York: Prometheus Books.

Allport, Gordon W. (1966). The religious context of prejudice. *Journal for the Scientific Study of Religion*, 5, 448–451.

Alwin, Duane F. (1986). Religion and parental child-rearing orientations: Evidence of a Catholic-Protestant convergence. *American Journal of Sociology*, 92, 412–440.

American Psychiatric Association (1994). *Diagnostic and Statistical Manual for Mental Disorders (4th ed.)*. Washington, D.C.: Author.

Amey, Cheryl H., Stan L. Albrecht, Michael K. Miller (1996). Racial differences in adolescent drug use: The impact of religion. *Substance Use and Misuse*, 31, 1311–1332.

Ammons, Linda L. (1999). What's God got to do with it? Church, state collaboration in the subordination of women and domestic violence. *Rutgers Law Review*, 51, 1207–1288.

Amoateng, Acheampong Y., & Stephen J. Bahr (1986). Religion, family, and adolescent drug use. *Sociological Perspectives*, 29, 53–76.

Appleby, R. Scott (2000). *The Ambivalence of the Sacred: Religion, Violence, and the Sacred*. New York: Rowan & Littlefield.

Ary, Dennis V., Elizabeth Tildesley, Hyman Hops, & Judy A. Andrews (1993). The influence of parent, sibling, and peer modeling and attitudes on adolescent use of alcohol. *International Journal of the Addictions*, 28, 853–880.

Asser, Seth M., & Rita Swan (1998). Child fatalities from religion-motivated medical neglect. *Pediatrics*, 101, 625–629.

Bahr, Stephen J., & Ricky D. Hawks (1995). Religious organizations. In Robert H. Coombs & Douglas Ziedonis (Eds.), *Handbook on Drug Abuse Prevention: A Comprehensive Strategy to Prevent the Abuse of Alcohol and Other Drugs* (pp. 159–179). Boston: Allyn and Bacon.

Bahr, Stephen J., Ricky D. Hawks, & Gabe Wang (1993). Family and religious influences on adolescent substance abuse. *Youth & Society*, 24, 443–465.

Bahr, Stephen J., Anastasios C. Marcos, & Suzanne L. Maughan (1995). Family, educational, and peer influences on the alcohol use of female and male adolescents. *Journal of Studies on Alcohol*, 56, 457–469.

Bahr, Stephen J., Suzanne L. Maughan, Anastasios C. Marcos, & Bingdan Li (1998). Family, religiosity, and the risk of adolescent drug use. *Journal of Marriage & the Family*, 60, 979–992.

Baier, Colin J., & Bradley R. E. Wright (2001). "If you love me, keep my commandments": A meta-analysis of the effect of religion on crime. *Journal of Research in Crime & Delinquency*, 38, 3–21.

Balk, David E. (1983). How teenagers cope with sibling death: Some implications for school counselors. *School Counselor*, 31, 150–158.

Balk, David E. (1999). Bereavement and spiritual change. *Death Studies*, 23, 485–493.

Balk, David E., & Nancy S. Hogan (1995). Religion, spirituality, and bereaved adolescents. In David Walker Adams & Eleanor J. Deveau (Eds.), *Beyond the Innocence of Childhood: Helping Children and Adolescents Cope with Threat to Their Lives, Dying, Death and Bereavement* (Vol. 3, pp. 61–88). Amityville, NY: Baywood.

Bane, Mary Jo, Brent Coffin, & Ronald Theimann (2000). *Who Will Provide? The Changing Role of Religion in American Social Welfare*. Boulder, CO: Westview Press.

Bankston, Carl L., III., & Min Zhou (1995). Religious participation, ethnic identification, and the adaptation of Vietnamese adolescents in an immigrant community. *Sociological Quarterly*, 36, 523–534.

Bankston, Carl L., III, & Min Zhou (1996). The ethnic church, ethnic identification, and the social adjustment of Vietnamese adolescents. *Review of Religious Research*, 38, 18–37.

Barnes, Grace M., & Michael P. Farrell (1992). Parental support and control as predictors of adolescent drinking, delinquency, and related problem behaviors. *Journal of Marriage and the Family*, 54, 763–776.

Barnes, Grace M., Michale P. Farrell, & Sarbani Banerjee (1994). Family influences on alcohol abuse and other problem behaviors among black and white adolescents in a general population sample. *Journal of Research on Adolescence*, 4, 183–201.

Bartowski, John P., & W. Bradford Wilcox (2000). Conservative Protestant child discipline: The case of parental yelling. *Social Forces*, 79, 865–891.

Bassett, William W. (2001). Private religious hospitals: Limitations upon autonomous moral choices in reproductive medicine. *Journal of Contemporary Health Law & Policy*, 17, 455–583.

Batson, C. Daniel, Patricia Schoenrade, & Larry W. Ventis (1993). *Religion and the Individual: A Social-Psychological Perspective*. New York: Oxford University Press.

Batten, Michelle, & Kevin Ann Oltjenbruns (1999). Adolescent sibling bereavement as a catalyst for spiritual development: A model for understanding. *Death Studies*, 23, 529–546.

Beck, Scott H., Bettie S. Cole, & Judith A. Hammond (1991). Religious heritage and premarital sex: Evidence from a national sample of adults. *Journal for the Scientific Study of Religion*, 30, 173–180.

Beisecker, Analee E., & Thomas D. Beisecker (1990). Patient information-seeking behaviors when communicating with doctors. *Medical Care*, 28, 19–28.

Bellotti v. Baird, 443 U.S. 622 (1979).

Benda, Brent B. (1995). The effect of religion on adolescent delinquency revisited. *Journal of Research in Crime & Delinquency*, 32, 446–466.

Benson, Peter L., Michael J. Donahue, & Joseph A. Erickson (1989). Adolescence and religion: A review of the literature from 1970 to 1986. *Research in the Social Scientific Study Religion*, 1, 153–181.

Berg, Thomas C. (2001). Anti-Catholicism and modern church-state relations. *Loyola University Chicago Law Journal*, 33, 121–172.

Berkowitz, Marvin W., & John H. Grych (1998). Fostering goodness: Teaching parents to facilitate children's moral development. *Journal of Moral Education*, 27, 371–392.

Bersoff, Donald N., & David J. Glass (1995). The not-so Weisman: The Supreme Court's continuing misuse of social science research. *Roundtable: A Journal of Interdisciplinary Legal Studies*, 2, 279–302.

Billingsley, Andrew (1999). *Mighty Like a River: The Black Church and Social Reform*. New York: Oxford University Press.

Bjarnason, Thoroddur (1998). Parents, religion, and perceived social coherence: A Durkheimian framework of adolescent anomie. *Journal for the Scientific Study of Religion*, 37, 742–754.

Blaine, Bruce Evan, Pamela Trivedi, & Amy Eshleman (1998). Religious belief and the self-concept: Evaluating the implications for psychological adjustment. *Personality & Social Psychology Bulletin*, 24, 1040–1052.

Blazak, Randy (2001). White boys to terrorist men: Target recruitment of Nazi skinheads. *American Behavioral Scientist*, 44, 982–1000.

Blum, Robert W., Trisha Beuhring, Marcia L. Shew, Linda H. Bearinger, Renee E. Sieving, & Michael D. Resnick (2000). The effects of race/ethnicity, income, and family structure on adolescent risk behaviors. *American Journal of Public Health*, 90, 1879–1885.

Board of Education of Kiryas Joel Village School District v. Grumet, 512 U.S. 687 (1994).

Board of Education of Westside Community Schools v. Mergens, 496 U.S. 226 (1990).

Bob Jones University v. United States, 461 U.S. 574 (1983).

Boggess, Scott, & Carolyn Bradner (2000). Trends in adolescent males' abortion attitudes, 1988–1995: Differences by race and ethnicity. *Family Planning Perspectives*, 32, 118–123.

Bowen v. Kendrick, 487 U.S. 589 (1988).

Bowlby, John (1988). *A Secure Base: Clinical Applications of Attachment Theory*. London: Routledge.

Bradfield v. Roberts, 175 U.S. 291 (1899).

Braunfeld v. Brown, 366 U.S. 599 (1961).

Breakey, William R. (2001). Psychiatry, spirituality, and religion. *International Review of Psychiatry*, 13, 61–66.

Brega, Angela G., & Lerita M. Coleman (1999). Effects of religiosity and racial socialization on subjective stigmatization in African-American adolescents. *Journal of Adolescence*, 22, 223–242.

Brendgen, Mara, Francois Bowen, Normand Rondeau, & Frank Vitaro (1999). Effects of friends' characteristics on children's social cognitions. *Social Development,* 8, 41–51.

Brendgen, Mara, Frank Vitaro, & William M. Bukowski (1998). Affiliation with delinquent friends: Contributions of parents, self-esteem, delinquent behavior, and peer rejection. *Journal of Early Adolescence*, 18, 244–265.

Bretherton, Inge (1985). Attachment theory: Retrospect and prospect. In Inge Bretherton & Everett Waters (Eds.), *Growing Points of Attachment Theory and Research* (pp. 3–35). *Monographs of the Society for Research in Child Development*, 50(1–2, Serial No. 209). Chicago: University of Chicago Press.

Brewster, Karin L. (1994). Neighborhood context and the transition to sexual activity among young black women. *Demography*, 31, 603–614.

Brewster, Karin L., John O. G. Billy, & William R. Grady (1993). Social context and adolescent behavior: The impact of community on the transition to sexual activity. *Social Forces*, 71, 713–740.

Brewster, Karin L., Elizabeth C. Cooksey, David K. Guilkey, & Ronald R. Rindfuss (1998). The changing impact of religion on the sexual and contraceptive behavior of adolescent women in the United States. *Journal of Marriage & the Family*, 60, 493–504.

Brody, Gene H., Zollina Stoneman, & Douglas Flor (1996). Parental religiosity, family processes, and youth competence in rural, two-parent African-American families. *Developmental Psychology*, 32, 696–706.

Brody, Gene H., Zollina Stoneman, Douglas Flor, & Chris McCrary (1994). Religion's role in organizing family relationships: Family processes in rural, two-parent African-American families. *Journal of Marriage and the Family*, 56, 878–888.

Brown v. Gilmore, 258 F.3d 265 (4th Cir.), cert. denied, 70 U.S.L.W. 3315 (2001).

Brown, Bradford (1993). Peer groups and peer cultures. In S. Shirley Feldman & Glen R. Elliot (Eds.), *At the Threshold: The Developing Adolescent* (pp. 171–196). Cambridge: Harvard University Press.

Brown, Brian Edward (1999). *Religion, Law, and the Land: Native Americans and the Judicial Interpretation of Sacred Land.* Westport, CT: Greenwood Press.

Brownfield, David, & Ann M. Sorenson (1991). Religion and drug use among adolescents: A social support conceptualization and interpretation. *Deviant Behavior*, 12, 259–276.

Bruno, Robert J. (1999). Child protection as a judicial burden on religion and the impact of religious liberty legislation. *Cardozo Law Review*, 21, 677–688.

Bryk, Anthony S., Valerie Lee, & Peter B. Holland (1993). *Catholic Schools and the Common Good.* Cambridge: Harvard University Press.

Bureau of Justice Statistics (2001). *Hate Crimes Reported in NIBRS, 1997–1999*, NCJ 186765. Washington, D.C.: United States Department of Justice.

Burris, Christopher T., & Lynne M. Jackson (1999). Hate the sin/love the sinner, or love the hater? Intrinsic religion and responses to partner abuse. *Journal for the Scientific Study of Religion*, 38, 160–174.

Burket, Roger C., Wade C. Myers, William L. Bradford, & Frank Carrera III (1994). Emotional and behavioral disturbances in adolescents involved in witchcraft and Satanism. *Journal of Adolescence*, 17, 41–52.

Burkett, Steven R. (1977). Religion, parental influence, and adolescent alcohol and marijuana use. *Journal of Drug Issues*, 7, 263–273.

Buss, Emily (1999). What does Frieda Yoder believe? *University of Pennsylvania Journal of Constitutional Law*, 2, 53–76.

Buss, Emily (2000). The adolescent's stake in the allocation of educational control between parent and state. *University of Chicago Law Review*, 67, 1233–1291.

Butman, Ricard E. (1990). The assessment of religious development: Some possible options. *Journal of Psychology and Christianity*, 9(2), 14–26.

Cairns, Robert B., Beverley D. Cairns, Holly J. Neckerman, Scott Gest, & Jean-Louis Gariepy (1988). Social networks and aggressive behavior: Peer support or social rejection? *Developmental Psychology*, 24, 815–823.

Cantwell v. Connecticut, 310 U.S. 296 (1940).

Capps, Donald (1992). Religion and child abuse: Perfect together. *Journal for the Scientific Study of Religion*, 31, 1–14.

Carr, David (1996). Rival conceptions of spiritual education. *Journal of Philosophy of Education*, 30, 159–178.

Casanova, José (1994). *Public Religions in the Modern World*. Chicago: University of Chicago Press.

Charitable Choice Expansion Act of 1999 (1999). S. 1113, 106th Cong.

Chassin, Laurie, David R. Pillow, Patrick J. Curran, Brooke S. G. Molina, & Manuel Barrera Jr. (1993). Relation of parental alcoholism to early adolescent substance use: A test of three mediating mechanisms. *Journal of Abnormal Psychology*, 102, 3–19.

Choper, Jesse H. (2000). Century of religious freedom. *California Law Review*, 88, 1709–1741.

Church of the Lukumi Babalu Aye, Inc. v. City of Hialeah, 508 U.S. 520 (1993).

City of Boerne v. Flores, 521 U.S. 507 (1996).

Civil Rights Act of 1964. Pub. L. No. 88-352 (1964), Title VII 701, 78 Stat. 253 (codified as amended at 42 U.S.C. §§ 2000e-17 (1999)).

Cline v. Catholic Diocese of Toledo, 206 F.3d 651 (6th Cir. 2000).

Cochran, John K. (1991). The effects of religiosity on adolescent self-reported frequency of drug and alcohol use. *Journal of Drug Issues*, 22, 91–104.

Cochran, John K. (1993). The variable effects of religiosity and denomination on adolescent self-reported alcohol use by beverage type. *Journal of Drug Issues*, 23, 479–491.

Coll, Cynthia Garcia, Gontran Lamberty, Renee Jenkins, Harriet Pipes McAdoo, Keith Crnic, Barbara Hanna Wasik, & Heidie Vazquez Garcia (1996). An integrative model for the study of developmental competencies in minority children. *Child Development*, 67, 1891–1914.

Committee for Public Education and Religious Liberty v. Nyquist, 413 U.S. 756 (1973).

Consequences for Juvenile Offenders Act of 1999 (1999). H.R. 1501, 106th Congress.

Cooksey, Elizabeth C., Ronald R. Rindfuss, & David K. Guilkey (1996). The initiation of adolescent sexual and contraceptive behavior during changing times. *Journal of Health and Social Behavior*, 3, 59–71.

Coombs, Robert H., Morris J. Paulson, & Mark A. Richardson (1991). Peer vs. parental influence in substance use among Hispanic and Anglo children and adolescents. *Journal of Youth and Adolescence*, 20, 73–89.

Cornwall, Marie (1988). The influence of three agents of religious socialization: Family, church, and peers. In Darwin L. Thomas (Ed.), *The Religion and Family Connection: Social Science Perspectives* (pp. 207–231). Provo, UT: Religious Studies Center, Brigham Young University.

Craycraft, Kenneth R., Jr. (1999). *The American Myth of Religious Freedom*. Dallas: Spence.

Csikszentmihalyi, Mihaly (1984). *Being Adolescent: Conflict and Growth in the Teenage Years*. New York: Basic Books.

Csikszentmihalyi, Mihaly, & Kevin Rathunde (1993). The measurement of flow in everyday life: Toward a theory of emergent motivation. In Janis E. Jacobs (Ed.), *Developmental Perspectives on Motivation* (pp. 57–97). Lincoln: University of Nebraska Press.

Curtis, John M., & Mimi J. Curtis (1993). Factors related to susceptibility and recruitment by cults. *Psychological Reports*, 73, 451–460.

Damphousse, Kelley R., & Ben M. Crouch (1992). Did the Devil make them do it? *Youth and Society*, 24, 204–227.

Danso, Henry, Bruce Hunsberger, & Michael Pratt (1997). The role of religious fundamentalism and right-wing authoritarianism in child-rearing goals and practices. *Journal for the Scientific Study of Religion*, 36, 496–511.

D'Antonio, William V., William M. Newman, & Stuart A. Wright (1982). Religion and family life: How social scientists view the relationship. *Journal for the Scientific Study of Religion*, 21, 218–225.

Darnell, Alfred, & Darren E. Sherkat (1997). The effect of Protestant fundamentalism on educational attainment. *American Sociological Review*, 62, 306–315.

Day, James M. (2001). From structuralism to eternity? Re-imagining the psychology of religious development after the cognitive-developmental paradigm. *International Journal for the Psychology of Religion*, 11, 173–183.

Day, James M., & Myriam H. L. Naedts (1995). Convergence and conflict in the development of moral judgment and religious judgment. *Journal of Education*, 177, 1–30.

Detwiler, Fritz (1999). *Standing on the Premises of God: The Christian Right's Fight to Redefine America's Public Schools*. New York: New York University Press.

DiCamillo, LaDonna (1998). Caught between the clauses and the branches: When parents deny their child nonemergency medical treatment for religious reasons. *Journal of Juvenile Law*, 19, 123–157.

DiIulio, John, Jr., David Larson, & Bryon Johnson (1999). *Religion: The Forgotten Factor in Cutting Youth Crime and Saving At-Risk Urban Youth. TJP Report 98-2*. New York: Manhattan Institute.

Dobratz, Betty A. (2001). The role of religion in the collective identity of the white racialist movement. *Journal for the Scientific Study of Religion*, 40, 287–301.

Donahue, Michael J. (1995). Religion and the well-being of adolescents. *Journal of Social Issues*, 51, 145–160.

Donelson, Elaine (1999). Psychology of religion and adolescents in the United States: Past to present. *Journal of Adolescence*, 22, 187–204.

Donnelly, Joseph, David F. Duncan, Eva Goldfarb, & Carolyn Eadie (1999). Sexuality attitudes and behaviors of self-described very religious urban students in middle school. *Psychological Reports*, 85, 607–610.

Drobac, Jennifer Ann (1998). For the sake of the children: Court consideration of religion in child custody cases. *Stanford Law Review*, 50, 1609–1663.

Dubow, Eric F., Kenneth I. Pargament, Paul Boxer, & Nalini Tarakeshwar (2000). Initial investigation of Jewish early adolescents' ethnic identity, stress, and coping. *Journal of Early Adolescence*, 20, 418–441.

Dudley, Roger L. (1999). Youth religious commitment over time: A longitudinal study of retention. *Review of Religious Research*, 41, 109–120.

Dudley, Roger L., & H. Philip Muthersbaugh (1996). Social attachment to the Seventh-Day Adventist Church among young adults. *Review of Religious Research*, 38, 38–50.

Dwyer, James G. (1998). *Religious Schools versus Children's Rights*. Ithaca: Cornell University Press.

Education Consolidation and Improvement Act of 1981, 20 U.S.C. §§ 7301–7373 (1981 & Supp. 2000).

Edwards v. Aguillard, 482 U.S. 578 (1987).

Eisenberg, Nancy (1991). Values, sympathy, and individual differences: Toward a pluralism of factors influencing altruism and empathy. *Psychological Inquiry*, 2, 128–131.

Eisgruber, Christopher L., & Lawrence G. Sager (1997). Congressional power and religious liberty after *City of Boerne v. Flores*. *Supreme Court Review*, 1997, 79–139.

Elkind, David (1971). The development of religious understanding in children and adolescents. In Merton P. Strommen (Ed.), *Research on Religious Development* (pp. 655–685). New York: Hawthorn Books.

Elliott, Delbert S., David Huizinga, & Suzanne S. Ageton (1985). *Explaining Delinquency and Drug Use*. Newbury Park, CA: Sage.

Ellison, Christopher G. (1992). Are religious people nice people? Evidence from the national survey of black Americans. *Social Forces*, 71, 411–30.

Ellison, Christopher G. (1994). Religion, the life stress paradigm, and the study of depression. In Jeffrey S. Levin (Ed.), *Religion in Aging and Health: Theoretical Foundations and Methodological Frontiers* (pp. 78–121). Thousand Oaks, CA: Sage.

Ellison, Christopher G., & John P. Bartowski (1995). 'Babies were being beaten': Exploring child abuse at ranch apocalypse. In Stuart A. Wright (Ed.), *Armageddon in Waco* (pp. 111–149). Chicago: University of Chicago Press.

Ellison, Christopher G., & Linda K. George (1994). Religious involvement, social ties, and social support in a Southeastern community. *Journal for the Scientific Study Religion*, 33, 46–61.

Ellison, Christopher G., & Marc A. Musick (1995). Conservative Protestantism and public opinion toward science. *Review of Religious Research*, 36, 245–262.

Ellison, Christopher G., & Darren E. Sherkat (1993a). Obedience and autonomy: Religion and parental values reconsidered. *Journal for the Scientific Study of Religion*, 32, 313–329.

Ellison, Christopher G., & Darren E. Sherkat (1993b). Conservative Protestantism and support for corporal punishment. *American Sociological Review*, 58, 131–144.

Ellison, Christopher G., John P. Bartowski, & Kristin L. Anderson (1999). Are there religious variations in domestic violence? *Journal of Family Issues*, 20, 87–113.

Ellison, Christopher G., John P. Bartkowski, & Michelle L. Segal (1996). Do conservative Protestant parents spank more often? Further evidence from the National Survey of Families and Households. *Social Science Quarterly*, 77, 663–673.

Emerson, Shirley, & Yvonne Syron (1995). Adolescent Satanism: Rebellion masquerading as religion. *Counseling & Values*, 39, 145–160.

Employment Division, Oregon Department of Human Resources v. Smith, 494 U.S. 872 (1990).

Erikson, Erik H. (1959/1980). *Identity and the Life Cycle*. New York: Norton.

Erickson, Joseph A. (1992). Adolescent religious development and commitment: A structural equation model of the role of family, peer group, and educational influences. *Journal for the Scientific Study of Religion*, 31, 131–152.

Equal Access Act (EAA), 20 U.S.C. §§ 4071–4074 (1984, 1994).

Evans, T. David, Francis T. Cullen, Velmer S. Burton, Jr., R. Gregory Dunaway, Gary L. Payne, & Sesha R. Kethineni (1996). Religion, social bonds, and delinquency. *Deviant Behavior*, 17, 43–70.

Evans, T. David, Francis T. Cullen, R. Gregory Dunaway, & Velmer S. Burton, Jr. (1995). Religion and crime reexamined: The impact of religion, secular controls, and social ecology on adult criminality. *Criminology*, 33, 195–217.

Everson v. Board of Education of Ewing, 330 U.S. 1 (1947).

Fathers Count Act of 1999 (1999). H.R. 3073, 106th Congress.

Fasick, Frank A. (1984). Parents, peers, youth culture, and autonomy in adolescence. *Adolescence*, 19, 143–157.

Filsinger, Erik E., & Margaret R. Wilson (1984). Religiosity, socioeconomic rewards, and family development: Predictors of marital adjustment. *Journal of Marriage and the Family*, 46, 663–670.

Forliti, John E., & Peter L. Benson (1986). Young adolescents: A national study. *Religious Education*, 81, 199–224.

Forste, Reneta T., & Tim B. Heaton (1988). Initiation of sexual activity among female adolescents. *Youth and Society*, 19, 250–268.

Fowler, James W. (1981). *Stages of Faith*. San Francisco: HarperCollins.

Fowler, James W. (1991). Stages in faith consciousness. In Fritz Oser & W. George Scarlett (Eds.), *Religious Development in Childhood and Adolescence. New Directions for Child Development* (Vol. 52, pp. 21–45). San Francisco: Jossey-Bass.

Fowler, James W. (2001). Faith development theory and the postmodern challenges. *International Journal for the Psychology of Religion*, 11, 159–172.

Francis, Leslie J., & Harry M. Gibson (1993). Parental influence and adolescent religiosity: A study of church attendance and attitude toward Christianity among adolescents 11 to 12 and 15 to 16 years old. *International Journal for the Psychology of Religion*, 3, 241–253.

Francis, Leslie J., Susan H. Jones, & Carolyn Wilcox (2000). Religiosity and happiness during adolescence, young adulthood, and later life. *Journal of Psychology and Christianity*, 19, 245–257.

Franklin, Karen (2000). Antigay behaviors among young adults: Prevalence, patterns, and motivators in a noncriminal population. *Journal of Interpersonal Violence*, 15, 339–362.

Frazee v. Illinois Department of Employment Security, 489 U.S. 829 (1989).

Free, Marvin D. (1994). Religiosity, religious conservatism, bonds to school, and juvenile delinquency among three categories of drug users. *Deviant Behavior*, 15, 151–170.

Freedman, David J. (2001). Wielding the ax of neutrality: The constitutional status of Charitable Choice in the wake of *Mitchell v. Helms*. *University of Richmond Law Review*, 35, 313–363.

Fulton, Aubyn S. (1997). Identity status, religious orientation, and prejudice. *Journal of Youth and Adolescence*, 26, 1–11.

Furrow, James L., & Linda Mans Wagener (2000). Lessons learned: The role of religion in the development of wisdom in adolescence. In Warren S. Brown (Ed.), *Understanding Wisdom: Sources, Science, & Society* (pp. 361–391). Philadelphia: Templeton Foundation Press.

Gallup, George H., Jr. (1999). *The Spiritual Life of Young Americans: Approaching the Year 2000*. Princeton: George H. Gallup International Institute.

Garbarino, James (1999). *Lost Boys: Why Our Sons Turn Violent and How We Can Save Them*. New York: Free Press.

Gay, Brenda (2000). Fostering spiritual development through the religious dimension of schools: The report of a pilot study in 17 independent schools. *International Journal of Children's Spirituality*, 5, 61–74.

Gershoff, Elizabeth Thompson, Pamela C. Miller, & George W. Holden (1999). Parenting influences from the pulpit: Religious affiliation as a determinant of parental corporal punishment. *Journal of Family Psychology*, 13, 307–320.

Gecas, Viktor, & Monica F. Serf (1990). Families and adolescents: A review of the 1980s. *Journal of Marriage and the Family*, 52, 941–958.

Gildin, Gary S. (2000). A blessing in disguise: Protecting minority faiths through state religious freedom nonrestoration acts. *Harvard Journal of Law & Public Policy*, 23, 411–486.

Gilligan, Carol (1982). *In a Different Voice: Psychological Theory and Women's Development*. Cambridge: Harvard University Press.

Ginsberg v. New York, 390 U.S. 629 (1968).

Glendon, Mary Ann, & Paul F. Yanes (1991). Structural Free Exercise. *Michigan Law Review*, 90, 477–550.

Goldman v. Weinberger, 475 U.S. 503, 504 (1986).

Good News Club v. Milford Central School, 121 S. Ct. 2093 (2001).

Greven, Philip (1991). *Spare the Child: The Religious Roots of Punishment and the Psychological Impact of Physical Abuse*. New York: Knopf.

Grotevant, Harold D. (1992). Assigned and chosen identity components: A process perspective on their integration. In Gerald R. Adams, Thomas P. Gullotta, & Raymond Montemayor (Eds.), *Adolescent Identity Formation: Advances in Adolescent Development* (pp. 73–90). Newbury Park, CA: Sage.

Gunnoe, Marjorie Lindner, E. Mavis Hetherington, & David Reiss (1999). Parental religiosity, parenting style, and adolescent social responsibility. *Journal of Early Adolescence*, 19, 199–225.

Gustafson, Carrie, & Peter Juviler (Eds.) (1999). *Religion and Human Rights: Competing Claims?* Armonk, NY: M. E. Sharpe.

Hackerman, Ann E., & Paul King (1998). Adolescent spirituality: A foundation for recovery from drug dependency. *Alcoholism Treatment Quarterly*, 16, 89–99.

Hafen, Bruce C., & Jonathan O. Hafen (1996). Abandoning children to their autonomy: The United Nations Convention on the Rights of the Child. *Harvard International Law Journal*, 6, 449–491.

Hall v. Baptist Memorial Health Care Corporation, 215 F.3d 618 (6th Cir. 2000).

Hamilton, Claire E. (2000). Continuity and discontinuity of attachment from infancy through adolescence. *Child Development*, 71, 690–695.

Hamilton, Marci A. (2001). Religion in the public square: Free? Exercise. *William & Mary Law Review*, 42, 823–881.

Hamilton, Marci A. (2000). Religion and the law in the Clinton era: An anti-Madison legacy. *Law & Contemporary Problems*, 63, 359–389.

Haney, Craig (1980). Psychology and legal change: On the limits of a factual jurisprudence. *Law and Human Behavior*, 4, 147–200.

Harris, Judith (1995). Where is the child's environment? A group socialization theory of development. *Psychological Review*, 102, 458–489

Hartsell, Jennifer L. (1999). Mother may I . . . live? Parental refusal of life-sustaining medical treatment for children based on religious objections. *Tennessee Law Review*, 66, 499–530.

Hawkins, J. David, Richard F. Catalano, & Janet Y. Miller (1992). Risk and protective factors for alcohol and other drug problems in adolescence and early adulthood: Implications for prevention. *Psychological Bulletin*, 112, 64–105.

Hazelwood School District v. Kuhlmeier, 484 U.S. 260 (1988).

Herdt, Gilbert H. (1981). *Guardians of the Flutes: Idioms of Masculinity*. New York: McGraw-Hill.

Hertel, Bradley R., & Michael J. Donahue (1995). Parental influences on God images among children: Testing Durkheim's metaphoric parallelism. *Journal for the Scientific Study of Religion*, 34, 186–199.

Hill, Hope M., Fernando I. Soriano, S. Andrew Chen, & Teresa D. LaFromboise (1994). Sociocultural factors in the etiology and prevention of violence among ethnic minority youth. In Leonard D. Eron, Jaquelyn H. Gentry, & Peggy Schlegal (Eds.), *Reason to Hope: A Psychosocial Perspective on Violence and Youth* (pp. 59–67). Washington, D.C.: American Psychological Association.

Hill, Peter C., Kenneth I. Pargament, Ralph W. Hood, Jr., Michael E. McCullough, James P. Swyers, David B. Larson, & Brian J. Zinnbauer (2000). Conceptualizing religion and spirituality: Points of commonality, points of departure. *Journal for the Theory of Social Behavior*, 30, 51–77.

Hirschi, Travis (1969). *Causes of Delinquency*. Berkeley: University of California Press.

Hohman, Melinda, & Craig Winston LeCroy (1996). Predictors of adolescent A.A. affiliation. *Adolescence*, 31, 339–353.

Holder, David W., Robert H. Durant, Treniece L. Harris, Jessica Henderson Daniel, Dawn Obeidallah, & Elizabeth Goodman (1999). The association between adolescent spirituality and voluntary sexual activity. *Journal of Adolescent Health*, 26, 295–302.

Holman, Thomas B., & John R. Harding (1996). The teaching of nonmarital sexual abstinence and members' sexual attitudes and behaviors: The case of the Latter-Day Saints. *Review of Religious Research*, 38, 51–60.

Huizinga, David, Rolf Loeber, & Terrence P. Thornberry (1995). *Urban Delinquency and Substance Abuse (OJJDP Research Summary, NCJ 143454)*. Washington, D.C.: Office of Juvenile Justice and Delinquency Prevention, Office of Justice Programs.

Hunsberger, Bruce, Michael Pratt, & S. Mark Pancer (2001). Adolescent identity formation: Religious exploration and commitment. *Identity: An International Journal of Theory and Research*, 1, 365–386.

Illinois ex rel. McCollum v. Board of Education, 333 U.S. 203 (1948).

In Re Gault, 387 U.S. 1 (1967).

Indian Child Welfare Act (ICWA). Pub. L. No. 95–608, 92 Stat. 3069 (1978) (codified as amended at 25 U.S.C. §§ 1901–1963 (1994)).

Individuals with Disabilities Education Act (IDEA). 20 U.S.C. §§ 1400 *et seq.* (1994 & Supp. 2000).

Jackson v. Benson, 578 N.W.2d 602, 608 (Wis.), cert. denied, 119 S. Ct. 466 (1998).

Jackson, Lynne M., & Bruce Hunsberger (1999). An intergroup perspective on religion and prejudice. *Journal for the Scientific Study of Religion*, 38, 509–523.

Jenness, Valerie, & Kendal Broad (1997). *Hate Crimes: New Social Movements and the Politics of Violence*. New York: Aldine de Gruyter.

Jeynes, William H. (1999). The effects of religious commitment on the academic achievement of black and Hispanic children. *Urban Education*, 34, 458–479.

Johnson, Byron R., Sung Joon Jang, David B. Larson, & Spencer de Li (2001). Does adolescent religious commitment matter? A reexamination of the effects of religiosity on delinquency. *Journal of Research in Crime and Delinquency*, 38, 22–44.

Johnson, Byron R., Sung Joon Jang, Spencer de Li, & David Larson (2000). The 'invisible institution' and black youth crime: The church as an agency of local social control. *Journal of Youth & Adolescence*, 29, 479–498.

Kahn, Joan R., Ronald R. Rindfuss, & David K. Guilkey (1990). Adolescent contraceptive method choices. *Demography*, 27, 323–335.

Kandel, Denise B. (1980). Drug and drinking behavior among youth. *Annual Review of Sociology*, 6, 235–285.

Kaplan, Howard B., Steven S. Martin, & Cynthia Robbins (1984). Pathways to adolescent drug use: Self-derogation, peer influence, weakening of social controls, and early substance use. *Journal of Health and Social Behavior*, 25, 270–289.

Kedem, Reri, & Debbie W. Cohen (1987). The effects of religious education on moral judgment. *Journal of Psychology and Judaism*, 11, 4–14.

Kennedy, Margaret (2000). Christianity and child sexual abuse: The survivors' voice leading to change. *Child Abuse Review*, 9, 124–141.

Khavari, Khalil A. (1993). Interpersonal influences in college students' initial use of alcohol and drugs—the role of friends, self, parents, doctors, and dealers. *International Journal of the Addictions*, 28, 377–388.

Killen, Melanie, & Daniel Hart (Eds.) (1995). *Morality in Everyday Life: Developmental Perspectives*. New York: Cambridge University Press.

Koenig, Harold G., Micahel E. McCullough, & Davbid B. Larson (2001). *Handbook of Religion and Health*. New York: Oxford University Press.

Kohlberg, Lawrence (1981). *The Philosophy of Moral Development*. San Francisco: Harper & Row.

Kole, Tara (2001). Juvenile offenders. *Harvard Journal on Legislation*, 38, 231–324.

Kooistra, William P., & Kenneth I. Pargament (1999). Religious doubting in parochial school adolescents. *Journal of Psychology & Theology*, 27, 33–42.

Koubek, Richard J. (1984). Correlation between religious commitment and students' achievement. *Psychological Reports*, 54, 262.

Kroger, Jane (1996). *Identity in Adolescence: The Balance between Self and Other*. New York: Routledge.

Lamb's Chapel v. Center Moriches Union Free School District, 508 U.S. 384 (1993).

Laible, Deborah J., Gustavo Carlo, & Marcela Raffaelli (2000). The differential relations of parent and peer attachment to adolescent adjustment. *Journal of Youth and Adolescence*, 29, 45–59.

Lee v. Weisman, 505 U.S. 577 (1992).

Leicester, Mal, Celia Modgil, & Sohan Modgil (Eds.) (2000). *Spiritual and Religious Education*. New York: Falmer Press.

Lemon v. Kurtzman, 403 U.S. 602 (1971).

Lerner, Natan (2000). *Religion, Beliefs, and International Human Rights*. Maryknoll, NY: Orbis Books.

Levesque, Roger J. R. (1995). The failures of foster care reform: Revolutionizing the most radical blueprint. *Maryland Journal of Contemporary Legal Issues*, 6, 1–35.

Levesque, Roger J. R. (1997). *Adolescents, Society, and the Law: Interpretive Essays and Bibliographic Guide*. American Bar Association, Chicago, IL.

Levesque, Roger J. R. (2000a). *Adolescents, Sex, and the Law: Preparing Adolescents for Responsible Citizenship*. Washington, D.C.: American Psychological Association.

Levesque, Roger J. R. (2000b). Cultural evidence, child maltreatment and the law. *Child Maltreatment*, 5, 146–160.

Levesque, Roger J. R.(2000c). Sexuality education: What adolescents' educational rights require. *Psychology, Public Policy, & Law*, 6, 953–988.

Levesque, Roger J. R. (2001). *Culture and Family Violence: Fostering Change through Human Rights Law*. Washington, D.C.: American Psychological Association.

Levesque, Roger J. R. (2002). *Dangerous Adolescents, Model Adolescents: Shaping the Role and Promise of Education*. New York: Plenum/Kluwer.

Levin, Jack, & Jack McDevitt (1993). *Hate Crimes: The Rising Tide of Bigotry and Bloodshed*. New York: Plenum.

Litchfield, Alan, Darwin L. Thomas, & Bing Dao Li (1997). Dimensions of religiosity as mediators of the relations between parenting and adolescent deviant behavior. *Journal of Adolescent Research*, 12, 199–226.

Lock, Sharon E., & Murray Vincent (1995). Sexual decision-making among rural adolescent females. *Health Values: The Journal of Health Behavior, Education, & Promotion*, 19, 47–58.

Lowney, Kathleen S. (1995). Teenage Satanism and oppositional youth subculture. *Journal of Contemporary Ethnography*, 23, 453–484.

Lugg, Catherine A. (2000). Reading, writing, and reconstructionism: The Christian Right and the politics of public education. *Educational Policy*, 14, 622–637.

Lynch v. Donnelly, 465 U.S. 668 (1984).

Lyng v. Northwest Indian Cemetery Protective Association, 485 U.S. 439 (1988).

Magen, Zipora (1998). *Exploring Adolescent Happiness: Commitment, Purpose, and Fulfillment*. Thousand Oaks, CA: Sage.

Mahrer, Alvin R. (1996). Existential-humanistic psychotherapy and the religious person. In Edward P. Shafranske (Ed.), *Religion and the Clinical Practice of Psychology* (pp. 433–460). Washington, D.C.: American Psychological Association.

Marcia, James E. (1966). Development and validation of identity status. *Journal of Personality and Social Psychology*, 3, 551–558.

Markstrom, Carol A. (1999). Religious involvement and adolescent psychosocial development. *Journal of Adolescence*, 22, 205–221.

Markstrom-Adams, Carol (1992). A consideration of intervening factors in adolescent identity formation. In Harold D. Grotevant & Catherine R. Cooper (Eds.), *Adolescent Identity Formation* (pp. 173–192). Newbury Park, CA: Sage.

Markstrom-Adams, Carol, & Melanie Smith (1996). Identity formation and religious orientation among high school students from the United States and Canada. *Journal of Adolescence*, 19, 247–261.

Marsiglio, William, & Constance L. Shehan (1993). Adolescent males' abortion attitudes: Data from a national survey. *Family Planning Perspectives*, 25, 162–169.

Masten, Ann S., & J. Douglas Coatsworth (1998). The development of competence in favorable and unfavorable environments: Lessons from research on successful children. *American Psychologist*, 53, 205–220.

Maton, Kenneth I., & Elizabeth A. Wells (1995). Religion as a community resource for well-being: Prevention, healing, and empowerment pathways. *Journal of Social Issues*, 51, 177–193.

Mawdsley, Ralph D. (1998). *Agostini v. Felton*: Is establishment of religion moving toward a nondiscrimination model? *Education Law Report*, 127, 13–35.

Maynard, Douglas W. (1991). Interaction and asymmetry in clinical discourse. *American Journal of Sociology*, 97, 448–495.

McIntosh, Daniel N., & Bernard Spilka (1990). Religion and physical health: The role of personal faith and control beliefs. *Research on the Social Scientific Study of Religion*, 2, 167–194.

McClain, Linda C. (2001). Care as a public value: Linking responsibility, resources, and Republicanism. *Chicago Kent Law Review*, 76, 1673–1731.

McDargh, John (2001). Faith development theory and the postmodern problem of foundations. *International Journal for the Psychology of Religion*, 11, 185–199.

McPhail, Beverly A. (2000). Hating hate: Policy implications of hate crime legislation. *Social Service Review*, 74, 635–654.

Merrill, Ray M., Richard D. Salazar, & Nicole W. Gardner (2001). Relationship between family religiosity and drug use behavior among youth. *Social Behavior & Personality*, 29, 347–358.

Meyer v. Nebraska, 262 U.S. 390 (1923).

Miller, Alan S., & John P. Hoffmann (1995). Risk and religion: An explanation of gender differences in religiosity. *Journal for the Scientific Study of Religion*, 34, 63–75.

Miller, Brent C., & Terrence D. Olson (1988). Sexual attitudes and behavior of high school students relation to background and contextual factors. *Journal of Sex Research*, 24, 194–200.

Miller, Lisa, Mark Davies, & Steven Greenwald (2000). Religiosity and substance use and abuse among adolescents in the National Comorbidity Survey. *Journal of the American Academy of Child and Adolescent Psychiatry*, 39, 1190–1197.

Milligan, Jeffrey Ayala (2000). Mapping the road toward Bethlehem: Parameters for discourse on the relationship between religion and public education. *Educational Policy*, 14, 685–702.

Minow, Martha (1999). Choice or commonality: Welfare and schooling after the end of welfare as we knew it. *Duke Law Journal*, 49, 493–559.

Mississippi Band of Choctaw Indians v. Holyfield, 490 U.S. 30 (1989).

Mitchell v. Helms, 120 S. Ct. 2530 (2000).

Montemayor, Raymond, Gerald R. Adams, & Thomas P. Gullotta (2000). *Adolescent Diversity in Ethnic, Economic, and Cultural Contexts*. Thousand Oaks: Sage.

Montgomery, Alice, & Leslie J. Francis (1996). Relationship between personal prayer and school-related attitudes among 11–16-year-old-girls. *Psychological Reports*, 78, 787–793.

Mueller v. Allen, 463 U.S. 388 (1983).

Murry-McBride, Velma (1996). An ecological analysis of coital timing among middle-class African-American adolescent females. *Journal of Adolescent Research*, 11, 261–279.

Needle, Richard, Hamilton McCubbin, Marc Wilson, Robert Reineck, Amonn Lazar, & Helen J. Mederer (1986). Interpersonal influences in adolescent drug use—the role of older siblings, parents, and peers. *International Journal of the Addictions*, 21, 739–766.

Newcomb, Michael D., Ebrahim Maddahian, Rodney Skager, & Peter M. Bentler (1987). Substance abuse and psychosocial risk factors among teenagers: Asso-

ciations with sex, age, ethnicity, and type of school. *American Journal of Drug and Alcohol Abuse*, 13, 413–433.
Nigosian, Solomon A. (1994). *World Faiths*. New York: St. Martin's Press.
Orcutt, James D. (1987). Differential association and marijuana use: A closer look at Sutherland (with a little help from Becker). *Criminology*, 25, 341–358.
Oser, Fritz (1991). The development of religious judgment. In Fritz Oser & W. George Scarlett (Eds.), *Religious Development in Childhood and Adolescence. New Directions for Child Development* (Vol. 52, pp. 5–15). San Francisco: Jossey-Bass.
Oser, Fritz, & K. Helmut Reich (1990). Moral judgment, religious judgment, world view, and logical thought: A review of their relationship. Part 1. *British Journal of Religious Education*, 12, 94–101.
Oser, Fritz, & K. Helmut Reich (1996). Psychological perspectives on religious development. *World Psychology*, 2, 365–396.
Ozorak, Elizabeth Weiss (1989). Social and cognitive influences on the development of religious beliefs and commitment in adolescence. *Journal for the Scientific Study of Religion*, 28, 448–463.
Palmer, C. Eddie, & Dorinda N. Noble (1986). Premature death: Dilemmas of infant mortality. *Social Casework*, 67, 332–339.
Palmer, Susan J., & Charlotte E. Hardman (Eds.) (1999). *Children in New Religions*. New Brunswick: Rutgers University Press.
Paloutzian, Raymond F. (1996). *Invitation to the Psychology of Religion*. Boston: Allyn & Bacon.
Pargament, Kenneth I. (1990). God help me. Toward a theoretical framework for the psychology of religion. In Monty L. Lynn & David O. Moberg (Eds.), *Research in the Scientific Study of Religion* (Vol. 2, pp. 195–224). Greenwich, CT: JAI.
Pargament, Kenneth I. (1997). *The Psychology of Religion and Coping: Theory, Research, and Practice*. New York: Guilford.
Pargament, Kenneth I., David S. Ensing, Kathryn Falgout, Hannah Olsen, Barbara Reilly, Kimberly Van Haitsma, & Richard K. Warren (1990). God help me: (I) Religious coping efforts as predictors of the outcomes to significant negative life events. *American Journal of Community Psychology*, 18, 793–824.
Parham v. J.R., 442 U.S. 584 (1979).
Parke, Ross, & Navaz P. Bhavnageri (1989). Parents as managers of children's peer relationships. In Deborah Belle (Ed.), *Children's Social Networks and Social Support* (pp. 241–259). New York: Wiley.
Payne, I. Reed, Allen E. Bergin, Kimberly A. Beilema, & Paul H. Jenkins (1991). Review of religion and mental health: Prevention and the enhancement of psychological functioning. In Robert Hess (Ed.), *Religion and Prevention in Mental Health: Conceptual and Empirical Foundations* (pp. 11–40). New York: Haworth.

Peach, Lucinda (2002). *Legislating Morality: Pluralism and Religious Identity in Lawmaking*. New York: Oxford University Press.

Pealver, Eduardo (1997). The concept of religion. *Yale Law Journal*, 107, 791–822.

Peek, Charles W., George D. Lowe, & Susan L. Williams (1991). Gender and God's word: Another look at religious fundamentalism and sexism. *Social Forces*, 69, 1205–1221.

Pellebon, Dwain A. (2000). Influences of ethnicity, interracial climate, and racial majority in school on adolescent ethnic identity. *Social Work in Education*, 22, 9–19.

Perry, Constance M., & Walter G. McIntire (1994). High school seniors' concern for others: Predictors and implications. *High School Journal*, 77, 199–205.

Personal Responsibility and Work Opportunities Reconciliation Act of 1996, Pub. L. No. 104-193, 110 Stat. 2501 (codified as amended in scattered sections of 8 U.S.C. & 42 U.S.C.).

Peshkin, Alan (1986). *God's Choice: The Total World of a Fundamentalist School*. Chicago: University of Chicago Press.

Pierce v. Society of Sisters, 268 U.S. 510 (1925).

Planned Parenthood v. Casey, 505 U.S. 833 (1992).

Plante, Thomas G., & Allen C. Sherman (Eds.) (2001). *Faith and Health: Psychological Perspectives*. New York: Guilford.

Pollner, Melvin (1989). Divine relations, social relations, and well-being. *Journal of Health & Social Behavior*, 30, 92–104.

Prince v. Massachusetts, 321 U.S. 158 (1944).

Queen, Edward L. (2000). *Serving Those in Need: A Handbook for Managing Faith-Based Human Services Organizations*. San Francisco: Jossey-Bass.

R.A.V. v. City of St. Paul, 112 S. Ct. 2538 (1992).

Ravitch, Frank S. (1999). *School Prayer and Discrimination: The Civil Rights of Religious Minorities and Dissenters*. Boston: Northeastern University Press.

Reed, Leslie A., & Lawrence S. Myers (1991). A structural analysis of religious orientation and its relation to sexual attitudes. *Educational and Psychological Measurement*, 51, 943–952.

Rees, Rebecca G. (1999). "If we recant, would we qualify?" Exclusion of religious providers from state social service voucher programs. *Washington & Lee Law Review*, 59, 1291–1339.

Regnerus, Mark D. (2000). Shaping school success: Religious socialization and educational outcomes in metropolitan public schools. *Journal for the Scientific Study of Religion*, 39, 363–370.

Regnerus, Mark D., & Christian Smith (1998). Selective deprivatization among American religious traditions: The reversal of the great reversal. *Social Forces*, 76, 1347–1372.

Reich, K. Helmut (1993). Cognitive-developmental approaches to religiousness: Which version for which purpose? *International Journal for the Psychology of Religion*, 3, 145–171.

Reimer, Sam, & Jerry Z. Park (2001). Tolerant (in)civility? A longitudinal analysis of white conservative Protestants' willingness to grant civil liberties. *Journal for the Scientific Study of Religion*, 40, 735–745.

Religious Freedom Restoration Act (RFRA), 42 U.S.C. §§ 2000bb *et seq.* (1994).

Reynolds v. United States, 98 U.S. 145 (1878).

Richardson, James T. (1999). New religions: From 'brainwashing' claims to chid sex abuse accusations. In Susan J. Palmer & Charlotte E. Hardman (Eds.), *Children in New Religions* (pp. 172–186). New Brunswick: Rutgers University Press.

Roberts, Robert E., Jean S. Phinney, Louise C. Masse, Y. Richard Chen, Catherine R. Roberts, & Andrea Romero (1999). The structure of ethnic identity of young adolescents from diverse ethnocultural groups. *Journal of Early Adolescence*, 19, 301–322.

Robinson, Beth, & Ellen M. Frye (1997). Cult affiliation and disaffiliation: Implications for counseling. *Counseling & Values*, 41, 166–173.

Rodell, Daniel E., & Brent B. Benda (1999). Alcohol and crime among religious youth. *Alcohol Treatment Quarterly*, 17, 53–66.

Rodger, Alex (2000). Moral, spiritual, religious—Are they synonymous? In Mal Leicester, Cecilia Modgil, & Sohan Modgil (Eds.), *Spiritual and Religious Education* (pp. 3–14). New York: Falmer Press.

Roof, Wase Clark (1994). *A Generation of Seekers*. San Francisco: HarperCollins.

Rosenberger v. Rector and Visitors of the University of Virginia, 515 U.S. 819 (1995).

Rotheram-Borus, Mary Jane, & Karen F. Wyche (1994). Ethnic differences in identity development in the United States. In Sally L. Archer (Ed.), *Interventions for Adolescent Identity Development* (pp. 62–83). Thousand Oaks, CA: Sage.

Rust v. Sullivan, 500 U.S. 173 (1991).

Ryan, Richard M., & Edward L. Deci (2000). Self-determination theory and the facilitation of intrinsic motivation, social development, and well-being. *American Psychologist, 55*, 68–78.

Ryan, Richard M., Scott Rigby, & Kristi King (1993). Two types of religious internalization and their relations to religious orientations and mental health. *Journal of Personality and Social Psychology*, 65, 586–596.

Salamon, Lester M. (1995). *Partners in Public Service: Government–Nonprofit Relations in the Modern Welfare State*. Baltimore: Johns Hopkins University Press.

Sander, William (1996). Catholic grade schools and academic achievement. *Journal of Human Resources*, 31, 540–548.

Santa Fe Independent School District v. Doe, 120 S. Ct. 2266 (2000).

Schaefer, Axel R. (1999). Evangelicalism, social reform, and the U.S. welfare state, 1970–1996. In David K. Adams & Cornelis A. van Minnen (Eds.), *Religious and Secular Reform in America: Ideas, Beliefs, and Social Change* (pp. 249–273). New York: New York University Press.

Scott, Elizabeth S., N. Dickon Reppucci, & Jennifer L. Woolard (1995). Evaluating adolescent decision making in legal contexts. *Law and Human Behavior*, 19, 221–244.

Seligman, Martin E. (1991). *Learned Optimism*. New York: Simon & Schuster.

Sharpe, Tanya Telfair (2000). The Identity Christian Movement: Ideology of domestic terrorism. *Journal of Black Studies*, 30, 604–623.

Sherbert v. Verner, 374 U.S. 398 (1963).

Sherkat, Darren E., & Alfred Darnell (1999). The effect of parents' fundamentalism on children's educational attainment: Examining differences by gender and children's fundamentalism. *Journal for the Scientific Study of Religion*, 38, 23–35.

Shrum, Wesley (1980). Religious and marital stability: Change in the 1970s? *Review of Religious Research*, 21, 135–147.

Simmons, Roberta G., & Dale A. Blyth (1987). *Moving into Adolescence: The Impact of Pubertal Change and School Context*. New York: Aldine de Gruyter.

Simson, Gary J., & Erika A. Sussman (2000). Keeping the sex in sex education debate: The First Amendment's Religion Clauses and the sex education debate. *Southern California Review of Law & Women's Studies*, 9, 265–297.

Singer, Margaret T. (1996). *Cults in Our Midst: The Hidden Menace in Our Everyday Lives*. San Francisco: Jossey-Bass.

Smith, Christian (2000). *Christian America? What Evangelicals Really Want*. Berkeley: University of California Press.

Smith, Steven D. (1995). *Foreordained Failure: The Quest for a Constitutional Principle of Religious Freedom*. New York: Oxford University Press.

Smith, Elizabeth S. (1999). The effects of investments in the social capital of youth on political and civic behavior in young adulthood. *Political Psychology*, 20, 553–580.

Sorenson, Ann Marie, Carl F. Grindstaff, & R. Jay Turner (1995). Religious involvement among unmarried adolescent mothers: A source of emotional support? *Sociology of Religion*, 56, 71–81.

Sperry, Len, & Paul Giblin (1996). Marital and family therapy with religious persons. In Edward P. Shafranske (Ed.), *Religion and the Clinical Practice of Psychology* (pp. 551–532). Washington, D.C.: American Psychological Association.

Stark, Rodney (1996). Religion as context: Hellfire and delinquency one more time. *Sociology of Religion*, 57, 163–173.

Steinberg, Laurence (2000). The family at adolescence: Transition and transformation. *Journal of Adolescent Health*, 27, 170–178.

Steinberg, Laurence (2001). We know some things: Parent-adolescent relationships in retrospect and prospect. *Journal of Research on Adolescence*, 11, 1–19.

Steinberg, Laurence, & Elizabeth Cauffman (1996). Maturity of judgment in adolescence: Psychosocial factors in adolescent decision making. *Law and Human Behavior*, 20, 249–272.

Steinberg, Matt (1995). Free exercise of religion: The conflict between a parent's rights and a minor child's right in determining the religion of the child. *University of Louisville Journal of Family Law*, 34, 219–235.

Stone v. Graham, 449 U.S. 39 (1980).

Straus, Murray A. (1994). *Beating the Devil Out of Them: Corporal Punishment in American Families*. San Francisco: Jossey-Bass.

Straus, Murray A., David B. Sugarman, & Jean Giles-Sims (1997). Spanking by parents and subsequent antisocial behavior of children. *Archives of Pediatric and Adolescent Medicine*, 151, 761–767.

Strayhorn, Joseph M., Carla S. Weidman, & David Larson (1990). A measure of religiousness and its relation to parent and child mental health variables. *American Journal of Community Psychology*, 18, 34–43.

Streib, Heinz (2001). Faith development theory revisited: The religious styles perspective. *International Journal for the Psychology of Religion*, 11, 143–158.

Studer, Marlana, & Arland Thornton (1987). Adolescent religiosity and contraceptive usage. *Journal of Marriage and the Family*, 49, 117–128.

Sullivan, Winnifred Fallers (2001). Reflections on Catholic lobbying in the Supreme Court of the United States. *Notre Dame Law Review*, 76, 993–1018.

Tamminen, Kalevi (1994). Comparing Oser's and Fowler's developmental stages. *Journal of Empirical Theology*, 7, 75–112.

Tamney, Joseph B., & Stephen D. Johnson (1997). Christianity and public book banning. *Review of Religious Research*, 38, 263–271.

Taylor, Robert Joseph, Christopher G. Ellison, Linda M. Chatters, Jeffrey S. Levin, & Karen D. Lincoln (2000). Mental health services in faith communities: The role of clergy in black church. *Social Work*, 45, 73–87.

Thierstein, Joel, & Yahya R. Kamalipour (2000). *Religion, Law, and Freedom: A Global Perspective*. Westport, CT: Praeger.

Thomas, Darwin L., & Craig Carver (1990). Religion and adolescent social competence. In Thomas P. Gullotta, Gerald R. Adams, & Raymond Montemayor (Eds.), *Developing Social Competency in Adolescence. Advances in Adolescent Development* (Vol. 3, pp. 195–219). Newbury Park, CA: Sage.

Thompson, Edward H. (1991). Beneath the status characteristic: Gender variations in religiousness. *Journal for the Scientific Study of Religion*, 30, 381–394.

Thornton, Arland (1985). Reciprocal influences of family and religion in a changing world. *Journal of Marriage and the Family*, 47, 381–394.

Thornton, Arland, & Donald Camburn (1989). Religious participation and adolescent sexual behavior and attitudes. *Journal of Marriage and the Family*, 51, 641–653.

Tinker v. Des Moines Independent Community School District, 393 U.S. 503 (1969) .

Todd, Greg (2000). "Fully participating" voucher programs and the Wisconsin template: A brick or a breach in the wall of church-state separation? *University of Pennsylvania Journal of Constitutional Law*, 2, 710–739.

Tribe, Lawrence H. (1988). *American Constitutional Law*. 2nd ed. Mineola, NY: Foundation Press.

Trusty, Jerry, & Richard E. Watts (1999). Relationship of high school seniors' religious perceptions and behavior to educational, career, and leisure variables. *Counseling & Values*, 44, 30–39.

United States v. Lee, 455 U.S. 252 (1982).

United States v. Seeger, 380 U.S.163 (1965).

United States Department of Education (1995). *Religious Expression in Public Schools*. http://www.ed.gov./Speeches/08-1995/religion. html.

Van Ausdale, Debra, & Joe R. Feagin (2001). *The First R: How Children Learn about Race and Racism*. New York: Rowan & Littlefield.

de Vaus, David A. (1983). The relative importance of parents and peers for adolescent religious orientation: An Australian study. *Adolescence*, 69, 147–159.

Violato, Claudio, & LeRoy Travis (1995). *Advances in Adolescent Psychology*. Alberta, Canada: Detselig.

Violent and Repeat Juvenile Offender Accountability and Rehabilitation Act of 1999 (1999). H.R. 1501, 106th Cong. (1999).

Wagner, Melinda Bollar (1990). *God's Schools: Choice and Compromise in American Society*. New Brunswick: Rutgers University Press.

Wallace v. Jaffree, 472 U.S. 38 (1985).

Wallace, John M., Jr., & Tyrone A. Forman (1998). Religion's role in promoting health and reducing risk among American youth. *Health Education & Behavior*, 25, 721–741.

Wallace, John M., Jr., & David R. Williams (1997). Religion and adolescent health-compromising behavior. In John Schulenberg, Jenniffer L. Maggs, & Klaus Hurrelmann (Eds.), *Health Risks and Developmental Transitions during Adolescence* (pp. 444–468). Cambridge: Cambridge University Press.

Wardle, Lynn D. (1993). Protecting the rights of conscience of health care providers. *Journal of Legal Medicine* 14, 177–230.

Watson, Paul J., Ronald J. Morris, & Ralph W. Hood (1988). Sin and self-functioning. Part l: Grace, guilt, and self-consciousness. *Journal of Psychology and Theology*, 16, 254–269.

Weihe, Vernon R. (1990). Religious influence on parental attitudes toward the use of corporal punishment. *Journal of Family Violence*, 5, 173–186.

Weissbourd, Richard (2000). Beyond "villages": New community building strategies for disadvantaged families. In Mary Jo Bane, Brent Coffin, & Ronald Theimann (Eds.), *Who Will Provide? The Changing Role of Religion in American Social Welfare* (pp. 219–237). Boulder, CO: Westview Press.

Weishaar, Peter J. (1994). School choice vouchers and the establishment clause. *Alabama Law Review*, 58, 543–574.

Welch, Michael R., Charles R. Tittle, & Thomas Petee (1991). Religion and deviance among adult Catholics: A test of the "moral communities" hypothesis. *Journal for the Scientific Study of Religion*, 30, 159–172.

Welsh v. United States, 398 U.S. 333 (1970).

West Virginia State Board of Education v. Barnette, 319 U.S. 624 (1943).

White, Kathleen A. (1999). Crisis of conscience: Reconciling religious faith care providers' beliefs and patients' rights. *Stanford Law Review*, 51, 1703–1749.

Widmar v. Vincent, 454 U.S. 263 (1981).

Williams, David R. (1994). The measurement of religion in epidemiologic studies. In Jeffrey S. Levin (Ed.), *Religion in Aging and Health: Theoretical Foundations and Methodological Frontiers* (pp. 125–148). Thousand Oaks, CA, Sage.

Wilson, John, & Thomas Janoski (1995). The contribution of religion to volunteer work. *Sociology of Religion*, 56, 137–152.

Wisconsin v. Mitchell, 113 S. Ct. 2194 (1993).

Wisconsin v. Yoder, 406 U.S. 205 (1972).

Witte, John, Jr. (2001). A Dickensian era of religious rights: An update on religious human rights in global perspective. *William and Mary Law Review*, 42, 707–770.

Witte, John, Jr., & Richard C. Martin (1999). *Sharing the Book: Religious Perspectives on the Rights and Wrongs of Proselytism.* Maryknoll, NY: Orbis Books.

Witters v. Washington State Department of Services for Blind, 474 U.S. 481 (1986).

Woolard, Jennifer L., N. Dickon Reppucci, & Richard E. Redding (1996). Theoretical and methodological issues in studying children's capacities in legal contexts. *Law & Human Behavior*, 20, 219–228.

Worthington, Everett L. (1989). Religious faith across the life span: Implications for counseling and research. *Counseling Psychologist*, 17, 555–612.

Wright, Loyd S., Christopher J. Frost, & Stephen J. Wisecarver (1993). Church attendance, meaningfulness of religion, and depressive symptomatology among adolescents. *Journal of Youth & Adolescence*, 22, 559–568.

Wybraniec, John, & Roger Finke (2001). Religious regulation and the courts: The judiciary's changing role in protecting minority religions from majoritarian rule. *Journal for the Scientific Study of Religion*, 40, 427–444.

Youniss, James, Jeffrey A. McLellan, & Miranda Yates (1999). Religion, community service, and identity in American youth. *Journal of Adolescence*, 22, 243–253.

Youth Drug and Mental Health Services Act (1999). S. 976, 106th Cong.

Zaleski, Ellen H., & Kathleen M. Schiaffino (2000). Religiosity and sexual risk-taking behavior during the transition to college. *Journal of Adolescence*, 23, 223–227.

Zern, David S. (1985). Positive links among obedience pressure, religiosity, and measures of cognitive accomplishment: Evidence for the secular value of being religious. *Journal of Psychology and Theology*, 15, 31–39.

Zern, David S. (1987). The relationship of religious involvement to a variety of indicators of cognitive ability and achievement in college students. *Adolescence*, 22, 883–895.

Zern, David S. (1989). Some connections between increasing religious-ness and academic accomplishment in a college population. *Adolescence*, 24, 141–154.

Zinnbauer, Brian J., Kenneth I. Pargament, Brenda Cole, Mark S. Rye, Eric M. Butter, Timothy G. Belavich, Kathleen M. Hipp, Allie B. Scott, & Jill L. Kadar (1997). Religion and spirituality: Unfuzzying the fuzzy. *Journal for the Scientific Study of Religion*, 36, 549–564.

Zobrest v. Catalina Foothills School District, 509 U.S. 1 (1993).

Zorach v. Clauson, 343 U.S. 306 (1952).

Index

About the Author

Roger J. R. Levesque received his J.D. from the Columbia University School of Law and his Ph.D. in cultural psychology from the University of Chicago. Prior to his current appointment as professor of criminal justice at Indiana University, he was professor of psychology and law at the University of Arizona and a fellow in the Law and Psychology Program at the University of Nebraska. He has published more than fifty research and scholarly articles and book chapters that deal with adolescence, family life, and the law. His most recent books include *Child Sexual Abuse: A Human Rights Perspective* (1999); *Adolescents, Sex, and the Law: Preparing Adolescents for Responsible Citizenship* (2000); *Culture and Family Violence: Fostering Change through Human Rights Law* (2001); and *Dangerous Adolescents, Model Adolescents: Shaping the Role and Promise of Education* (2002).